Biochemical Applications of Nonlinear Optical Spectroscopy

T0225604

OPTICAL SCIENCE AND ENGINEERING

Founding Editor
Brian J. Thompson
University of Rochester
Rochester, New York

Biochemical Applications of Nonlinear Optical Spectroscopy

Edited by

Vladislav V. Yakovlev

CRC Press
Taylor & Francis Group
Boca Raton London New York

CRC Press is an imprint of the
Taylor & Francis Group, an **informa** business

Taylor & Francis Group
6000 Broken Sound Parkway NW, Suite 300
Boca Raton, FL 33487-2742

First issued in paperback 2017

ISBN 13: 978-1-138-11287-2 (pbk)
ISBN 13: 978-1-4200-6859-7 (hbk)

Library of Congress Cataloging-in-Publication Data

Biochemical applications of nonlinear optical spectroscopy / editor, Vladislav Yakovlev.
    p. ; cm. -- (Optical science and engineering ; 138)
    Includes bibliographical references and index.
    ISBN 978-1-4200-6859-7 (hardcover : alk. paper)
    1. Laser spectroscopy. 2. Nonlinear optical spectroscopy. 3. Lasers in biology. 4. Lasers in biophysics. 5. Lasers in chemistry. 6. Biochemistry--Technological innovations. I. Yakovlev, Vladislav. II. Title. III. Series: Optical science and engineering (Boca Raton, Fla.) ; 138
    [DNLM: 1. Microscopy--methods. 2. Spectrum Analysis. 3. Optics. QD272.S6 B6155 2008]

QP519.9.L37B56 2008
621.36'6--dc22
                                              2008022656

Visit the Taylor & Francis Web site at
http://www.taylorandfrancis.com

and the CRC Press Web site at
http://www.crcpress.com

Contents

Contributors

Yair Andegeko
Department of Chemistry
Michigan State University
East Lansing, Michigan, USA

Rajan Arora
Department of Physics
University of Wisconsin–Milwaukee
Milwaukee, Wisconsin, USA

Christopher J. Bardeen
Department of Chemistry
University of California at Riverside
Riverside, California, USA

Virginijus Barzda
Department of Physics
Department of Chemical and
  Physical Sciences
Institute for Optical Sciences
University of Toronto
Mississauga, Ontario, Canada

Richard Cisek
Department of Physics
Department of Chemical and
  Physical Sciences
Institute for Optical Sciences
University of Toronto
Mississauga, Ontario, Canada

Marcos Dantus
Department of Chemistry
Michigan State University
East Lansing, Michigan, USA

Tatsuya Fujino
Department of Chemistry
Graduate School of Science and
  Engineering
Tokyo Metropolitan University
Hachioji-Shi, Tokyo, Japan

Feruz Ganikhanov
Department of Physics
West Virginia University
Morgantown, West Virginia, USA

Catherine Greenhalgh
Department of Physics
Department of Chemical and
  Physical Sciences
Institute for Optical Sciences
University of Toronto
Mississauga, Ontario, Canada

Kerry M. Hanson
Department of Chemistry
University of California at Riverside
Riverside, California, USA

Norihiko Hayazawa
RIKEN
Nanophotonics Laboratory
Wako, Saitama, Japan
and
CREST
Japan Science and Technology Corp.
Kawaguchi, Saitama, Japan

Taro Ichimura
CREST
Japan Science and Technology Corp.
Kawaguchi, Saitama, Japan
and
Department of Applied Physics
Osaka University
Suita, Osaka, Japan

Katsuyoshi Ikeda
RIKEN
Nanophotonics Laboratory
Wako, Saitama, Japan
and
Graduate School of Science
Hokkaido University
Hokkaido, Japan

Satoshi Kawata
RIKEN
Nanophotonics Laboratory
Wako, Saitama, Japan
and
CREST
Japan Science and Technology Corp.
Kawaguchi, Saitama, Japan
and
Department of Applied Physics
Osaka University
Suita, Osaka, Japan

Vishnu Vardhan Krishnamachari
Department of Chemistry
University of California–Irvine
Irvine, California, USA

Arkady Major
Department of Electrical
  and Computer Engineering
University of Manitoba
Winnipeg, Manitoba, Canada

Marcus Motzkus
Fachbereich Chemie
  Physikalische Chemie
Phillipps–Universität
Marburg, Germany

Dmitry Pestov
Department of Chemistry
Michigan State University
East Lansing, Michigan, USA

Georgi I. Petrov
Department of Physics
University of Wisconsin
Milwaukee, Wisconsin, USA

Eric Olaf Potma
Department of Chemistry
University of California–Irvine
Irvine, California, USA

Nicole Prent
Department of Physics
Department of Chemical and
  Physical Sciences
Institute for Optical Sciences
University of Toronto
Mississauga, Ontario, Canada

Valerica Raicu
Department of Physics
Department of Biological Sciences
University of Wisconsin–Milwaukee
Milwaukee, Wisconsin, USA

Daaf Sandkuijl
Department of Physics
Department of Chemical and
  Physical Sciences
Institute for Optical Sciences
University of Toronto
Mississauga, Ontario, Canada

Marius Schmidt
Department of Physics
Department of Biological Sciences
University of Wisconsin–Milwaukee
Milwaukee, Wisconsin, USA

Vladislav Shcheslavskiy
Laboratory of Biomedical Optics
EPFL
Lausanne, Switzerland

Michael Stoneman
Department of Physics
Department of Biological
  Sciences
University of Wisconsin–Milwaukee
Milwaukee, Wisconsin, USA

Tahei Tahara
RIKEN
Molecular Spectroscopy Laboratory
Wako, Saitama, Japan

Adam Tuer
Department of Physics
Department of Chemical and
  Physical Sciences
Institute for Optical Sciences
University of Toronto
Mississauga, Ontario, Canada

Bernhard von Vacano
BASF SE
Ludwigshafen, Germany

Vladislav V. Yakovlev
Department of Physics
University of Wisconsin–Milwaukee
Wisconsin, USA

Introduction

Nonlinear optical spectroscopy originates from the pioneering work of Marija Göppert-Mayer, who in her Ph.D. thesis described the concept of two-photon excitation (Göppert-Mayer 1931). In the 1960s, with the invention of lasers and the development of ultra-short, high-peak-power coherent light sources, a number of optical techniques using nonlinear response of dielectric materials to the electric field of light were created. While all the major concepts were developed in 1960s and 1970s, the real applications had to wait until the late 1990s, when further progress of short-pulsed laser systems and low-noise detectors sparked a variety of applications of nonlinear optical spectroscopy to real-time structural and chemical analysis.

What makes nonlinear optical spectroscopy so valuable for biochemical applications? Nonlinear optical spectroscopy is governed by the higher order susceptibility tensor, whose properties are often significantly affected by a small variation of material structure and/or composition (Boyd 2003). In some examples, such as vibrational spectroscopy, nonlinear effects often generate signals orders of magnitude stronger than their linear counterparts. Last but not least, nonlinear optical spectroscopy often uses detection at a different frequency, allowing near-infrared (IR) excitation, which is considered safer for biological matter. These advantages of nonlinear optical spectroscopy make it indispensable for biochemical applications, where minimally invasive diagnostics is needed.

In Chapter 1, Raicu et al. make a unique comparison of nonlinear optical spectroscopy and x-ray crystallography in terms of the fundamental understanding of the time-resolved structural dynamics of myoglobin.

In Chapter 2, Hanson and Bardeen show a great use of multiphoton fluorescence to study chemical microenvironment in the skin.

Traditional optical spectroscopy, such as fluorescence imaging, often gets a substantial boost from nonlinear optics, which, as demonstrated by Fujino and Tahara (Chapter 3), can provide temporal resolution and the ability to detect transient chemical species.

One of the most common trends in the recent development of nonlinear optical spectroscopies for biological imaging is the multimodal approach, which incorporates several spectroscopic techniques in one instrument. This approach allows the employment of several contrast mechanisms, such as fluorescence, second harmonic, and third harmonic (Cisek et al., Chapter 4), significantly increasing the informational content of microscopic imaging.

Similar approaches were adopted by Ganikhanov (Chapter 5), who developed a state-of-the-art laser system, benefiting simultaneous third-harmonic and nonlinear Raman microscopy, and Yakovlev et al. (Chapter 6), who applied third-harmonic generation microscopy and nonlinear Raman microspectroscopy for biochemical analysis in microfluidic devices.

Using the broadband laser pulse generated through continuum generation and an optical pulse shaper, Vacano and Motzkus (Chapter 7) and Dantus et al. (Chapter 8) demonstrate the substantial untapped potential of this approach for microscopic biological imaging and biochemical analysis.

The ability to shape the incident beams in time (Chapters 7 and 8) and space (Chapters 9 and 10) is one of the most promising improvements that modern optical technology can make for nonlinear optical spectroscopy.

Potma and Krishnamachari (Chapter 9) explore the effects of spatial beam shaping on the generated nonlinear Raman signals in a tightly focused geometry, demonstrating its enhanced ability to image chemical interfaces.

Finally, Hayazawa et al. (Chapter 10), using a plasmonic tip-enhanced arrangement, extend nonlinear optical spectroscopy to the nanoscale, providing it with the capacity to resolve individual molecules.

REFERENCES

Boyd, R. W. 2003. *Nonlinear Optics*. 2nd ed. San Diego, CA: Academic Press.
Göppert-Mayer, M. 1931. Über Elementarakte mit zwei Quantensprüngen. *Ann. Phys.* 9:273–95.

1 Structural Dynamics and Kinetics of Myoglobin-CO Binding: Lessons from Time-Resolved X-Ray Diffraction and Four-Wave Mixing Spectroscopy

Valerica Raicu, Marius Schmidt,
and Michael Stoneman

CONTENTS

I. INTRODUCTION

Structure, function, and dynamics of biomolecules are coupled closely together. It is not only the structure but also the dynamics of the structural properties that enable a protein to perform its function. Consider, for instance, the case of enzymes, which are catalytically active proteins that assist biochemical reactions. By mixing substrate with an active, functional enzyme, the progress of the reaction toward its end can be monitored by using adequate physical methods. Time becomes an important parameter in such experiments. Measuring how the distribution of molecules over discrete states varies as a function of time is equivalent to observing the kinetics of the reaction. There is a fundamental difference between protein dynamics and kinetics. Protein dynamics lies at the base of kinetics and represents a set of structural changes required for a reaction to occur, whereas kinetics is intrinsically statistical. Without protein dynamics there are no structural changes, no catalytic activity, and hence no kinetics.

In the study of protein dynamics and reaction kinetics, heme proteins are of special interest, primarily because much is known about them and also because ligand dissociation and rebinding may be studied in the time domain following excitation with light pulses. Myoglobin (Mb) is most commonly used as a model heme protein, since it presents all the features required for kinetic studies and yet is structurally and functionally simpler than other proteins, such as hemoglobin. In 1958, myoglobin became the first protein whose structure was determined (Kendrew et al. 1958), and almost 40 years later it became also the first protein to be investigated by nanosecond time-resolved crystallography (Srajer et al. 1996). Its biological role is to bind O_2 and CO_2 in the muscles, but in most dynamic studies of Mb-ligand recombination, CO is used as a ligand, since it can be easily dissociated using light. Also, heme Fe easily oxidizes *in vitro* in MbO_2, which is thus unstable, unlike MbCO.

Figure 1.1 shows the molecular structure of myoglobin consisting of eight α-helical segments wrapped around the functional heme chromophore. Here, carbon-monoxide (CO) is bound to the heme iron. However, this bond can be ruptured by an intense laser flash and the structure starts to relax to the deoxy form.

Obvious questions, such as what the timescales and the structural bases of the relaxations are, where the CO is located, and what the intermediate states of the protein as it progresses toward ligand rebinding are, can be answered by linear and nonlinear laser spectroscopy and time-resolved crystallography.

In this chapter, we first review some of the basic theory of reaction kinetics as well as knowledge accumulated over the past three decades on MbCO recombination kinetics and associated protein motions. Then we present a nonlinear optical technique used in studies of MbCO dynamics and kinetics, the *transient phase grating*. In discussing the results, we focus our attention on the kinetics of protein-ligand recombination, while leaving aside most of the details of segmental motion in proteins during ligand

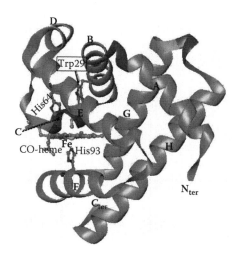

FIGURE 1.1 The architecture of myoglobin. 8 α-helices (A–H) accommodate a heme group, which is attached to the protein via the proximal His93. CO is bound to the iron (Fe) on the distal side of the heme. Here the L29W mutant (Nienhaus et al., 2005) is depicted with the bulky Trp29 bound to helix B on the distal heme side. Figure prepared with "ribbons" (Carson et al. 1997).

dissociation and rebinding. While taking note of the existence of various interpretations of experimental data and competing theoretical models, we only adopt particular interpretations wherever the discussion of experimental data seems to demand it.

II. THEORY OF TRANSIENT STATE KINETICS AND DYNAMICS

A. SYSTEMS WITH A SMALL NUMBER OF DISCRETE STATES

Since proteins are essentially polymers, they may adopt many, almost iso-energetic and very similar structures called protein substates (Austin et al. 1975; Frauenfelder et al. 1988). At ambient temperatures, proteins switch between their substates; hence, they are constantly changing their structure (Parak et al. 1982; Parak 2003). By knowing the coordinates of each of the atoms and their velocity vectors at any instant of time, the dynamics of the structural changes, i.e., the trajectories of the molecule's atoms, can be followed exactly. However, this has been achieved only in computer simulations so far.

At thermal equilibrium, a protein may reach a state of minimal energy. If a ligand binds to it, another structure is energetically favored and the protein starts to relax toward it. It is the protein dynamics that couples to the heat bath and enables the molecules to surmount the barriers of energy, which otherwise would prevent the reaction from progressing. Since kinetics is the result of the underlying dynamics, to observe kinetics, an ensemble of reacting molecules is required.

Consider the simple two-state diagram of protein molecules reacting from state A to state B sketched in Figure 1.2. State A is occupied by an ensemble of molecules

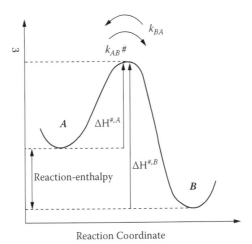

FIGURE 1.2 Two-state diagram (states A and B) to describe chemical kinetics. The reaction coordinate can be considered as the pathway with lowest energy through the multidimensional energy hyper-surface of the reaction. # represents the transition state, i.e., the state with highest free energy on the reaction coordinate. The reaction enthalpy and the enthalpy differences between the two states and the transition state are depicted by arrows. Note that for the rate coefficients k_{AB} and k_{BA} the free energy differences, including enthalpy and entropy differences, have to be considered.

that undergo complex internal motions individually and independently. There is a finite probability that they will reach the top of the barrier of activation, which is the transition state number (Eyring 1935; Cornish-Bowden 1999), and may relax to state B. Measuring the progressively smaller occupancy of state A is equivalent to monitoring the kinetics of that reaction. On the other hand, measuring the trajectory of each of the molecules within the states and across the barrier is equivalent to probing the dynamics. At this point, the conceptual difference between kinetics and dynamics becomes clearer.

The reaction rate (or the velocity), which is defined as the change of concentration of a particular species (or protein state), may also be defined as the number of molecules per unit time crossing the barrier from state A to B. It depends on the height of the barrier, the number of molecules available in state A, and the temperature T. The height of the barrier and the temperature dependence are parameterized by the rate coefficient, k, defined as

$$k = \Omega \cdot e^{\frac{-\Delta H^{\#} + T\Delta S^{\#}}{k_B T}} = \Omega \cdot e^{\frac{-\Delta G^{\#}}{k_B T}} \tag{1.1}$$

where $\Delta G = \Delta H - T\Delta S$ is the Gibbs free energy, Ω may be interpreted as the number of trials per unit time the molecules attempt to jump over the barrier, and the exponential term is the probability that they succeed. This probability depends on the enthalpy of activation, $\Delta H^{\#}$, the entropy difference, $\Delta S^{\#}$, between the states A (or B) and the transition state #, the Boltzmann constant, k_B, and, of course, the temperature.

It should be pointed out at this juncture that a reaction with no enthalpy of activation ($\Delta H^{\#} = 0$) can nevertheless have a very small rate coefficient, since the entropy of the transition state can be very small compared to that in the state A or B. Accordingly, $\Delta S^{\#}$ becomes negative. In other words, there is a possibility that only a small loophole or a very narrow path in conformational space exists through which molecules may migrate from state A to B or vice versa.

If the free energy of the reaction from A to B has a large negative value, the reaction is irreversible, and the backwards reaction from B to A can be neglected, since k_{BA} is extremely small. The kinetics is then described by the following simple differential equations:

$$\frac{dA}{dt} = -k_{AB}A, \tag{1.2a}$$

$$\frac{dB}{dt} = k_{AB}A. \tag{1.2b}$$

The temporal derivatives on the right-hand sides represent the changes in concentration of molecules in state A or B, which are the rates or the velocities of the reaction(s). The second equation follows immediately from Equation (1.2a) and the independent requirement that $A(t) + B(t) = $ const (i.e., the conservation of mass). The rate is negative when molecules leave a state and positive when they populate a state. In addition, the rate is considered proportional to the concentration of molecules in state A. Hence, the larger the number of molecules in state A, the larger the velocity of the reaction. Integration of Equation (1.2a) shows that the molecules in A vanish according to simple exponential law,

$$A(t) = A_0 e^{-k_{AB}t}, \tag{1.3a}$$

while the population in state B increases as

$$B(t) = A_0(1 - e^{-k_{AB}t}), \tag{1.3b}$$

where A_0 is the concentration of molecules in state A at the beginning of the reaction, when the concentration of B is zero (i.e., $B_0 = 0$). The functions $A(t)$ and $B(t)$ are called "the concentration profiles of the reaction."

If the free energy of the reaction is small, the reaction is reversible. Then the change in the concentration of molecules in state A depends on the efflux of molecules from A to B *and* the influx of molecules from B into A. Similar arguments hold for the flux of molecules into and from state B. The differential equations are then coupled,

$$\frac{dA}{dt} = -k_{AB}A + k_{BA}B, \tag{1.4a}$$

$$\frac{dB}{dt} = k_{AB}A - k_{BA}B. \tag{1.4b}$$

Equations (1.4a) and (1.4b) can be written using matrix notation as

$$
\begin{pmatrix} \dfrac{dA}{dt} \\[2ex] \dfrac{dB}{dt} \end{pmatrix} = \begin{pmatrix} -k_{AB} & k_{BA} \\[1ex] k_{AB} & -k_{BA} \end{pmatrix} \begin{pmatrix} A \\[1ex] B \end{pmatrix},
\tag{1.5}
$$

or, in a compressed form, as

$$
\frac{dA}{dt} = KA ,
\tag{1.6}
$$

where A is a vector with the components A and B, dA/dt is its derivative with respect to time, and K is the coefficient matrix. It may be shown that the coupled differential equations can be integrated by finding the eigenvalues and eigenvectors of the coefficient matrix K (Steinfeld et al. 1989). The results again take exponential forms:

$$
A(t) = \frac{A_0}{k_{AB} + k_{BA}} \left[k_{BA} + k_{AB} e^{-(k_{AB}+k_{BA})t} \right],
\tag{1.7a}
$$

$$
B(t) = \frac{k_{AB} A_0}{k_{AB} + k_{BA}} \left[1 - e^{-(k_{AB}+k_{BA})t} \right].
\tag{1.7b}
$$

Equation (1.6) was solved by assuming that at the beginning of the reaction only state A is occupied with the concentration A_0 and state B is empty. The eigenvalues of this problem are 0 and $-(k_{AB} + k_{BA})$. Note, that the exponents in Equations (1.7a) and (1.7b) both have the same relaxation time, $\tau = 1/(k_{AB} + k_{BA})$. This can be generalized. The observed relaxation times are the inverse of the eigenvalues of the coefficient matrix describing the mechanism (Steinfeld et al. 1989). The number of observable relaxation times is equal to the number of states minus 1 (Matsen and Franklin 1950; Fleck 1971; Henry and Hofrichter 1992). Concentration profiles can be easily determined by numerical methods for any form of the coefficient matrix derived from any mechanism. Figure 1.3 shows concentration profiles for a reaction mechanism employing four states, the initial state and three intermediate states, during a reaction. As expected, three relaxation times are observable.

Any change in the catalytic properties of the protein will change the rate coefficients, k, which in turn will change the concentration profile and the relaxation times. Transient state kinetics becomes a tool to investigate the catalyst. In extreme cases the reaction will stop at some stage and the protein ceases to work. Then, a potent inhibitor might have been found that blocks one of the four states of this reaction. This lies at the base of any kinetic investigation.

B. Distribution of Barrier Heights

Oftentimes, the concentration profile follows neither a simple exponential in time nor a discrete combination of exponentials, but a stretched exponential, which is also

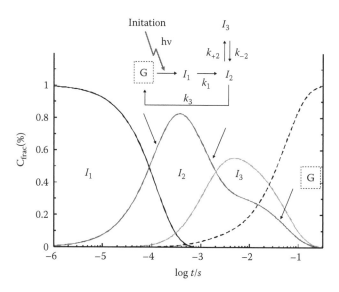

FIGURE 1.3 Concentration profile from a mechanism employing three intermediates, I_1, I_2, I_3, plus the initial (dark) state I_0. $I_1(0) = 1$ is assumed. The cyclic reaction is started (initiated) by a laser flash. Molecules relax through the intermediate states back to the initial state. The concentration profile of intermediate I_2 shows all three relaxation times (arrows).

called the Kohlrausch-Williams-Watt (KWW) equation,

$$A(t) = A_0 e^{-(k_{AB}t)^\beta} = A_0 e^{-\left(\frac{t}{\tau_{AB}}\right)^\beta}, \tag{1.8}$$

which can cover several decades in time. The power, β, of the exponent is the stretching constant and takes values over the interval [0,1]. If the concentration profile traces a straight line in a log-log representation, it is described by a "power-law" function,

$$A(t) = A_0(1 + k_{AB}t)^{\alpha-1} = A_0\left(1 + \frac{t}{\tau_{AB}}\right)^{\alpha-1}. \tag{1.9}$$

To retain the analogy with a simple exponential function, it is considered in the cases described by Equations (1.8) and (1.9) that there is a distribution of barrier heights, $g(G)$, each height corresponding to an exponential relaxation (Austin et al. 1975; Nagy et al. 2005). The concentration profile is in this case described by

$$A(t) = \int_0^\infty g(G_{AB})e^{-k_{AB}t}dG_{AB}, \tag{1.10}$$

where $g(G_{AB})$ represents the probability of finding a molecule with activation energy between G_{AB} and $G_{AB} + dG_{AB}$. If we assume that Ω is independent of G, Equation (1.1) gives

$$dG_{AB} = -k_B T \frac{dk_{AB}}{k_{AB}}, \tag{1.11}$$

and Equation (1.10) becomes

$$A(t) = k_B T \int_0^\Omega \frac{g(k_{AB})}{k_{AB}} e^{-k_{AB}t} dk_{AB}.$$ (1.12)

where $g(k_{AB})$ is the distribution of rate constants.

By using the normalization condition for the distribution of relaxation times,

$$k_B T \int_0^\Omega g(k_{AB}) dk_{AB} = 1$$

Equation (1.12) may be rewritten as

$$A(t) = \frac{\int_0^\Omega \frac{g(k_{AB})}{k_{AB}} e^{-k_{AB}t} dk_{AB}}{\int_0^\Omega \frac{g(k_{AB})}{k_{AB}} dk_{AB}} = \frac{\int_0^\Omega g(k_{AB}) e^{-k_{AB}t} d(\ln k_{AB})}{\int_0^\Omega g(k_{AB}) d(\ln k_{AB})},$$ (1.13)

which expresses the concentration as a fraction of a maximum value.

Expressed in terms of the distribution of relaxation times, $g(\tau_{AB})$, Equation (1.13) reads

$$A(t) = \frac{\int_{\tau_{min}}^\infty \frac{g(\tau_{AB})}{\tau_{AB}} e^{-\frac{t}{\tau_{AB}}} d\tau_{AB}}{\int_{\tau_{min}}^\infty \frac{g(\tau_{AB})}{\tau_{AB}} d\tau_{AB}} = \frac{\int_{\tau_{min}}^\infty g(\tau_{AB}) e^{-\frac{t}{\tau_{AB}}} d(\ln \tau_{AB})}{\int_{\tau_{min}}^\infty g(\tau_{AB}) d(\ln \tau_{AB})},$$ (1.14)

where $\tau_{min} = 1/\Omega$. Since in most practical circumstances, the largest relaxation time has to be a finite number, the upper limit in Equation (1.13) may be replaced by some τ_{max}. Hence,

$$A(t) = \frac{\int_{\tau_{min}}^{\tau_{max}} \frac{g(\tau_{AB})}{\tau_{AB}} e^{-\frac{t}{\tau_{AB}}} d\tau_{AB}}{\int_{\tau_{min}}^{\tau_{max}} \frac{g(\tau_{AB})}{\tau_{AB}} d\tau_{AB}} = \frac{\int_{\tau_{min}}^{\tau_{max}} g(\tau_{AB}) e^{-\frac{t}{\tau_{AB}}} d(\ln \tau_{AB})}{\int_{\tau_{min}}^{\tau_{max}} g(\tau_{AB}) d(\ln \tau_{AB})}.$$ (1.15)

If entropy changes are negligible, one usually speaks of enthalpic barrier heights that the reaction has to overcome. The distribution of relaxation times or rate constants in this case may be equivalently considered a distribution of reaction barrier heights.

A distribution of relaxation times has been proposed (Raicu 1999) that encompasses all known types of relaxation functions (i.e., concentration profiles) when introduced into Equation (1.15). A convenient particular form of this distribution reads

$$g(\tau_{AB}) = \frac{1}{2\pi} \frac{\left(\dfrac{\tau_{AB}}{\tau_p}\right)^{\alpha} \sin[\pi(1-\beta)] + \left(\dfrac{\tau_{AB}}{\tau_p}\right)^{1-\beta} \sin[\pi\alpha]}{\cosh\left[(1-\alpha-\beta)\ln\left(\dfrac{\tau_{AB}}{\tau_p}\right)\right] + \cos[\pi(1-\alpha-\beta)]} \tag{1.16}$$

where τ_p is the most probable relaxation time, and $0 \leq \alpha, \beta \leq 1$ are real constants.

When introduced into Equation (1.15), the distribution given by Equation (1.16) leads, e.g., to the simple exponential form of Equations (1.3a) and (1.3b) for $\tau_{AB} \rightarrow \tau_p$ and $\alpha = \beta = 0$, the fractional power-law of Equation (1.9) for $\alpha = 1 - \beta$, as well as to other known functions; this distribution is especially useful when the relaxation follows fractal or self-similar pathways (Raicu et al. 2001), and it allows for almost any conceivable relaxation function to be introduced without a need to formulate complicated theoretical models.

C. WHAT CAN BE MEASURED

The ultimate goal of kinetics studies is the identification of a (unique) chemical kinetic mechanism, which consists of a reaction scheme such as the one shown in Figure 1.3 and the corresponding numerical values of the rate coefficients, k, which incorporate entropy and enthalpy differences. This is an inverse problem, since only the concentration profile or, in less favorable conditions, only the relaxation times can be observed, and the reaction mechanism must be deduced from this information. Any experimental method that establishes a connection between the signal and the concentration of molecules can be used to investigate kinetics. However, it is necessary that the method has sufficient time resolution since time is the crucial parameter in kinetic experiments.

Many methods of investigation of protein-ligand binding kinetics that are based on linear processes are of a "pump-probe" type. In this approach an optical pulse, called a "pump," starts a photoreaction (such as dissociation of MbCO into Mb and CO), and its progress is probed a time Δt later. The probe could be, for example, a weak laser pulse, which detects the spectral changes in the heme during the protein-ligand recombination, or an x-ray pulse, which allows determination of the protein structure at a particular instant in time.

Both in linear and nonlinear methods, the minimum time delay accessible to the experimenter is the time resolution, and it is determined by either the duration of the pump or the probe pulse, whichever is longer. Two linear methods are discussed in section II, while a nonlinear method is presented in section IV. Typical timescales for protein catalyzed reactions range in the nanosecond (ns) to millisecond (ms) time range and the time resolution must be much better in order to sample the time range sufficiently. However, there are processes in proteins that are much faster, often occurring at femtosecond (fs) timescales (Franzen et al. 1995; Lim et al. 1993; Jackson et al. 1994; Armstrong et al. 2003; Nagy et al. 2005). To observe these processes, the time resolution must be exquisite (Norrish et al. 1965). This adds tremendous complication to the experiments.

In the next section, we describe two pump-probe approaches and review their use in experiments with MbCO. This places the nonlinear optical techniques of transient phase grating introduced in the subsequent section in their proper context.

III. PUMP-PROBE STUDIES OF PROTEIN-LIGAND BINDING

A. Experimental Methods

1. Optical Pump-Probe Spectroscopy

In the typical setup, excitation light is provided by a pulsed (e.g., nanosecond) laser (emitting in the visible range, e.g., at 532 nm, if Mb is investigated), while the probe is delivered by a continuous-wave (cw) laser. The two beams are spatially overlapped in the sample, and the temporal changes in the optical properties (such as optical absorption or frequency shift) that follow the passage of the pump pulse are registered by a detector with short response time (relative to time scale of the processes monitored), such as a fast photodiode.

If the processes of interest occur on the nanosecond or longer time scale, one can usually visualize the evolution of the parameter of interest using an oscilloscope. For shorter timescales, standard electronics is too slow, and temporal information is determined by passing the pulsed probe light through a delay line with adjustable length (Nagy et al. 2005). Then, the time coordinate is obtained by dividing the length of the beam path in the delay line by the speed of light.

2. Time-Resolved Crystallography

X-ray crystallography is the method of choice for determination of structures of large macromolecules such as proteins. Nowadays, roughly 48,000 x-ray structures are stored in the Protein Data Bank (http://www.rcsb.org; Berman et al. 2000). X-ray crystallography is traditionally a static method, i.e., without time resolution. In order to follow the kinetics and to determine the structure of the transiently occupied intermediate states of proteins, time-resolved crystallography has to be used (Moffat 1989). The time resolution, t_{min}, has to be as good as for any other method employed to follow reaction kinetics. This implies that x-ray data must be collected as fast as possible.

In third-generation synchrotrons, the x-rays are generated in intense flashes of ~100 picoseconds (ps) duration. If, during this time, an entire diffraction pattern is recorded, the time resolution t_{min} equals 100 ps (Szebenyi et al. 1992; Srajer et al. 1996, Schotte et al. 2003). However, the traditional monochromatic oscillation diffraction method cannot be used since there is no way to rotate the crystal during this 100 ps timeframe to collect the integrated intensity of a Bragg reflection. Still exposures, therefore, have to be used.

Although other methods could be used to determine the integrated intensity, the Laue method (Amoros et al. 1975; Bartunik et al. 1992; Ren et al. 1999) has been the method of choice so far. In this method, the crystal is subjected to a spectrum of x-ray radiation. Each reflection accepts a small fraction of this bandwidth, which covers the entire reflection range of that particular reflection. Hence, the integrated

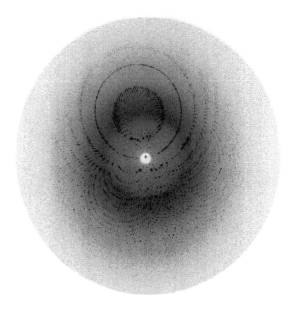

FIGURE 1.4 A protein Laue diffraction pattern.

intensity can be recorded instantaneously without rotating the crystal. This is the sole physical reason for the use of the Laue method. Of practical value is that a substantial fraction of the reciprocal space is sampled with one diffraction pattern (Figure 1.4). Consequently, an entire Laue data set can be collected rapidly. With optimized, polychromatic, synchrotron-based undulator x-ray sources, a number of different crystal settings spaced approximately 2–3 degrees apart cover the unique volume of the reciprocal space.

Using time-resolved crystallographic experiments, molecular structure is eventually linked to kinetics in an elegant fashion. The experiments are of the pump-probe type. Preferentially, the reaction is initiated by an intense laser flash impinging on the crystal and the structure is probed a time delay, Δt, later by the x-ray pulse. Time-dependent data sets need to be measured at increasing time delays to probe the entire reaction. A time series of structure factor amplitudes, $|F_t|$, is obtained, where the measured amplitudes correspond to a vectorial sum of structure factors of all intermediate states, with time-dependent fractional occupancies of these states as coefficients in the summation. Difference electron densities are typically obtained from the time series of structure factor amplitudes using the difference Fourier approximation (Henderson and Moffatt 1971). Difference maps are correct representations of the electron density distribution. The linear relation to concentration of states is restored in these maps. To calculate difference maps, a data set is also collected in the dark as a reference. Structure factor amplitudes from the dark data set, $|F_D|$, are subtracted from those of the time-dependent data sets, $|F_t|$, to get difference structure factor amplitudes, ΔF_t. Using phases from the known, precise reference model (i.e., the structure in the absence of the photoreaction, which may be determined from static x-ray diffraction), difference structure factors, and corresponding difference maps, $\Delta\rho_t$, are calculated for each time point, t.

Difference electron densities can be globally analyzed by methods from linear algebra, as has been successfully demonstrated in the literature (Schmidt et al. 2003; Schmidt et al. 2004; Rajagopal et al. 2004, 2005; Ihee et al. 2005). At the base of the analysis is a component analysis, the singular value decomposition (*SVD*). It takes a time series of difference maps (matrix *A*) and decomposes it into a set of time-independent, singular difference maps, the left singular vectors *U* and their respective time courses, the right singular vectors *V*. Matrices *U* and *V* are connected by a diagonal matrix *S*, which contains the singular values

$$A = U\,S\,V^{T}. \tag{1.17}$$

How matrix *A* is set up and details of the *SVD* analysis of time-resolved x-ray data are described elsewhere (Schmidt et al. 2003, 2005b; Schmidt 2008). Essentially the relaxation times of the kinetics are found in a global way in the right singular vectors *V*. By interpreting the kinetics with a suitable chemical kinetic model, it becomes possible to project the left singular vectors onto the intermediate states (Henry and Hofrichter 1992; Schmidt et al. 2003). Basically, this operation linearly combines the content of the singular vectors in a concentration- dependent way. As a result, the time-independent difference electron densities of the intermediates can be found, from which the structures of the intermediates are determined (Schmidt et al. 2004; Rajagopal et al. 2005; Ihee et al. 2005).

B. PUMP-PROBE STUDIES OF MB-CO RECOMBINATION

The interest in studying the structures of intermediates in myoglobin can be traced back to the groundbreaking work of Frauenfelder and coworkers (Austin et al. 1975). They flashed away the CO and observed the rebinding kinetics at various temperatures spectroscopically. Steps or phases in the kinetics were assumed to correspond to different intermediate states. What do these intermediate states look like? Several possible positions of the CO in the heme pocket were determined independently by three groups in the 1990s using cryo-crystallographic techniques (Schlichting et al. 1994; Teng et al. 1994, 1997; Hartmann et al. 1996). From those experiments, it appeared that the CO molecule can be driven by extended illumination to a distance of up to 3.5 Å from the heme iron (Figure 1.5, B-site). Other positions could not be populated at low temperatures, which required other experimental schemes to be developed (Ostermann et al. 2000; Nienhaus et al. 2005). Results from one of these schemes are described in Figure 1.6.

MbCO L29W mutant crystals are cooled from 180 K to 105 K during laser illumination. The CO is found on the proximal side of the heme. This site is identical to one of the four sites previously characterized by binding xenon (Tilton et al. 1984). They are called Xe1…Xe4 sites (see Figure 1.5 for their approximate positions). Only Xe1 is populated in this experiment. No electron density of the CO is found at the iron-binding site (Figure 1.5, B-site). Subsequent heating followed by structure determination at 105 K show the rebinding of the CO from Xe1 back to the iron.

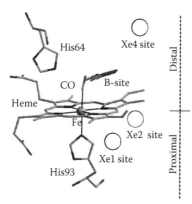

FIGURE 1.5 The heme pocket of myoglobin. B-site: CO found here also at cryogenic temperatures. Xe1…Xe4 sites: identified in xenon binding experiments.

Effective ligand rebinding from the Xe1 site is only observed when the temperature has risen above a characteristic temperature $T_c \sim 180$ K (Figure 1.6), which is the temperature where protein dynamics sets in. However, the time-related information is lost in experiments at cryogenic temperatures. Time-resolved crystallography was applied to restore the time scale and observe undisturbed relaxations.

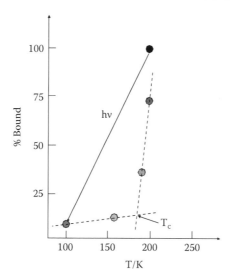

FIGURE 1.6 Photoflash experiments with temperature cycling on L29W MbCO. Amount of CO (occupancy of CO) bound to the iron determined from integrating the electron density at the iron-binding site of the CO. Black circle (a): initial species, not photolyzed. Dark blue circle (b): directly after the crystal was cooled to 105 K under continuous illumination. Light blue circle (c): after warming to 160 K. Green circle (d): after 180 K. Red circle (e) after 200 K. (From Nienhaus, K., Ostermann, A., Nienhaus, U., Parak, F., and Schmidt, M., *Biochemistry* 44: 5095–5205, 2005. With permission.)

Currently, up to four different species of sperm whale myoglobin have been investigated by time-resolved crystallography: the wild-type (Srajer et al. 1996, 2001;
Schotte et al. 2004), the L29W mutant (Schmidt et al., 2005a), the L29F mutant
(Schotte et al. 2003, 2004) and the YQR triple mutant (Bourgeois et al. 2003, 2006).
Table 1.1 shows the positions and timescales on which the CO was discovered in
their respective sites. The L29W mutant is unique, since it is the slowest rebinder of
all species. Other mutants like the L29F mutant are interesting, since the CO resides
a much shorter time interval in the primary docking B as compared to the wild-type.
Other cavities such as Xe4 and Xe2 are populated rapidly and the subsequent occu-
pation of these cavities can be followed (Schotte et al. 2003).

In the L29W mutant, the CO could be observed in the Xe1 site 300 ns after the
photoflash. No other binding site could be identified even on the fastest times around 1 ns.
Xe4 site was most likely dynamically occluded by a bulky tryptophan residue. Even on
very fast timescales, the CO stays in Xe1 for an exceptionally long time. Comparison
of relevant relaxation times derived from the time-resolved crystallographic data on the
L29W mutant and the wild-type made it possible to determine the pathway of the CO
out of the myoglobin. Figure 1.7 shows the relaxations observed in the L29W mutant
and marks relevant timescales observed in the wild-type. In the wild-type, depopulation of the distal pocket binding site B follows a stretched exponential with $\tau \sim 70$ ns.
At the same time, the proximal site Xe1 becomes populated. Xe1 depopulates biphasically on timescales of a few hundred nanoseconds and 100 μs. Rebinding of CO to the
iron occurs in concert with the depopulation of Xe1. However, in the L29W the CO
stays for 1.5 ms in the proximal Xe1 site. This time scale is much larger than those
observed in the wild-type. In addition, in the L29W mutant the CO molecules first

---

TABLE 1.1
Sites Where CO can be Observed in Photoflash Experiments in Different Species of Sperm Whale Myoglobin (References in the Text)

Species/Site	B	Xe4	Xe2	Xe1
Wild-type[a]	Up to 70 ns	—	—	30 ns to 100 μs
L29W[b]	—	—	—	300 ns to 1.5 ms
L29F[c]	Up to 1 ns	1 ns to 3 ns	3 ns to 30 ns	30 ns to >3 μs
YQR[d]	—	Up to 20 ns	—	20 ns to >3 μs

Notes: Relaxation times are available for the wild-type and L29W mutant. For the L29F and YQR Mb
mutants, approximate relaxation times were derived by visual inspection of the difference maps and
by inspecting the time course of integrated difference electron densities, respectively.

[a] Data from Srajer, V., Ren, Z., Teng, T. Y., Schmidt, M., Ursby, T., Bourgeois, D., Praderv and, C.,
Schildkamp, W., Wulff, M., and Moffat, K. 2001. *Biochemistry* 40:13802–15.

[b] Data from Schmidt, M., Nienhaus, K., Pahl, R., Krasselt, A., Nienhaus, U., Parak, F., and Srajer, V.
2005. *Proc. Natl. Acad. Sci. USA* 13:11704–9.

[c] Data from Schotte, F., Lim, M., Jackson, T. A., Smirnov, A. V., Soman, J., Olson, J. S., Phillips, G.
N. Jr., Wulff, M., and Anfinrud, P. A. 2003. *Science* 300:1944–47.

[d] Data from Bourgeois, D., Vallone, B., Arcovito, A., Sciara, G., Schotte, F., Anfinrud, P. A., and
Brunori, M. 2006. *Proc. Natl. Acad. Sci.* USA 103:4924–29.

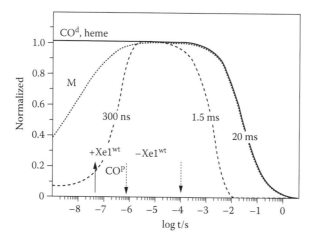

FIGURE 1.7 Rebinding kinetics in the L29W myoglobin. CO^d, heme: instantaneous displacement of bound CO and heme, final relaxation follows CO rebinding with $\tau = 20$ ms. M: protein moieties, initial stretched relaxation, final relaxation follows CO rebinding with $\tau = 20$ ms. CO^p: CO in Xe1 site. In L29W Mb Xe1 is populated with $\tau = 300$ ns and depopulated with $\tau = 1.5$ ms. In the wild-type Xe1 is populated with ~30 ns (solid arrow at $+Xe1^{wt}$) and depopulated biphasically on timescales of ~800 ns and ~100 μs (dashed arrows at $-Xe1^{wt}$). Note the very large time window accessible in the L29W Mb. (From *Proc. Natl. Acad. Sci. USA* 101:4799–4804. Used with permission.)

leave the Xe1 site, most likely to the solvent, before they rebind to the iron with $\tau \sim$ 20 ms. Hence, rebinding is two orders of magnitude slower than in the wild-type. The only structural difference between the myoglobin species is that a leucine at position 29 in the wild-type is replaced by a bulky tryptophan in the L29W mutant. Although Trp29 is on the distal side of the heme, it has a dramatic effect on the migration of CO. Obviously, Trp29 blocks the most important migration pathway. Hence in the L29W the CO has to find another way. As suggested by molecular dynamics simulations, several pathways are conceivable that take advantage of the network of cavities in myoglobin (Cohen et al. 2006). These pathways, however, must be much slower than the pathway on which CO leaves Xe1 in the wild-type. Hence, these pathways are not important for the protein's function. In the wild-type, CO migrates to the solvent through the distal side of the heme, which is not blocked there.

On timescales faster than a few nanoseconds, the relaxation kinetics of the protein structure is non-exponential (Lim et al. 1993; Jackson et al. 1994). This has also been observed with time-resolved crystallography for the YQR triple mutant (Bourgeois et al. 2003, 2006) and for the L29W mutant (Figure 1.7, protein moieties M) (Schmidt et al. 2005a). In this case, the picture of simple kinetics drawn by the two-state model in Figure 1.2 most likely fails. These fast timescales coincide with characteristic timescales of protein-specific motions that range from 10 ps to a few ns in myoglobin (Parak 2003). On these timescales, equilibration between all modes of motion (degrees of freedom) is not achieved and the relaxation of the ensemble is likely observed as diffusive motion along a rough energy surface (Hagen and Eaton 1996). However, at this time, structural models do not exist on ultrafast timescales. To develop structural and kinetic descriptions for the fastest reactions in proteins, heme proteins will remain one of the future prime targets for time-resolved methods.

IV. NONLINEAR TRANSIENT PHASE GRATING SPECTROSCOPY

A. EXPERIMENTAL

1. General Principles

Two pulsed coherent beams with parallel polarization intersect in the sample at an angle θ_{ex} to form an interference pattern in the material (Figure 1.8). The spatial variation of the intensity along the spatial coordinate x is given by

$$I(x) = \frac{I_m}{2}\left(1 + \cos\frac{2\pi x}{\Lambda}\right), \tag{1.18}$$

where I_m is the maximum intensity and Λ is the fringe spacing of the grating, given by (Nagy et al. 2005; Eichler et al. 1986; Nelson et al. 1982):

$$\Lambda = \frac{\lambda_{ex}}{2\sin(\theta_{ex}/2)}. \tag{1.19}$$

In this equation, $\Lambda = 2\pi/|k_{ex1} - k_{ex2}|$, and λ_{ex} is the wavelength of both excitation beams.

The bright fringes of the interference pattern modulate the material optical properties through photoinduced processes to form a diffraction grating in the sample. The probe pulse, incident on the sample at the Bragg angle for diffraction (Θ), diffracts off the grating with an efficiency given by the expression (Eichler et al. 1986; Kogelnik 1969):

$$\eta(\lambda, t) = \zeta(\lambda_p, t)[\Delta n(\lambda, t)^2 + \Delta\kappa(\lambda, t)^2], \tag{1.20}$$

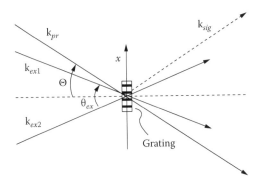

FIGURE 1.8 Phase matching diagram for transient grating experiments. Significance of the symbols: k_{ex1}, k_{ex2} = wave vectors of the excitation beams; k_{pr} = wave vector of the probe beam; k_{sig} = wave vector of the signal (diffracted) beam.

where $\Delta n(\lambda,t)$ and $\Delta\kappa(\lambda,t)$ are changes in the real and imaginary parts of the complex index of refraction ($n^* = n + i\kappa$) of the sample, respectively, and $\zeta(\lambda_p,t)$ is given by

$$\zeta(\lambda_p,t) = G(w,\theta_{ex})\left(\frac{\pi d}{\lambda_p \cos\Theta}\right)^2 \exp\left[\frac{-2.3\alpha(\lambda_p,t)}{\cos\Theta}\right]^2, \qquad (1.21)$$

with $G(w,\theta_{ex})$ being a constant factor that depends on the waist (w) and the incidence angle (θ_{ex}) of the Gaussian excitation beams, α the average density at the probe wavelength (λ_p), d the thickness of the grating in the sample, and Θ the Bragg angle of diffraction.

Since the beams are pulsed (usually, at nanosecond level), the grating forms only transiently, i.e., it dissipates within a short time after the passage of the two pulses through the sample. Therefore, the probe senses the ensuing transient dynamics in the sample following formation of the grating and before its dissipation.

As mentioned above, the diffracted signal reports on excitation-induced changes in both the real and the imaginary parts of the index of refraction. Changes in the imaginary part ($\Delta\kappa$) reflect changes in absorption by the sample, while changes in the real part are due to a superposition of mainly three effects, as given by

$$\Delta n = \Delta n_{protein} + \Delta n_{ex} + \Delta n_{th}, \qquad (1.22)$$

where $\Delta n_{protein}$ are changes due to the protein dynamics (which is the information sought after in studies of protein dynamics). The third term (Δn_{th}) reflects contributions to the signal arising from density changes due to heating of the sample; it consists of a fast oscillating component with a period of ~1 ns (for liquid samples) and an exponentially decaying part due to thermal diffusion. Both of these components can be identified and/or corrected for (Nagy et al. 2005; Walther et al. 2005). Finally, the second term (Δn_{ex}) and also $\Delta\kappa$ mentioned above are electronic contributions that arise from the absorption differences between the protein and its dissociated state (i.e., between MbCO and deoxy-Mb, in the case of myoglobin) (Ogilvie et al. 2002). These contributions can be dramatically reduced by using an off-resonant probe beam (Ogilvie et al. 2002; Deak et al. 1998). In this way, the transient grating method detects only changes in the real part of the complex index of refraction, which incorporate important information on the protein dynamics.

2. Diffractive Optics-Based Four-Wave Mixing with Heterodyne Detection

The already high sensitivity of the transient grating spectroscopy may be further improved by using optical heterodyne detection. In this method, a relatively weak signal field is mixed with a much more intense coherent field, called a *local oscillator* or *reference field* (such as the beam denoted by "ref1" in Figure 1.9). The experimental setup requires in this case two probe beams, crossing each other in the transient grating at twice the Bragg angle (Figure 1.9). The signal is represented by a small fraction of the probe intensity that is diffracted by the transient grating, such as the

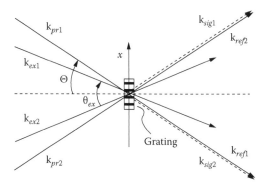

FIGURE 1.9 Schematic representation of the excitation, probe, signal, and reference beams used for heterodyne detection in transient phase grating experiments.

beam characterized by k_{sig1} in Figure 1.9, while the local oscillator is constituted by an intense beam, such as the beam characterized by k_{ref2} in the figure, which is the portion of probe 1 that passes undiffracted through the grating.

The mixing of the signal and reference beams results in a measured intensity that incorporates a strong constant contribution from the reference field $[E_{ref2}(t)]$ and a term that combines the strong reference with the weak signal field (Goodno et al. 1999), as expressed by

$$I_{OHD} = |E_{ref2}(t)|^2 + 2|E_{sig1}||E_{ref2}|\cos(\varphi_{sig1} - \varphi_{ref2}), \qquad (1.23)$$

where φ_{sig1} and φ_{ref2} are the phases of the signal and reference fields, respectively, and a term corresponding to the intensity of the signal field ($|E_{sig1}(t)|^2$) has been considered negligibly small. The mixed term retains the phase of E_{sig1} through modulation of the more intense reference field (E_{ref2}), and thereby allows detection of E_{sig1} with a higher signal-to-noise ratio (Levenson and Eesley 1979) compared to the method that measures the signal field directly.

As seen from Figure 1.9, exactly the same procedure may be used for detection of E_{sig2} from mixing the fields of k_{sig2} and k_{ref1}.

It can be shown that, under phase-matching conditions, I_{OHD} is related to Δn and $\Delta \kappa$ by the expression (Nagy et al. 2005):

$$I_{OHD} \cong 4n|E_{sig}||E_{ref}|(\Delta n \sin\phi - \Delta \kappa \cos\phi), \qquad (1.24)$$

in which the constant term $|E_{ref}(t)|^2$ has been ignored and ϕ has replaced $\varphi_{sig} - \varphi_{ref}$.

Rogers and Nelson outlined the use of a transmission grating for easy alignment of all the beams in transient grating experiments (Rogers and Nelson 1996). Critical refinements, introduced by the group of Miller (Miller 2002; Goodno et al. 1998, 1999) has led to the development of diffractive-optics-based transient grating experiments with heterodyne detection (or diffractive-optics-based four-wave mixing). Besides its inherent simplicity, the technique features very high sensitivity and stability.

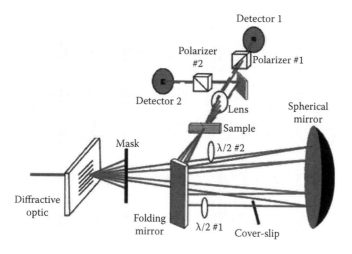

FIGURE 1.10 Schematic representation of the experimental setup for diffractive optics-based four-wave mixing with heterodyne detection. (From Ogilvie, J. P., Plazanet, M., Dadusc, G., and Miller, R. J. D. 2002. *J. Phys. Chem.* 109:10460–67. With permission)

In the typical setup (Figure 1.10), excitation light is provided by a pulsed (e.g., nanosecond) laser emitting at 532 nm, while the off-resonant probe is delivered by a cw laser (for instance an Nd:YVO laser operating at 1064 nm). The two coherent beams are overlapped using a dichroic mirror and then passed together through a surface relief diffractive optics. A spatial filter (i.e., mask) blocks all but the $\pm$ 1-order diffracted beams, which are focused onto the sample by a spherical mirror. The focused excitation beams (532 nm) form a transient grating in the sample, which diffracts the probe beams. The two diffracted beams and the residual probe beams that cross the sample are collimated by a lens and directed in pairs of one diffracted (i.e., signal) and one undiffracted (i.e., reference) beam to two separate fast photodiodes (with rise-time of ~1 ns), which are connected to an oscilloscope. The time-course of the heterodyne signals from each beam pair is recorded separately or differentially. The relative phase [ϕ in Equation (1.22)] between the diffracted and undiffracted beams is adjusted by tilting a cover slip in the path of one of the probe beams before entering the sample. This ensures easy separation of the real part of the signal (i.e., Δn) from the imaginary part ($\Delta \kappa$) (Ogilvie et al. 2002, Walther et al. 2005), and a further increase in the signal-to-noise ratio.

B. Overview of Transient Grating Results

1. MbCO in Aqueous Solutions

In aqueous solutions at room temperature, the bond-breaking event is usually followed by ligand escape from the protein. Rebinding requires that the ligand overcomes a series of barrier heights beginning with the first step of repenetrating the protein. The ligand motion through the protein requires that the protein as a whole undergoes a series of structural changes that facilitate motion of CO away from the heme following photoflash and then back until the final event of rebinding occurs (Ogilvie et al. 2001; Dadusc et al. 2001; Sakakura et al. 2001). As mentioned above, x-ray crystallography suggests that ligand escape is guided through a few specific internal cavities; these are presumably open and closed by fluctuations in the protein conformations. Ligand motion through the protein must therefore be accompanied by protein volume changes (ΔV) or development of material strain ($\Delta V/V$) due to protein deformation to accommodate the unbound ligand (Dadusc et al. 2001)

Phase grating spectroscopy provides exquisite sensitivity for monitoring volume changes in the protein, which are reflected by changes in the index of refraction [see Equations (1.21) and (1.22)]. Figure 1.11 shows the time changes in the real part of the refractive index obtained from transient phase grating measurements on MbCO following photo-induced bond breaking (Ogilvie et al. 2001; Dadusc et al. 2001).

Several features can be distinguished in Figure 1.11. First, a peak is observed within the first tens of nanoseconds (region I, in Figure 1.11), which has been ascribed to structural birefringence (Dadusc et al. 2001). The increase in the index of refraction suggests protein contraction. The barely perceptible plateau followed by the large decay (region II) that induces a sign change in the signal were assigned to CO motions inside the protein; region II may be divided into two subregions, corresponding to CO transition between two intermediate cavities inside the protein and between one of the cavities and the solvent, respectively (Dadusc et al. 2001). However, the time scale of phase II coincides with the time scale identified in time-resolved crystallographic experiments on various myoglobin species (Schmidt et al. 2005a; Bourgeois et al. 2003, 2006). Processes occurring at this time scale have been associated with structural relaxation from an energetically unfavorable deoxy species formed shortly after Fe-CO bond breaking toward the equilibrium of the deoxy-Mb species. Since the transient grating experiment is sensitive to structural

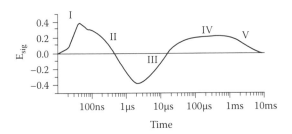

FIGURE 1.11 Changes in the real part of the refractive index following photodissociation of MbCO (in arbitrary units). Recall that the real part of the index of refraction probe's volume changes in the whole protein and is insensitive to the optical absorption of the heme. Various regions are marked in the figure by Roman numerals and the underlying physical mechanisms are identified in the text. The transient grating has been created in the sample by using a 20-ns laser pulse at 527 nm, and the volume changes have been probed by a cw beam at 1064 nm. (From Ogilvie, J. P., Armstrong, M. R., Plazanet, M., Dadusc, G., and Miller, R. J. D. 2001. *J. Luminesc.* 94–95:489–92. With permission.)

changes of the entire protein, it is conceivable that most of the signal of phase II in Figure 1.11 is due to this structural relaxation. Changes in region III are due to slower relaxation dynamics and are dominated by the thermal diffusion of the grating, while the further slower increase in the signal in region IV has been assigned to CO diffusion out of the grating (Dadusc et al. 2001). Finally, the bimolecular recombination (i.e., recombination of CO with a different Mb molecule) is responsible for the decrease in signal in region V toward zero, and the protein is ready for a new dissociation-recombination cycle induced by a new transient grating.

For further discussion of the above results, and for correlations with existing pump-probe data as well as with other techniques, the reader is referred to original papers (e.g., Ogilvie et al. 2001, 2002 Dadusc et al. 2001; Sakakura et al. 2001) as well as to a recent review by Nagy et al. (2005).

2. Non-Exponential Kinetics and Dynamics of MbCO in a Glass

Low-temperature measurements, especially of flash photolysis, have revealed very complicated kinetics, especially non-exponential dependence on time, of Mb-CO recombination when the temperature goes below the glass transition temperature of the surrounding medium (Austin et al. 1975; Srajer et al. 1988; Tian et al. 1992; Lim et al. 1993; Hagen and Eaton, 1996). This behavior deviates markedly from simple exponential forms observed at room temperature. A possible explanation has been proposed that below the glass-transition temperature proteins are "frozen" in various conformational substates, characterized by a broad distribution of enthalpic barrier heights (Austin et al. 1975; Frauenfelder et al. 1988; Steinbach et al. 1991), each height corresponding to an individual substate. Accordingly, this leads to the non-exponential time dependence of the rebinding kinetics observed experimentally. At room temperature, proteins are assumed to change their conformations on timescales much faster than binding occurs, so that these fluctuations are averaged out. An average reaction rate constant is obtained, which leads to the observed exponential behavior. We note that general agreement on the detailed interpretation of non-exponential kinetics has yet to be reached (see, e.g., Frauenfelder et al. 1988; Hagen and Eaton 1996; Parak 2003; Ye et al. 2007), for various views regarding this issue).

Much of the work on Mb-CO recombination following photodissociation has been devoted to the study of *geminate* recombination (i.e., rebinding to the same heme of the CO molecules trapped inside protein pockets), and in particular to whether rebinding kinetics couples to the outer protein motions. In their papers on the effect of viscosity on the rapid conformational changes in the protein following photodissociation, Ansari et al. suggested that protein relaxation may slow down the geminate recombination and increase barrier heights for ligand rebinding to Mb in high-viscosity solvents at room temperature (Ansari et al. 1992, 1994). Based on an extrapolation of their results to high viscosities, those authors suggested that a sufficiently viscous solvent should suppress protein relaxation as well as the interconversion of conformational substates. Similar to what happens at low temperatures in aqueous solvents, this viscosity effect should then lead to non-exponential rebinding kinetics. This prediction has been confirmed experimentally in studies of MbCO embedded in a room-temperature trehalose glass (Hagen et al. 1995, 1996), although interpretation of those results in terms of coupling between heme-protein motions has been questioned by other authors (Sastry and Agmon 1997).

In a recent paper by Walther et al. (2005), the ability of the trehalose glass to prevent large protein relaxations at room temperature has been used to determine the binding energy of the CO group to the heme of Mb. A binding energy of ~34 kcal/mol has been observed, which agrees with molecular dynamics simulations (Rovira and Parrinello 2000).

Analysis of pump-probe absorption data of Walther et al. (2005; not shown here) indicated two stretched exponentials characterizing the geminate recombination. In agreement with pump-probe results, their transient grating data (reproduced in Figure 1.12) have been fitted by an equation,

$$\text{Re}(I_{OHD}) = \left\{ A_{th} + A_{th1} \left[1 - e^{(k_1 t)^{\beta_1}} \right] + A_{th2} \left[1 - e^{(k_2 t)^{\beta_2}} \right] \right\} e^{-t/\tau_{th}}, \qquad (1.25)$$

which incorporates two stretched exponentials corresponding to two distinct geminate processes (subscripts "1" and "2") and a simple exponential (without subscript) accounting for the thermal grating decay. Very similar values for τ_{th} have been

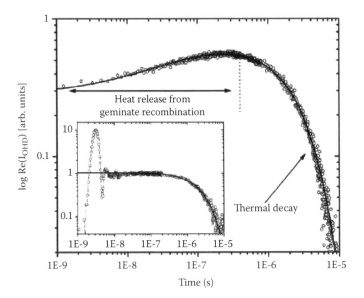

FIGURE 1.12 Measured diffraction grating signal (in arbitrary units) of MbCO embedded in a trehalose glass at room temperature (*open circles*) and theoretical simulation using equation (25) (*solid line*). The fitting parameters corresponding to the best-fit curve are $A_{th} = 0.27 \pm 0.01$, $A_{th1} = 0.21 \pm 0.01$, $A_{th2} = 0.25 \pm 0.01$, $k_1 = (2.29 \pm 0.13) \times 10^7$ s^{-1}, $k_2 = (2.80 \pm 0.40) \times 10^5$ s^{-1}, $\beta_1 = 0.67 \pm 0.04$, and $\beta_2 = 0.40 \pm 0.02$. The time constant [$\tau_{th} = (2.40 \pm 0.02)$ µs] of the thermal decay part of the curve has been calibrated from diffraction grating measurements on malachite green (data shown in the inset). (From Walther, M., Raicu, V., Ogilvie, J. P., Phillips, R., Kluger, R., and Miller, R. J. D. 2005. *J. Phys. Chem. B* 109:20605–11. Used with permission.)

obtained both from fitting of the data with Equation (1.25) and from measurements on malachite green, which presents a pure thermal grating decay and thereby serves as a good reference for the analysis of phase grating data from MbCO. The transient-phase grating as well as the pump-probe absorption data of Walter et al. will be reanalyzed in the next subsection to extract physical information about the protein undergoing ligand photodissociation and geminate recombination.

C. REANALYSIS OF DATA FROM TREHALOSE-EMBEDDED MBCO

1. Extraction of the Distribution of Relaxation Functions from Experimental Data

One of the widely used methods of analysis of kinetic data is based on extraction of the distribution of relaxation times or, equivalently, enthalpic barrier heights. In this section, we show that this may be done easily by using the distribution function introduced by Raicu (1999; see Equation [1.16] above). To this end, we use the data reported by Walther and coworkers (Walther et al. 2005) from pump-probe as well as the transient phase grating measurements on trehalose-embedded MbCO. Their pump-probe data have been used without modification herein, while the phase grating data (also reproduced in Figure 1.12) have been corrected for thermal diffusion of the grating using the relaxation time reported above, τ_{th}, and Equation (1.25).

Equation (1.15) has been integrated numerically on a logarithmic scale to determine the dummy variable τ_{AB}. The logarithmic scale weighs more uniformly the relaxation times, which may cover several orders of magnitude. A change of variables, $y = \ln(k_p / k_{AB}) = \ln(\tau_{AB} / \tau_p)$, transformed the distribution function given by Equation (1.16) into the following function:

$$g(y) = \frac{1}{2\pi} \frac{e^{y\alpha}\sin[\pi(1-\beta)] + e^{y(1-\beta)}\sin[\pi\alpha]}{\cosh[(1-\alpha-\beta)y] + \cos[\pi(1-\alpha-\beta)]}. \tag{1.26}$$

The expression for the concentration profile given by Equation (1.15) has been accordingly replaced by

$$A(t) = \frac{\int_{y_{min}}^{y_{max}} g(y)e^{-\frac{t}{e^{y}\tau_p}}\,dy}{\int_{y_{min}}^{y_{max}} g(y)\,dy}. \tag{1.27}$$

where $y_{min} = \ln(\tau_{min} / \tau_p)$ and $y_{max} = \ln(\tau_{max} / \tau_p)$.

Equations (1.26) and (1.27) were used to fit the pump-probe and transient grating data from MbCO embedded in trehalose glass at room temperature, by assuming a symmetric cutoff for the relaxation times at low and high values (i.e., assuming that $y_{min} = -y_{max}$). This reduced the total number of fitting parameters to the following four: α, β, τ_p, and τ_{max} (or τ_{min}). The goodness-of-fit was first evaluated by visually

TABLE 1.2
Best-Fit Parameter Values Corresponding to the Solid Lines in Figure 1.13

Measurement Method	τ_p (s)	α	β	τ_{max} (s)
Pump-probe	3.5×10^{-8}	0.08	0.14	8.54×10^{-5}
Transient grating	6.0×10^{-8}	0.06	0.15	6.53×10^{-2}

comparing the theoretical line to the experimental data, and then refining the parameter values to minimize a fitting residual given by the following expression:

$$\text{Residual} = \sum_i \frac{(A_{\text{theo},i} - A_{\text{exp},i})^2}{(A_{\text{exp},i})^2},$$

where i is a summation index ranging from 1 to the total number of time values. Figure 1.13 shows comparatively the experimental data and the best-fit curve, using the parameter values collected in Table 1.2.

In Figure 1.13 two recombination steps are clearly distinguishable, especially in the pump-probe data. These have been fitted previously by two separate stretched exponentials. Here, we used a single distribution of relaxation times, which accounted for both recombination steps.

Once the best-fit parameters are obtained, the distribution of relaxation times or, equivalently, barrier heights may be easily computed from Equation (1.16). The distribution function corresponding to the data in Figure 1.13 is plotted for the two types of measurements in Figure 1.14.

It is apparent from the above analysis that the most probable relaxation time, τ_p, for the transient phase grating is larger than the one for the pump-probe. While pump-probe absorption measurements report on changes in optical absorption accompanying ligand rebinding to the heme, the transient grating probes the structural changes in the entire protein. The longer relaxation time derived from the transient grating experiment therefore implies that the structural changes in the whole protein lag behind the formation of the iron–CO bond.

However, little more can be said about these differences without a detailed physical model at hand. Some physical significance of the data may be extracted by using a simple model proposed by Srajer, et al. (1988), as discussed below.

2. Determination of Certain Physical Parameters from Pump-Probe and Transient Grating Measurements

Srajer, Reinisch, and Champion (SRC) have proposed a model (Srajer et al. 1988; Srajer and Champion 1991), which separates the enthalpic barrier for ligand binding to the protein into the following two terms:

$$H = H_p + H_D = \frac{1}{2} Ka^2 + H_D, \tag{1.28}$$

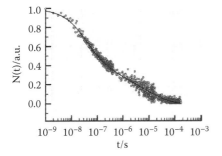
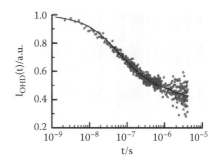

FIGURE 1.13 Experimental results (*points*) obtained from pump-probe (*left*) and transient phase grating (*right*) measurements of MbCO recombination as a function of time, *t*, following photodissociation. Solid lines were computed from equations (26) and (27), and the best-fit parameter values are listed in Table 1.2. (From Walther, M., Raicu, V., Ogilvie, J. P., Phillips, R., Kluger, R., and Miller, R. J. D. 2005. *J. Phys. Chem. B* 109:20605–11. With permission.)

where H_D takes into account the enthalpy barrier of the distal portions of the protein, and H_p describes the contributions to the enthalpic barrier from the heme distortions, which is necessary to bring the iron porphyrin system into the planar transition state configuration. The forces responsible for the displacement of the iron, *a*, to the in-plane geometry are modeled as a linear spring system, with *K* serving as an effective force constant (Srajer et al. 1988; Ye et al. 2007). In this model, the distribution of rate constants, or relaxation times, arises from the fact that there is a distribution of heme displacements, verified by x-ray studies. This distribution is assumed to be Gaussian in nature with a standard deviation σ_a around an equilibrium position a_0. Each enthalpic barrier of the distribution is associated with a different binding rate constant. The

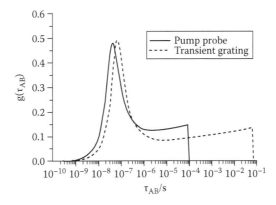

FIGURE 1.14 Distribution of relaxation times for the recombination of photodissociated Mb-CO.

displacement is related to the enthalpic barrier, as given by Equation (1.28), and the distribution of barrier heights can be written in the following form:

$$g(H) = \frac{1}{2\sigma_a [\pi K(H - H_D)]^{0.5}} \left[e^{-\frac{\left(\sqrt{H-H_D} - a_0 \sqrt{\frac{K}{2}} \right)^2}{K\sigma_a^2}} + e^{-\frac{\left(\sqrt{H-H_D} + a_0 \sqrt{\frac{K}{2}} \right)^2}{K\sigma_a^2}} \right], \quad (1.29)$$

where a_0 and σ_a^2 are the average and standard deviation of the assumed Gaussian distribution of the heme displacements. Similar to Equation (1.10), the concentration profile of bound ligands over time can be described by

$$A(t) = \int_{H_D}^{\infty} g(H) e^{\Omega t e^{\frac{-H}{k_B T}}} \, dH, \quad (1.30)$$

where it is assumed that $g(H)$ is zero when $H < H_D$.

For fitting purposes, a transformation of variables (Ye et al. 2007),

$$B = \sqrt{\frac{k_B T}{K\sigma_a^2}}, \ C = \frac{a_0}{\sqrt{2}\sigma_a}, \ t_0 = \frac{e^{\frac{H_D}{k_B T}}}{\Omega}, \text{ and } x = \frac{H - H_D}{k_B T}, \quad (1.31)$$

reduces the number of fitting parameters to just three. The concentration profile equation then becomes

$$A(t) = \int_0^{\infty} \frac{B}{2\sqrt{\pi x}} \left[e^{-\left(B\sqrt{x} - C \right)^2} + e^{-\left(B\sqrt{x} + C \right)^2} \right] e^{\frac{-t}{t_0} e^{-x}} \, dx. \quad (1.32)$$

We used this equation to fit the data obtained from the pump-probe and transient phase grating studies for MbCO in a trehalose glass discussed above. Plots of the experimental and simulated data are shown in Figure 1.15, while the parameters used to obtain the best-fit curve in the figure are listed in Table 1.3. Notice that, while the simulations based on the SRC model follow the general features of the curves in Figure 1.15, specific details (in particular the two steps, which were properly accounted for by Raicu's distribution function; see Figure 1.13) are not properly described by the SRC model. The precise reason for this is as yet unknown. The simplest, although admittedly the least spectacular, explanation would be the trapping of CO in different Xe pockets, wherefrom it recombines with the heme in two steps.

Equations (1.31) were solved to give

$$\sigma_a = \left(\frac{k_B T}{KB^2} \right)^{1/2}, \quad (1.33a)$$

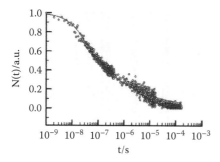

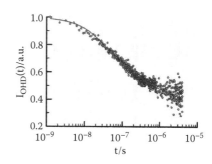

FIGURE 1.15 Reanalysis of the transient-grating and pump-probe data (*points*) of Walther et al. using the SRC model. Solid lines are theoretical simulations using equation (32) and the best-fit parameters given in Table 1.3. (From Walther, M., Raicu, V., Ogilvie, J. P., Phillips, R., Kluger, R., and Miller, R. J. D. 2005. *J. Phys. Chem. B.* 109:20605–11. With permission.)

$$a_0 = \left(\frac{2k_B T}{K} \right)^{1/2} \frac{C}{B},\qquad (1.33b)$$

$$\Omega = \frac{1}{t_0} \exp\left(\frac{H_D}{k_B T} \right),\qquad (1.33c)$$

while the average $<H_p>$ is determined from its definition and using Equation (1.33b):

$$<H_p> = \frac{1}{2} K a_0{}^2 = k_B T \left(\frac{C}{B} \right)^2 .\qquad (1.33d)$$

To determine the parameters a_0, σ_a, and H_p, we used the values for fitting parameters B and C (see above), the temperature $T = 293\ K$ and a spring constant, K, of 17 N/m (Srajer and Champion 1991). Further, by assuming a value of 18 kJ/mol for H_p + H_D (Ye et al. 2007; Srajer et al. 1988), and using Equation (1.33c), the value of Ω was determined. All those values are listed in Table 1.4 along with results reported in previous publications, as follows: The first two rows of Table 1.4 list the results obtained from pump-probe and grating data obtained from MbCO embedded in a glass at room temperature (i.e., data from Figure 1.15). The third row lists the

TABLE 1.3

Best-Fit Parameters Obtained from Simulations of the Data in Figure 1.15 with the SRC Model [Equation (1.32)]

Sample	Method	t_0 (s)	B	C
MbCO	Pump-probe	1.1×10^{-8}	0.87	1.50
MbCO	Phase grating	2.6×10^{-8}	0.35	0.62

TABLE 1.4
Parameters Extracted from Fitting SRC Model to Data Obtained from MbCO as well as from the Protoheme Alone Embedded in Various Solvents

Sample	Method	Medium	T (K)	a_0 (Å)	σ_a (Å)	H_p (kJ/mol)	Ω (s^{-1})
MbCO[a]	Pump-probe	Trehalose	293	0.38	0.18	7.2	7.7×10^9
MbCO[a]	Phase grating	Trehalose	293	0.39	0.44	7.6	2.7×10^9
MbCO[b]	Flash photolysis	Glycerol	160	0.20	0.10	2.0	2.8×10^9
H_2O-FePPIXCO[c]	Pump-probe	Glycerol	293	0.27	0.11	6.0	1.5×10^{11}
H_2O-FePPIXCO[c]	Pump-probe	Glycerol	150	0.12	0.06	1.0	1.5×10^{11}

Note: The significance of the parameters and the methods used for their computation are described in the text.

[a] Data from Walther, M., Raicu, V., Ogilvie, J. P., Phillips, R., Kluger, R., and Miller, R. J. D. 2005. *J. Phys. Chem. B* 109:20605–11, analyzed as described above.

[b] Results from Srajer, V., and Champion, P. M. 1991. *Biochemistry* 30:7390–7402.

[c] Results from Ye, X., Ionascu, D., Gruia, F., Yu, A., Benabbas, A., and Champion, P. M. 2007. *Proc. Nat. Acad. Sci. USA* 104:14682–87.

parameters obtained from MbCO recombination studies performed in mixtures of glycerol and water at low temperatures (Srajer and Champion 1991). Finally, the last two rows correspond to studies of CO recombination to the bare protoheme (H_2O-FePPIXCO), i.e., in the absence of the protein (Ye et al. 2007).

It is interesting to note that the values of a_0 for MbCO in a room-temperature glass (trehalose) and in water-glycerol solvent at room temperatures are similar. The a_0 values obtained from the room-temperature studies agree well with those obtained from standard and time-resolved crystallographic studies comparing CO and deoxy structures of wild-type and mutant myoglobin (Vojtechovski et al. 1999; Kachalova et al. 1999; Nienhaus et al. 2005; Schmidt et al. 2005a), which range from 0.29 Å to 0.36 Å. Photoflash experiments on crystals at low temperatures show lower Fe displacements on the order of 0.1 Å to 0.2 Å (Schlichting et al. 1994; Teng et al. 1994, 1997; Hartmann et al. 1996). Similar values are also listed for the low temperature measurements in Table 1.4.

We also note that the values obtained for the prefactor Ω are the same for room-temperature and low-temperature glass; also, in both glasses, the values of Ω are about two orders of magnitude lower than in the case of bare protoheme. This latter difference was interpreted by Ye et al. (2007) as entropic control in the protein. This idea may be rationalized upon glancing at Equation (1.1), which predicts that, if not used explicitly in data analysis, a possible entropic contribution to the change in Gibbs free energy would be collected in the prefactor Ω.

Finally, the values of the proximal enthalpic barrier, H_p, are the same for trehalose-embedded protein and for the heme alone at room temperature, while at low temperatures both the Mb and the protoheme alone present similar H_p values. The equality between the proximal enthalpic barriers for the protein in liquid solvent and in a room-temperature glass may imply the existence of water in the trehalose-embedded MbCO, which may preserve the mobility of the protein (Librizzi et al. 2002).

ACKNOWLEDGMENTS

We thank Vukica Srajer for reading and commenting on an earlier version of the manuscript. V. R. and M. St. are supported by a UWM-RGI grant (X066), and M. Sch. is supported by the Deutsche Forschungsgemeinschaft (grant SCHM 1423/2-1).

REFERENCES

Amoros, J. L., Buerger, M. J., Canut de Amoros, M. 1975. *The Laue method*. New York: Academic Press.

Ansari, A., Jones, C. M., Henri, E. R., Hofrichter, J., and Eaton, W. A. 1992. The role of solvent viscosity in the dynamics of protein conformational changes. *Science* 256:1796–98.

Ansari, A., Jones, C. M., Henri, E. R., Hofrichter, J., and Eaton, W. A. 1994. Conformational relaxation and ligand binding in myoglobin. *Biochemistry* 33:5128–45.

Armstrong, M. R., Ogilvie, J. P., Cowan, M. L., Nagy, A. M., and Miller, R. J. D. 2003. Observation of the cascaded atomic-to-global length scales driving protein motion. *Proc. Nat. Acad. Sci. USA* 100:4990–4994.

Austin, R. H., Beeson, K. W., Eisenstein, L., Frauenfelder, H., and Gunsalus, I. C. 1975. Dynamics of ligand binding to myoglobin. *Biochemistry* 14:5355–73.

Bartunik, H. D., Bartsch, H. H., and Qichen, H. 1992. Accuracy in Laue x-ray diffraction analysis of protein structures. *Acta Crystallogr. A* 48:180–88.

Berman, H. M., Westbrook, J., Feng, Z., Gilliland, G., Bhat, T. N., Weissig, H., Shindyalov, I. N., and Bourne, P. E. 2000. The protein data bank. *Nucleic Acid Res.* 28:235–42.

Bourgeois, D., Vallone, B., Arcovito, A., Sciara, G., Schotte, F., Anfinrud, P. A., and Brunori, M. 2006. Extended subnanosecond structural dynamics of myoglobin revealed by Laue crystallography. *Proc. Natl. Acad. Sci. USA* 103:4924–29.

Bourgeois, D., Vallone, B., Schotte, F., Arcovito, A., Miele, A. E., Sciara, G., Wulff, M., Anfinrud, P., and Brunori, M. 2003. Complex landscape of protein structural dynamics unveiled by nanosecond Laue crystallography. *Proc. Natl. Acad. Sci. USA* 100:8704–9.

Carson, M. 1997. Ribbons. *Methods Enzymol.* 277: 493–505.

Cohen, J., Arkhipov, A., Braun, R., and Schulten, K. 2006. Imaging the migration pathways for O_2, CO, NO, and Xe inside myoglobin. *Biophys. J.* 91:1844–57.

Cornish-Bowden, A. 1999. *Fundamentals of enzyme kinetics*, Revised Edition. London: Portland Press.

Dadusc, G., Ogilvie, J. P., Schulenberg, P., Marvet, U., and Miller, R. J. D. 2001. Diffractive optics-based heterodyne-detected four-wave mixing signals of protein motion: From "protein quakes" to ligand escape for myoglobin. *Proc. Nat. Acad. Sci. USA* 98:6110–6115.

Deak, J., Chiu, H. L., Lewis, C. M., and Miller, R. J. D. 1998. Ultrafast phase grating studies of heme proteins: Observation of the low-frequence modes directing functionally important protein motions, *J. Phys. Chem.* 102:6621–34.

Eichler, H. J., Gunter, P., and Pohl, D. W. 1986. *Laser-induced dynamic gratings*. Berlin: Springer-Verlag.

Eyring, H. 1935. The activated complex in chemical reactions. *J. Chem. Phys.* 3:107–15.

Fleck, G. M. 1971. *Chemical reaction mechanism*. New York: Holt, Rinehart and Winston.

Franzen, S., Bohn, B., Poyart, C., and Martin, J. L. 1995. Evidence for sub-picosecond heme doming in hemoglobin and myoglobin: A time-resolved resonance Raman comparison of carbonmonoxy and deoxy species. *Biochemistry* 34:1224–37.

Frauenfelder, H., Parak, F., and Young, R. D. 1988. Conformational substates in proteins. *Annu. Rev. Biophys. Biophys. Chem.* 17:451–79.

Goodno, G. D., Astinov, V., and Miller, R. J. D. 1999. Diffractive optics-based heterodyne-detected grating spectroscopy: Application to ultrafast protein dynamics. *J. Phys. Chem. A* 103:10619.

Goodno, G. D., Dadusc, G., and Miller, R. J. D. 1998. Ultrafast heterodyne-detected transient-grating spectroscopy using diffractive optics. *J. Opt. Soc. Am. B* 15:1791–94.

Hagen, S. J., and Eaton, W. A. 1996. Nonexponential structural relaxations in proteins. *J. Chem. Phys. B* 103:603–607.

Hagen, S. J., Hofrichter, W. A., and Eaton, W. A. 1995. Protein reaction-kinetics in a room-temperature glass. *Science* 269:959–62.

Hagen, S. J., Hofrichter, W. A., and Eaton, W. A. 1996. Geminate rebinding and conformational dynamics of myoglobin embedded in a glass at room temperature. *J. Phys. Chem.* 100:12008–21.

Hartmann, H., Zinser, S., Komninos, P., Schneider, R. T., Nienhaus, G. U., and Parak, F. 1996. X-ray structure determination of a metastable state of carbonmonoxy myoglobin after photodissociation. *Proc. Natl. Acad. Sci. USA* 93:7013–16.

Henderson, R., and Moffat, J. K. 1971. The difference Fourier technique in protein crystallography and their treatment. *Acta Crystallogr. B* 27:1414–20.

Henry, E. R., and Hofrichter, J. 1992. Singular value decomposition: Application to analysis of experimental data. *Meth. Enzymol.* 210:129–92.

Ihee, H., Rajagopal, S., Srajer, V., Pahl, R., Schmidt, M., Schotte, F., Anfinrud, P. A., Wulff, M., and Moffat, K. 2005. Visualizing chromophore isomerization in photoactive yellow protein from nanoseconds to seconds by time-resolved crystallography. *Proc. Natl. Acad. Sci., USA* 102:7145–50.

Jackson, T. A., Lim, M., and Anfinrud, P. A. 1994. Complex nonexponential relaxation in myoglobin after photodissociation of MbCO: Measurement and analysis from 2 ps to 56 ms. *Chemical Physics* 180:131–40.

Kachalova, G. S., Popov, A. N., and Bartunik, H. D. 1999. A steric mechanism for inhibition of CO binding to heme proteins. *Science* 284:473–76.

Kendrew, J. C., Bodo, G., Dintzis, H. M., Parrish, R. G., Wyckoff, H., and Phillips, D. C. 1958. A three-dimensional model of the myoglobin molecule obtained by X-ray analysis. *Nature* 181:662–66.

Kogelnik, H. 1969. Coupled wave theory for thick hologram gratings. *Bell Syst. Tech. J.* 48:2909–47.

Levenson, M. D., and Eesley, G. L. 1979. Polarization selective optical heterodyne detection for dramatically improved sensitivity in laser spectroscopy. *Appl. Phys.* 19:1–17.

Librizzi, F., Viapianni, C., Abbruzzetti, S., and Cordone, L. 2002. Residual water modulates the dynamics of the protein and of the external matrix in "trehalose-coated" MbCO: An infrared and flash-photolysis study. *J. Chem. Phys.* 116:1193–1200.

Lim, M., Jackson, T. A., and Anfinrud, P. A. 1993. Nonexponential protein relaxation: Dynamics of conformational change in myoglobin. *Proc. Natl. Acad. Sci. USA* 90:5801–4.

Matsen, F. A., and Franklin, J. L. 1950. A general theory of coupled sets of first order reactions. *J. Am. Chem. Soc.* 72:3337–41.

Miller, R. J. D. 2002. John C. Polany Award Lecture. Mother Nature and the molecular Big Bang. *Can. J. Chem.* 80:1–24.

Moffat, K. 1989. Time resolved macromolecular crystallography. *Annu. Rev. Biophys. Biophys. Chem.* 18:309–23.

Nagy, A., Raicu, V., and Miller, R. J. D. 2005. Nonlinear optical studies of heme protein dynamics: Implications for proteins as hybrid states of matter. *Biochim. Biophys. Acta* 1749:148–72.

Nelson, K. A., Miller, R. J. D., Lutz, D. R., and Fayer, M. D. 1982. Optical generation of tunable ultrasonic waves. *J. Appl. Phys.* 53:1144–49.

Nienhaus, K., Ostermann, A., Nienhaus, U., Parak, F., and Schmidt M. 2005. Ligand migration and protein fluctuations in myoglobin mutant L29W. *Biochemistry* 44:5095–5105.

Norrish, R. G. W. 1965. The kinetics and analysis of very fast chemical reactions. *Science* 149:1470–82.

Ogilvie, J. P., Armstrong, M. R., Plazanet, M., Dadusc, G., and Miller, R. J. D. 2001. Myoglobin dynamics: Evidence for a hybrid solid/fluid state of matter. *J. Luminesc.* 94-95:489–92.

Ogilvie, J. P., Plazanet, M., Dadusc, G., and Miller, R. J. D. 2002. Dynamics of ligand escape in myoglobin: Q-band transient absorption and four-wave mixing studies. *J. Phys. Chem.* 109:10460–67.

Ostermann, A., Waschipky, R., Parak, F. G., Nienhaus, G. U. 2000. Ligand binding and conformational motions in myoglobin. *Nature* 404:205–8.

Parak, F. G. 2003. Proteins in action: The physics of structural fluctuations and conformational changes. *Curr. Opin. Struct. Biol.* 13:552–57.

Parak, F., Knapp, E. W., and Kucheida, D. 1982. Protein dynamics. Mössbauer spectroscopy on deoxymyoglobin crystals. *J. Mol. Biol.* 161:177–94.

Raicu, V. 1999. Dielectric dispersion of biological matter: Model combining Debye-type and "universal" responses. *Phys. Rev. E* 60:4667–80.

Raicu, V., Sato, T., and Raicu, G. 2001. Non-Debye dielectric relaxation in biological structures arises from their fractal nature. *Phys. Rev. E* 64:021916.

Rajagopal, S., Anderson, S., Srajer, V., Schmidt, M., Pahl, R., and Moffat K. 2005. A structural pathway for signaling in the E46Q mutant of photoactive yellow protein. *Structure* 13:55–63.

Rajagopal, S., Schmidt, M., Anderson, S., and Moffat, K. 2004. Methodology for analysis of experimental time-resolved crystallographic data by singular value decomposition. *Acta Cryst.* D 60:860–71.

Ren, Z., Bourgeois, D., Helliwell, J. R., Moffat, K., Srajer, V., and Stoddard, B. L. 1999. Laue crystallography: Coming of age. *J. Synchrotron Rad.* 6:891–917.

Rogers, J. A., and Nelson, K. A. 1996. A new photoacoustic/photothermal device for real-time materials evaluation: An automated means for performing transient grating experiments. *Physica B* 56264:219–220.

Rovira, C., and Parrinello, M. 2000. First-principles molecular dynamics simulations of models of the myoglobin active center. *Int. J. Quantum Chem.* 80:1172–80.

Sakakura, M., Yamguchi, S., Horota, N., and Terazima, M. 2001. Dynamics of structure and energy of horse carboxymyoglobin after photodissociation of carbon monoxide. *J. Am. Chem. Soc.* 123:4286–94.

Sastry, G. M., and Agmon, N. 1997. Trehalose prevents myoglobin collapse and preserves its internal mobility. *Biochemistry* 36:7097–7108.

Schlichting, I., Berendzen, J., Phillips, G. N., and Sweet, R. M. 1994. Crystal structure of photolysed carbonmonoxy-myoglobin. *Nature* 371:808–12.

Schmidt, M. 2008. Structure based kinetics by time-resolved crystallography. In *Ultrashort laser pulses in biology and medicine*. M. Braun, P. Gilch, W. Zinth, Eds. New York: Springer Heidelberg.

Schmidt, M., Nienhaus, K., Pahl, R., Krasselt, A., Nienhaus, U., Parak, F., and Srajer, V. 2005a. Kinetic analysis of protein structural relaxations—A time-resolved crystallographic study. *Proc. Natl. Acad. Sci. USA* 13:11704–9.

Schmidt, M., Pahl, R., Ihee, H., and Srajer, V. 2005b. Protein ligand interactions probed by time-resolved X-ray structure determination. In *Methods in molecular biology, vol. 305: Protein-ligand interactions: Methods and Applications*. G. U. Nienhaus, Ed. Totowa, NJ: Humana Press.

Schmidt, M., Pahl, R., Srajer, V., Anderson, S., Ren, Z., Ihee, H., Rajagopal, S., and Moffat, K. 2004. Protein kinetics: Structures of intermediates and reaction mechanism from time-resolved X-ray data. *Proc. Natl. Acad. Sci. USA* 101:4799–4804.

Schmidt, M., Rajagopal, S., Ren, Z., and Moffat, K. 2003. Application of singular value decomposition to the analysis of time-resolved macromolecular x-ray data. *Biophys. J.* 84:2112–29.

Schotte, F., Lim, M., Jackson, T. A., Smirnov, A. V., Soman, J., Olson, J. S., Phillips, G. N. Jr., Wulff, M., and Anfinrud, P. A. 2003. Watching a protein as it functions with 150-ps time-resolved X-ray crystallography. *Science* 300:1944–47.

Schotte, F., Soman, J., Olson, J. S., Wulff, M., and Anfinrud, P. A. 2004. Picosecond time-resolved X-ray crystallography: Protein function in real time. *J. Struct. Biol.* 147:235–46.

Srajer, V., and Champion, P. M. 1991. Investigations of optical-line shapes and kinetic hole burning in myoglobin. *Biochemistry* 30:7390–7402.

Srajer, V., Reinisch, L., and Champion, P. M. 1988. Protein fluctuations, distributed coupling, and the binding of ligands to heme-proteins. *J. Am. Chem. Soc.* 110:6656–70.

Srajer, V., Ren, Z., Teng, T. Y., Schmidt, M., Ursby, T., Bourgeois, D., Pradervand, C., Schildkamp, W., Wulff, M., and Moffat, K. 2001. Protein conformational relaxation and ligand migration in myoglobin: Nanosecond to millisecond molecular movie from time-resolved Laue X-ray diffraction. *Biochemistry* 40:13802–15.

Srajer, V., Teng, T. Y., Ursby, T., Pradervand, C., Ren, Z., Adachi, S., Schildkamp, W., Bourgeois, D., Wulff, M., and Moffatt, K. 1996. Photolysis of the carbon monoxide complex of myoglobin: Nanosecond time-resolved crystallography. *Science* 274:1726–29.

Steinbach, P. J., Ansari, A., Berendzen, J., Braunstein, D., Chu K., Cowen, B. R. et al. 1991. Ligand binding to heme proteins: Connection between dynamics and function. *Biochemistry* 30:3988–4001.

Steinfeld, J. I., Francisco, J. S., and Hase, W. L. 1989. *Chemical kinetics and dynamics*. Englewood Cliffs, NJ: Prentice Hall.

Szebenyi, D. M. E., Bilderback, D. H., LeGrand, A., et al. 1992. Quantitative analysis of Laue diffraction pattern recorded with a 120 ps exposure from an X-ray undulator. *J. Appl. Cryst.* 25:414–23.

Teng, T. Y., Srajer, V., and Moffat, K. 1994. Photolysis-induced structural changes in single crystals of carbonmonoxy myoglobin at 40 K. *Struct. Biol.* 1:701–5.

Teng, T. Y., Srajer, V., and Moffat, K. 1997. Initial trajectory of carbon monoxide after photodissociation from myoglobin at cryogenic temperatures. *Biochemistry* 36:12087–12100.

Tian, W. D., Sage, J. T., Srajer, V., and Champion, P. M. 1992. Relaxation dynamics of myoglobin in solution. *Phys. Rev. Lett.* 68:408–11.

Tilton, R. F. Jr., Kuntz, I. D. Jr., and Petsko, G. A. 1984. Cavities in proteins: Structure of a metmyoglobin-xenon complex solved to 1.9 A. *Biochemistry* 23:2849–57.

Vojtechovsky, J., Chu, K., Berendzen, J., Sweet, R. M., and Schlichting, I. 1999. Crystal structures of myoglobin-ligand complexes at near-atomic resolution. *Biophys. J.* 77:2153–74.

Walther, M., Raicu, V., Ogilvie, J. P., Phillips, R., Kluger, R., and Miller, R. J. D. 2005. Determination of the Fe-CO bond energy in myoglobin using heterodyne-detected transient thermal phase grating spectroscopy. *J. Phys. Chem. B* 109:20605–11.

Ye, X., Ionascu, D., Gruia, F., Yu, A., Benabbas, A., and Champion, P. M. 2007. Temperature-dependent heme kinetics with nonexponential binding and barrier relaxation in the absence of protein conformational substates. *Proc. Nat. Acad. Sci. USA* 104:14682–87.

2 Using Two-Photon Fluorescence Microscopy to Study Chemical Phenomena in the Skin

Kerry M. Hanson and Christopher J. Bardeen

CONTENTS

I. INTRODUCTION

An outer layer of skin, roughly 1 mm thick, provides the only barrier between the body's soft tissues and the external world. It is the largest single organ of the body. From a medical standpoint, diseases of the skin, including cancer, psoriasis, acne, and eczema, cost the United States an estimated $38.6 billion in medical costs every year (Group 2004). In addition to disease, attempts to improve skin quality and counteract aging processes provide the foundation of the multi-billion-dollar cosmeceutical industry. Thus, there are strong medical, social, and commercial motivations for improving our understanding of the structure and function of skin.

Human skin is divided into two regions: the epidermis and dermis (Figure 2.1). The epidermis is composed of four stratified layers (in ascending order: strata basale,

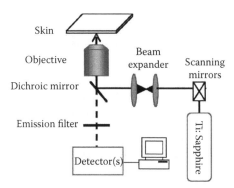

FIGURE 2.1 A diagram of a multi-photon microscope. For AF, SHG and exogenous probe fluorescence only one laser is used. The fluorescence or SHG signal passes through a dichroic mirror to the detector(s). One or two detectors and the appropriate filters can be used to collect multiple emission signals simultaneously. A spectrometer can be placed in the detector path to collect spectra, or a polarizer can be used to collect polarization data.

spinosum, granulosum, corneum) and typically ranges between 50 and 1500 µm thick depending upon body site (MacKenzie 1969). The stratum corneum, the outermost epidermal layer, is composed of anucleated keratinocytes and is a biochemically complicated region, having an active role in immunoregulation and barrier homeostasis. Its main functions are to limit physical trauma and penetration of topical agents in addition to inhibiting extrusion of bodily fluids. The three nucleated layers below the stratum corneum host Langerhans cells that are responsible for regulating immune response in the skin. Melanocytes are also present in the epidermis, residing primarily in the basale layer and extending upward into the upper epidermal layers, and give rise to the degree of skin pigmentation. The dermis is primarily composed of the structural proteins collagen and elastin and typically ranges between 100 and 500 µm thick.

Until recently, it has been difficult to probe the structural and chemical properties of the skin using photons (i.e., optical microscopy) because skin is a strongly scattering medium that is opaque at ultraviolet and visible wavelengths. However, this changed in 1990 when Denk et al. showed that the two-photon fluorescence excitation using near infrared (IR) femtosecond pulses could lead to sectioned, three-dimensional imaging of biological cells with submicron spatial resolution (Denk et al. 1990). Their work was followed by additional advances in nonlinear optical microscopy (NLOM) methods that have provided new ways to probe the structure and function of skin using near infrared light. Examples of NLOM methods include coherent anti-Stokes Raman spectroscopy (CARS), second harmonic generation (SHG) autofluorescence (AF), and two-photon fluorescence microscopy (TPM). In this chapter, we focus on the last technique.

In TPM, fluorescence emission of a fluorophore (either intrinsic to the sample or an exogenously applied chromophore) is detected. A fluorophore is used to obtain specific chemical information in the skin, as discussed in more detail below. To excite the fluorophore, two-photon excitation is used, which is achieved by the simultaneous interaction of two infrared photons and is a $\chi^{(3)}$ process. TPM has several advantages over UV-visible confocal fluorescence microscopy and optical coherence tomography methods, which rely on $\chi^{(1)}$ processes, i.e., one-photon absorption or scattering (Denk et al. 1990; Denk et al. 1995; Xu and Webb 1997; Masters et al. 1997; So et al. 2000; Ragan et al. 2003). First, the use of IR light affords greater depth penetration (<1000 μm) than visible or UV excitation (<50 μm) (Masters et al. 1997; Masters and So 1999; Ragan et al. 2003). This allows one to obtain physical and biochemical information deep within the skin with the submicron spatial resolution that TPM affords (Masters et al. 1997; Masters and So 1999; Ragan et al. 2003). Second, TPM is inherently confocal, allowing for 3D-sectioning without the use of a pinhole that reduces light collection and affects image quality (Xu and Webb 1997; Masters and So 1999; So et al. 2000; Ragan et al. 2003). Third, because near-IR light is not resonant with endogenous skin chromophores and the excitation is localized only in the focal region, photodamage to the tissue sample is minimized (Masters et al. 1997; Masters and So 1999; Ragan et al. 2003).

In this chapter we cover recent applications of TPM to probe chemical characteristics of live skin using exogenous chromophores. TPM has opened up the skin to optical study, providing a tool that allows the skin to remain intact, while still preserving its chemical environment as close to in vivo as possible. By combining TPM with the use of fluorescent probes that are sensitive to different types of molecular species, it is possible to observe chemical processes occurring at all depths within the skin. We begin with a general description of the instrumentation and sample preparation methods, followed by in-depth summary of recent work from our group on the measurement of reactive oxygen species in the skin, and the measurement of pH levels in response to various types of stimuli.

II. INSTRUMENTATION AND METHODS

TPM, like all NLOM methods, relies upon the interaction of more than one photon with a chromophore. It is a $\chi^{(3)}$ process. A detailed discussion of nonlinear optical processes can be found in the literature (Mukamel 1995; Shen 2003).

A. TWO-PHOTON FLUORESCENCE MICROSCOPE

Ragan et al. have published a comprehensive review of the instrumentation characteristics required for multiphoton microscopy, including TPM (Ragan et al. 2003). TPM uses a multiphoton microscope that requires a laser light source, a microscope, and detectors (Figure 2.2) (Denk et al. 1990, 1995; Xu and Webb 1997; Masters et al. 1997; Masters and So 1999; So et al. 2000; Ragan et al. 2003). In general, nonlinear optical processes can only be driven efficiently using short, intense pulses, and thus the light source with as output of femtosecond (10^{-15}) or picosecond (10^{-12}) pulses is needed. As mentioned above, the ideal wavelength range for skin imaging is the near IR. Finally, the laser should be tunable across a wide wavelength range in order to excite a variety of chromophores. These three requirements—short pulses, near-IR wavelength, and tunablility—have lead to Ti:sapphire laser oscillators becoming the

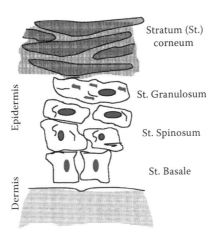

FIGURE 2.2 A diagram of human skin. Epidermal thickness depends upon body site being thickest on the palms and soles (~1500 μm) and thinnest around the eyes (~10 μm). The stratum corneum is the only layer composed of anucleated, terminally differentiated keratinocyte cells called corneocytes. All other epidermal layers contain nucleated keratinocytes. The dermis is composed primarily of the structural proteins collagen and elastin.

most widely used light sources for TPM. To generate femtosecond pulses, frequency-doubled continuous wave (CW) green laser light pumps a titanium:sapphire femtosecond laser for excitation between 720 and 900 nm. Most major laser manufacturers offer lasers that meet these criteria.

The laser light travels through the epifluorescence or side port of the microscope. A dichroic mirror reflects the laser light and passes the green fluorescence to either of the detectors. Detectors are positioned on the bottom port of the inverted microscope or the top port of the upright microscope. The choice of detector is discussed in more detail below. Broadband and band-pass filters placed in the detection path prevent residual IR from reaching either of the detectors.

B. EXCITATION OF FLUORESCENCE

Excitation of the probe molecule's fluorescence can be excited by selecting the appropriate wavelength (Denk et al. 1990, 1995; Xu and Webb 1997; Masters et al. 1997; Masters and So 1999; So et al. 2000; Ragan et al. 2003; Laiho et al. 2005) (Table 2.1). Tuning between 720 nm and 760 nm on a tunable Ti:sapphire laser can maximize autofluorescence (AF) (the fluorescence emitted by chromphores within the skin, emission between 435 nm and 700 nm), whereas tuning above 800 nm can reduce AF contributions to the overall signal, and depending upon the exogenous probe, enhance its fluorescence. The epidermis and dermis contain numerous chromophores that exhibit autofluorescence or even second harmonic generation (SHG) upon multiphoton excitation (Table 2.1), and care must be taken when selecting the excitation wavelength to minimize contributions from these phenomena to the overall signal.

TABLE 2.1
Endogenous Skin Chromophores

	Fluorescence	
Chromophore	Excitation κ_{ex} (nm)	Emission κ (nm)
Retinol[a]	700–830	450
NADH[b]	340 690–730	450–470
Vitamin D[a]	< 700	450
Flavins[c]	370, 350 700–730	430
Melanin[d]	280–450	440, 520, 575
Elastin[d]	300–340 700–740	420–460
Collagen fluorescence	300–340 700–740	420–460
CARS	**Excitation**	**Emission ω (cm^{-1})**
C-H stretch	See[e]	2845
Sebaceous glands		2845
Adipocytes		2956

[a] Data from Zipfel, W. R., Williams, R. M., Christie, R., Nikitin, A. Y., Hyman, B. T., and Webb, W. W. 2003. Live *Proc. Nat. Acad. Sci.* USA 100:7075–80.

[b] Data from Campagnola, P. J., Millard, A. C., Teraaki, M., Hoppe, P. E., Malone, C. J., Mohler, W. A., and One, C. T. 2002. *Biophys. J.* 81:493–508. Piston, D. W., Kirby, M. S., Cheng, H. P., Lederer, P., Webb, W. W., and Guy, M. A. 1994. *Appl. Opt.* 33:662–69. Patterson, G. H., Knobel, S. M., Arkhammar, P., Thastrup, O., and Piston, D. W. 2000. *Proc. Natl. Acad. Sci. USA* 97:5203–7. Huang, S. H., Heikal, A. A., and Webb, W. W. 2002. *Biophys. J.* 82:2811–25. Bennett, B. D., Jetton, T. L., Ying, G., Magnuson, M. A., and Piston, D. W. 1996. *J. Biol. Chem.* 271:3647–51.

[c] Data from Campagnola, P. J., Millard, A. C., Teraaki, M., Hoppe, P. E., Malone, C. J., Mohler, W. A., and One, C. T. 2002. *Biophys. J.* 81:493–508.

[d] Data from Schenke-Layland, K., Riemann, I., Damour, O., Stock, U. A., and Konig, K. 2006. *Adv. Drug Deliv. Rev.* 58:878–96.

[e] Data from Evans, C. L., Potma, E. O., Puoris'haag, M., Cote, D., Lin, C. P., and Xie, X. S. 2005. *Proc. Nat. Acad. Sci.* 102:16807–12. Djaker, N., Lenne, P. F., Marguet, D., Colonna, A., Hadjur, C., and Rigneault, H. 2007. *Nuc. Inst. Meth. Phys. Res. A* 571:177–81.

C. Scanning and Imaging

There are four ways to scan across the *x-y* plane to obtain an image in TPM. Point-by-point scanning can be accomplished by using either galvanometer-driver scanning mirrors (Cambridge Technology, Cambridge, Massachusetts) or a motorized stage (H101, Prior Scientific, UK). These two methods are relatively slow, yielding a typical frame rate of 0.5–10 s (Denk et al. 1995; Kim et al. 1999). To achieve video rate imaging (~30 frames/s), both line-scanning and multiphoton multifocal microscopy have proven successful (Brakenhoff et al. 1996; Guild and Webb 1995; Evans et al. 2005). The former uses a line focus and a CCD camera for detection and yields a reduced point-spread function (PSF, the impulse response of the laser), which in turn yields greater image resolution. The latter uses an array of lenses that when rotated focus on multiple spots uniformly over the x image plane and a synchronized galvanometric mirror to scan the y axis. The advantage of the latter technique is that video-rate (40 μs/line) is achieved without compromising the PSF. Achieving video-rate reduces data collection time by 100× over the typical point-scanning method, allowing for data from multiple skin samples and multiple areas to be collected. The ability to examine multiple areas is often required to draw accurate conclusions when imaging in the highly heterogeneous skin environment (Yu et al. 2002). To scan along the axial direction, a motorized piezo-driven Z-stage can be used to position the focal spot of the beam at different depths within the tissue.

There are some specific requirements for using TPM to image skin samples. When choosing an objective, the numerical aperture (NA) should be closest to the average of the index of refraction (n) of skin, ~1.4 (Yu et al. 2002). Because the skin is heterogeneous, the n values vary with depth (stratum corneum $n = 1.47$, stratum basale $n = 1.34$, and dermis $n = 1.41$), and thus it is impossible to match all n values of the skin (Ragan et al. 2003; Niesner et al. 2005). Both water and oil high NA (1.3) objectives have been found to yield high-quality images, with oil objectives collecting more fluorescence when imaging the dermis (Hsia et al. 2006; Dong et al. 2004). Typical objectives are 40× infinity corrected oil or water objectives (F Fluor, 1.3 NA; Zeiss, S Fluor NA 1.3, Nikon).

For two-photon fluorescence, typical average excitation powers are <5 mW for ex vivo tissues and 50 mW in vivo. Selecting an excitation power can be a trade-off between image quality and tissue damage. The TPM signal decays exponentially with depth due to scattering of the excitation photons, which reduces TPM image quality (Dunn et al. 2000). Scattering of the excitation light does not alter the PSF or focal volume. However, care must be taken to avoid photodamage to the sample by using excessive incident laser power (Ragan et al. 2003; Dunn et al. 2000). Contrast agents like glycerol, propylene glycol, and, to a lesser extent, glucose may help with this compromise by reducing excitation scatter and improve image contrast and penetration depth (Cicchi et al. 2005). They may do this by removing water through osmosis and reducing heterogeneity in the n for better matching with the NA. Tissue damage is a concern for higher laser powers. Masters et al. point out that photodamage in the skin from two-photon excitation can occur three ways: (i) by the absorption of intracellular chromophores, which may be similar to damage caused by UV radiation; (ii) from dielectric breakdown from EM radiation; and (iii) from one-photon absorption of IR (Masters et al. 2004; Pustovalov 1995; Zipfel et al. 2003). The dominant mechanism can depend sensitively on excitation conditions and sample. For example, care must be taken to minimize thermal mechanical damage during TPM optical biopsy of the skin. In this application, it was found that the majority of photodamage in skin occurs at the epidermal–dermal junction and results from the one-photon absorption of IR by melanin causing cavitation (aka explosive evaporation) at the focus. The damage was best minimized by reducing the laser repetition rate to reduce the average energy deposited in the sample, as opposed to simply reducing the pulse energy, which also compromised the fluorescence signal. For every TPM experiment, conditions must be found that optimize signal while avoiding sample damage.

D. Detection

By altering the detector setup, data can be collected from fluorescence intensity or fluorescence lifetime measurements. For TPM data acquisition for home-built systems, the SimFCS computer program (Laboratory for Fluorescence Dynamics, University of California at Irvine) is a free resource available to researchers.

1. Intensity Measurements

Often only simple intensity measurements are needed, which require a single PMT for single channel collection. A band-pass filter is placed before the PMT to collect the emission of interest and reduce or eliminate any contribution from unwanted background fluorescence. To collect both fluorescence (430–700 nm, depending upon fluorophore) and second harmonic generation (~370–410 nm, or $\lambda/2$, depending upon λ), two PMTs for dual channel (aka dual color) detection can be employed where a dichroic and filters are placed in the detection path to separate the two colors. For video-rate imaging provided by line-scanning or multifocal microscopy (described above), a CCD camera can be used. Emission spectra from fluorescence or SHG can be collected by placing a spectrometer before a CCD camera. For CARS microscopy, a red-sensitive PMT is needed (Evans et al. 2005). Instruments can be designed to collect different data simultaneously. For example, Yazdanfar et al. developed a trimodal instrument to image both linear (index of refraction contrast, absorption, scatter) and nonlinear events (two-photon fluorescence, SHG) combining optical coherence microscopy with spectrally separated SHG and two-photon fluorescence (Yazdanfar et al. 2007). They used a dichroic to separate backscattered IR light for the optical coherence microscopy from the visible light generated by two-photon fluorescence and SHG. The emission from the latter two was in turn separated by the placement of an additional dichroic before two PMT detectors.

2. Fluorescence Lifetime Imaging Microscopy

In many samples, variations in the fluorescence intensity of exogenous chromophores are able to provide information about chemical processes; however, caution must be used when interpreting intensity images of exogenous fluorescence from the skin. All probes inhomogeneously label the skin such that probe intensity in a pixel may appear more or less depending simply upon the amount of label present in that pixel. If one used a pH-sensitive dye to detect local variations in skin acidity, a region having higher intensity could indicate higher pH, or it could just indicate a greater local concentration of dye at the same pH. Intensity measurements are appropriate if one simply wants to detect the presence of fluorescence that is absent in a control (see ROS section below). But in cases where the intensity of the probe fluorescence cannot be directly compared to a control intensity image due to inhomogeneous labeling (see pH section below), then fluorescence lifetime imaging microscopy (FLIM) is necessary (Straub and Hell 1998; Dong et al. 2003; Clegg et al. 2003; Gadella et al. 1993; Owen et al. 2007; Margineanu et al. 2007; Cole et al. 2000). This is because FLIM data are independent of probe concentration and inhomogeneities in excitation and emission paths, and thus FLIM is a powerful method useful for probes whose inhomogeneous distribution in the skin compromises the intensity images, and where a comparative control image cannot be made with confidence.

There are two ways to collect FLIM data: frequency-domain or time-domain data acquisition (Alcala et al. 1985; Jameson et al. 1984). Briefly, in frequency domain FLIM, the fluorescence lifetime is determined by its different phase relative to a frequency modulated excitation signal using a fast Fourier transform algorithm. This method requires a frequency synthesizer phase-locked to the repetition frequency of the laser to drive an RF power amplifier that modulates the amplification of the detector photomultiplier at the master frequency plus an additional cross-correlation frequency. In contrast, time-domain FLIM directly measures τ using a photon counting PMT and card.

III. IMAGING FOR CHEMICAL INFORMATION IN THE SKIN WITH EXOGENOUS CHROMOPHORES

Exogenous fluorescence probes have proven to be the only effective means to obtain more specific biochemical information from the skin, although CARS microscopy may prove to be the exception to this rule in the future (Evans et al. 2005; Owen et al. 2007; Cole et al. 2000; Djaker et al. 2007). To study the biochemical environment of the skin, exogenous chromophores such as organic dyes and inorganic quantum dots are applied to the skin surface and incubated for a period of time, and then the skin is imaged (Larson et al. 2003; Hanson and Clegg 2002, 2003, 2005; Hanson et al. 2002, 2006; Malone et al. 2002; Tirlapur et al. 2006). When exogenous chromophores are used, their concentrations are typically much lower than that of the endogenous chromophores, and fluorescence is the only practical detection modality. Thus, two-photon fluorescence microscopy is the method of choice used for detection of these low concentration species.

A. CHARACTERIZATION OF THE pH OF THE STRATUM CORNEUM

Our first example of imaging the skin for chemical information focuses on the pH gradient within the stratum corneum, the outmost epidermal layer of the skin composed of anucleated keratinoctyes (Figure 2.1). Its brick-and-mortar-like structure of keratinocytes surrounded by a lipid-rich mortar crucially provides the sole barrier between the body and the world. Although it is only, on average, 10 μm thick, the pH changes from neutral (pH 7) at the SC basement to acidic (pH 4.5–5.6, depending upon body site and sex) at the skin surface. The acidic stratum corneum surface is referred to as the acid mantle. Regulation of the acid mantle is crucial for normal barrier function, where if absent and with the SC at neutral, pH barrier recovery is significantly impaired (Mauro et al. 1998). Thus, a first step in the study of the biology of the SC barrier is to measure its pH at different SC depths and under different conditions (Niesner et al. 2005; Hanson et al. 2002; Behne et al. 2002, 2003).

To measure pH in the SC, FLIM measurements of skin incubated with the lifetime-sensitive pH probe 2′,7′-bis-(2-carboxyethyl)-5-(and-6)-carboxyfluorescein (BCECF) (Figure 2.3) have been conducted. The lifetime (t_f) of BCECF changes with pH—at pH 4.5, $t_f = 2.75$ ns, and at pH 7.1, $t_f = 3.97$ ns (Hanson and Clegg 2002; Szmacinksi

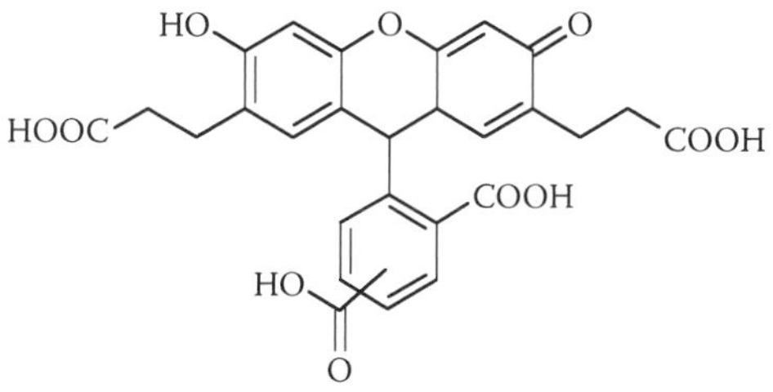

FIGURE 2.3 The fluorescence lifetime-sensitive pH probe 2′,7′-bis-(2-carboxyethyl)-5-(and-6)-carboxyfluorescein (BCECF). Because BCECF's fluorescence lifetime (t_f) changes with pH, which is not affected by inhomogenous labeling, it can be used to accurately monitor pH in the skin. At pH 4.5, $t_f = 2.75$ ns, and at pH 7.1, $t_f = 3.97$ ns.

and Lakowicz 1993). FLIM measurements identified the presence of ~1 μm diameter acidic microdomains in the lipid-rich extracellular matrix compared to the neutral intracellular space of the corneocytes (Figure 2.4) (Hanson and Clegg 2002). The changing ratio of acidic microdomains:neutral regions is the source of the change in pH over the short SC distance. This work showed that the pH variation in skin is not continuous, but rather is controlled by the creation of these microdomains, which provide a way to generate very large pH gradients that are not affected by molecular diffusion. The images in Figure 2.4 exemplify the importance of using FLIM when labeling is heterogeneous. In a homogeneous environment, the fluorescence intensity of BCECF is greater at neutral pH than at acidic pH (Szmacinksi and Lakowicz 1993). However, as Figure 2.4 shows, BCECF does not label uniformly; rather, its intensity is greatest in areas of acidic pH. Thus, although the intensity images show a bright

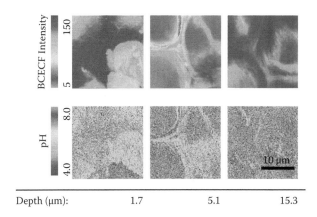

FIGURE 2.4 Fluorescence intensity of the lifetime-sensitive pH-probe BCECF and corresponding pH maps of mouse skin at different epidermal depths. The pH maps were calculated using BCECF's lifetime values, and not its intensity. The intensity images show the importance of taking lifetime measurements. BCECF's fluorescence intensity is greatest at high pH, but the areas in the skin with the greatest intensity are in reality at low pH. The intensity variations are due to inhomogenous labeling by the fluorophore.

fluorescence in some regions, the corresponding pH is not neutral, but rather is acidic. Previously, bulk methods were employed to determine pH as a function of SC depth, where skin layers were successively tape-stripped and the pH was measured with a pH probe (Dikstein and Zlotogorski 1994; Ohman and Vahlquist 1994). In contrast, FLIM allowed the pH to be characterized with submicron spatial resolution at different depths and without disrupting the sample.

Further work found that the formation of acidic microdomains occurs at the stratum granulosum–SC interface and is regulated by the sodium-proton exchanger NHE1. In addition, the acidic SC surface is not fully developed at birth, and rather acidic microdomains at the SC-SG interface develop postnatally (Behne et al. 2003). Niesner et al. also used TP FLIM on artificial skin constructs and found that an identical pH gradient to that found in mammalian skin exists, which could enable further research on barrier function without the need for human or animal tissues (Niesner et al. 2005).

B. DETECTION OF REACTIVE OXYGEN SPECIES IN THE SKIN

The second example of how TPM with probe fluorophores can provide information about chemical processes in the skin is provided by our work on the role of sunscreens in skin protection. The use of topical creams and oils to prevent ultraviolet-radiation-induced damage to the skin is a billion dollar industry. While there are a variety of mechanisms by which ultraviolet (UV) radiation can damage the skin, one important one is indirect, by promoting the generation of reactive oxygen species (ROS), which in turn damage various types of macromolocules in the skin, including DNA and collagen. Since the stratum corneum consists mainly of anucleated cells, we must be able to image deeper into the skin below the stratum corneum surface and into the nucleated keratinocyte cells of the epidermis ($z = 10$–100 μm). In particular, we are interested in how UV light influences the generation of ROS within the epidermal keratinocytes, and how the generation of ROS is affected by the application of commercial sunscreen products.

ROS are highly reactive derivatives of oxygen and include superoxide anion, hydroxyl radical, and singlet oxygen. They are formed naturally during cellular respiration, and through energy transfer to or reaction with O_2 following UVB (280–320 nm) and UVA (320–450 nm) absorption by skin chromophores including urocanic acid, NADH, riboflavin, and melanin (Cunningham et al. 1985; Peak and Peak 1986, 1989; Menon et al. 2003; Haralampus-Grynaviski et al. 2002; Nofsinger et al. 2002). Over-expression of ROS leads to oxidative stress, which can induce photoaging, immunomodulation, DNA damage, and actinic keratosis (skin cancer precursors) (Pathak and Caronare 1992; Chen et al. 1995; Wlaschek et al. 1997; Vile and Tyrrell 1995; Iwai et al. 1999). The damage done by direct UV absorption by DNA and its role in promoting skin cancer has been well studied, but recently more evidence is mounting that shows photodamage caused by ROS is of great concern to the health and integrity of skin. The first goal of our research was to develop a strategy to use TPM to study the effects of solar UV radiation on the generation of ROS in the skin (Hanson and Clegg 2005; Hanson and Clegg 2002, 2003; Hanson et al. 2006).

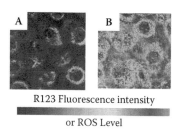

R123 Fluorescence intensity

or ROS Level

FIGURE 2.5 Fluorescence intensity images of skin ($z = 30$ μm) incubated with DHR before (a) and after (b) UVB irradiation. The fluorescence in (a) results from autofluorescence and DHR conversion to R123 due to mitochondrial respiration. The increase in fluorescence in (b) results from R123 that forms from the reaction of DHR with ROS. R123 fluorescence is detected primarily in the cytoplasm of the keratinocytes, which may result from inhomogenous labeling by DHR.

ROS in skin can be detected by exogenous chromophores like dihydrorhodamine (DHR) (Figure 2.5). DHR is nonfluorescent until reaction with ROS when it becomes fluorescent rhodamine-123 (R123, $\lambda_{em} = 535$ nm). The reaction scheme is given in Figure 2.6. By simply measuring the increase in R123 fluorescence, we can estimate how many ROS are generated by solar irradiation. Using this quantification method, we found that UVB irradiation (equivalent to 2 hr noonday summer sun in North America) of ex vivo skin samples generates 14.7 mm of ROS in the stratum corneum and 0.01 mm in all of the viable layers for the average adult-size face of 258 cm^{-2} (Hanson and Clegg 2002). Because DHR may not have labeled cell membranes, nuclei, and other cellular components, these experiments may underestimate the level of ROS that are truly generated; however, they do show that ROS are generated in significant amounts by a UVB dose often obtained on a summer day (Hanson and Clegg 2002).

After developing a method to quantitate the generation of ROS in skin, we could begin to study how the number of ROS can be reduced by the application of FDA-approved UV filters used in sunscreens (Hanson et al. 2006). Our measurements showed that octocrylene (OC), octylmethoxycinnamate (OMC), and benzophenone-3

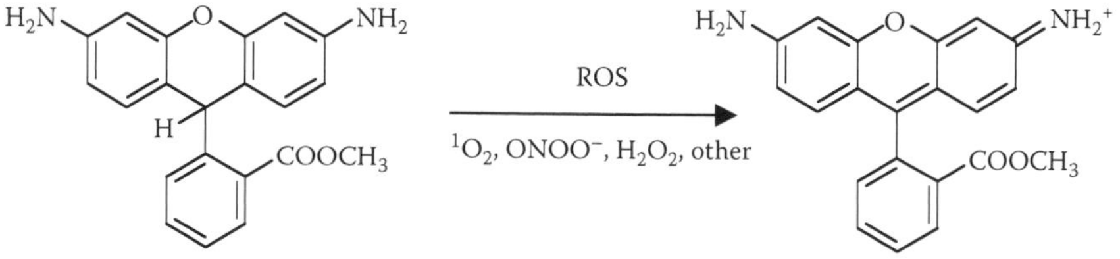

FIGURE 2.6 The ROS probe dihydrorhodamine (DHR) is non-fluorescent until it reacts with ROS to form fluorescent rhodamine-123 (R1230). DHR is not a selective reactant and may react with many other ROS than those listed above. It also does not localize in nuclei nor in cell membranes and cannot identify if ROS are generated in these regions on keratinocytes. Other ROS probes may prove to be useful to provide more data on these cellular locations.

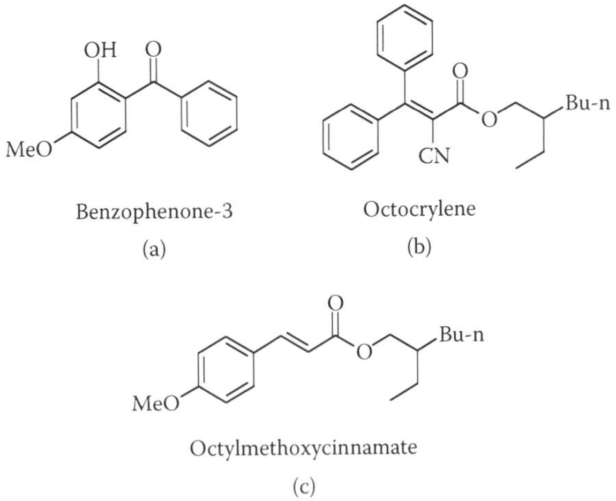

OH   O

MeO

Benzophenone-3

(a)

O

Bu-n

O

CN

Octocrylene

(b)

O

Bu-n

O

MeO

Octylmethoxycinnamate

(c)

FIGURE 2.7 Three FDA-approved UV filters commonly used in over-the-counter sunscreens: (a) benzophenone-3 (B3), (b) octocrylene (OC), and (c) octylmethoxycinnamate (OMC).

(B3) (Figure 2.7) all reduced the number of ROS generated in the epidermis following irradiation by solar-simulated UVB-UVA if they remained on the skin surface (Figure 2.8). However, as the skin was incubated for $t = 20$ or $t = 60$ min with OC, OMC, and B3 formulations, the UV filters penetrated the SC surface. These molecules then absorbed the solar-simulated UVB-UVA (20 mJ·cm^{-2} [~10 min summer sun in North America]) and generated ROS themselves deep within the nucleated

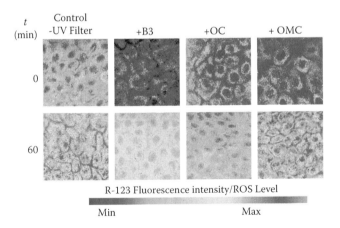

t
(min)   Control
-UV Filter   +B3   +OC   + OMC

0

60

R-123 Fluorescence intensity/ROS Level

Min   Max

FIGURE 2.8 R123 fluorescence intensity of epidermis ($z = 60$ μm) after 20 mJ cm^{-2} UVB-UVA radiation. Skin applied with cream containing B3, OC or OMC and incubated $t = 0$ min show a decrease in fluorescence compared to the placebo. After $t = 60$ min, the fluorescence of B3-, OC-, or OMC-applied skin is greater than the placebo fluorescence. Identical results were found for all nucleated epidermal layers at $t = 60$ min, indicating that the UV filters penetrated the skin surface and generated ROS themselves.

layers (Figure 2.8). These results show that if OC, OMC, and B3 penetrate the skin surface they can generate more ROS in the nucleated epidermis than if sunscreen wasn't used; however, a concomitant attenuation of UV at the skin surface (i.e., from reapplication of the sunscreen) should inhibit OC, OMC, and B3 from sensitizing ROS because no UV light could reach them to initialize the ROS sensitization. The data show that sunscreen molecules can act both as UV shields to suppress ROS generation, and also as sensitizers to enhance ROS generation, depending on where they are in the skin. Skin penetration and the role of surrounding molecules are thus key elements of the overall efficacy of a sunscreen.

For example, the vehicle (cream) plays a significant role in the degree of penetration of a UV filter or any topical agent. Ideally, one should formulate a vehicle to improve retention of a UV filter on the skin surface, so that it acts in a manner similar to latex paint (Hanson et al. 2006). In addition, topically applied antioxidant precursors (vitamin E aceate and sodium ascorbyl phosphpate) have been found to reduce ROS levels in the nucleated epidermis, although typically a large amount must be present in the formulation to significantly reduce the number of UVB-UVA–induced ROS (Hanson and Clegg 2003). Additionally, TPM has shown that dietary lutein reduces UV-induced ROS in mouse epidermis (Lee et al. 2004). These results illustrate that TPM can provide more detailed data on the efficacy of a sun protection product. Sunscreens do an excellent job at protecting against sunburn when used correctly. However, sunburn may not be the only risk factor for skin cancer, and reactions indistinguishable to the naked eye, such as those instigated by ROS, may play a significant role as well. Clearly, there appears to be room for more research in photoprotection science.

IV. SUMMARY

Before two-photon fluorescence microscopy, traditional optical experiments had limited success at acquiring relevant data on the chemical environment within the skin. Because TPM affords submicron spatial resolution, sectioned imaging, and little, if any, photodamage to the sample, the biochemical environment within the skin now is accessible for optical study. Its flexibility provides the opportunity for multiple experiments to be performed. For example, TPM could be used to image layers upon layers of cells up to 100 μm deep into the epidermis to determine its response to a stimuli, or it can be used to image a smaller region such as the nucleus of a keratinocyte to, for example, monitor nuclear response to UV light. Because imaging skin using TPM provides basic scientific information (i.e., barrier and biochemical properties) of the skin, which until now has been impossible to obtain in unfixed tissues, it is likely to play a key role in answering important questions in skin biology. Examples, to name just a few, include what is the role of calcium in barrier homeostasis, how does rosacea develop, how do hormones affect hair loss, and do ROS cause skin cancer? Currently there are a limited number of exogenous fluorophores that can be used on live subjects, and thus TPM is not used for in vivo biopsy or study; however, it is showing to be highly applicable for ex vivo dermatopathology whose traditional methods are time-consuming and require fixation of a sample. In addition, and principle, with more advanced probes, TPM could be coupled with other NLOM methods (autofluorescence, second harmonic, CARS) to form novel clinical tools that provide an alternative to traditional biopsy and histology methods. For example, with further advances, multiphoton endoscopes that can image noninvasively in vivo may become even more common (Chen 1995) that could correlate chemical information (pH, calcium concentration, protein presence) within skin via fluorescence while concurrently differentiating between normal, diseased, and cancerous tissues.

ACKNOWLEDGMENTS

This work was supported by the National Science Foundation, grant MCB-0344719.

REFERENCES

Alcala, J. R., Gratton, E., and Jameson, D. M. 1985. A multifrequency phase fluorometer using the harmoic content of a mode-locked laser. *Anal. Instr.* 14:225–50.

Behne, M. J., Barry, N. P., Hanson, K. M., Aronchik, I., Clegg, R. M., Gratton, E., Feingold, K., Holleran, W. M., Elias, P. M., and Mauro, T. M. 2003. Neonatal development of the stratum corneum pH gradient: Localization and mechanisms leading to the emergence of optimal barrier function. *J. Invest. Dermatol.* 120:998–1006.

Behne, M. J., Meyer, J. W., Hanson, K. M. Barry, N. P., Murata, S., Crumrine, D., Clegg, R. W., et al. 2002. NHE1 regulates the SC permeability barrier homeostasis. *J. Biol. Chem.* 277:47399–406.

Brakenhoff, G. J., Squier, J., Norris, T., Bliton, A. C., Wade, H., and Athey, B. 1996. Real-time two-photon confocal microscopy using a femtosecond, amplified Ti:sapphire system. *J. Micros.* 181:253–59.

Chen, Q., Fischer, A., Reagan, J. D., Yan, L. J., and Ames, B. N. 1995. Oxidative DNA damage and senescence of human diploid fibroblast cells. *Proc. Nat. Acad. Sci. USA* 92:4337–41.

Cicchi, R., Pavone, R. S., Massi, D., and Sampson, D. D. 2005. Contrast and depth enhancement in TPM of human skin ex vivo by use of optical clearing agents. *Opt. Exp.* 13:2337–44.

Clegg, R. M., Holub, O., and Gohlke, C. 2003. Fluorescence lifetime resolved imaging: Measuring lifetimes in an image. *Methods in Enzymology* 360:509–42.

Cole, M. J., Siegel, J., Webb, S. E., Jones, R., Dowling, K., French, P. M., Lever, M. J., et al. 2000. Whole-field optically sectioned fluorescence lifetime imaging. *Opt. Lett.* 25:1361–63.

Cunningham, M. L., Krinsky, N. I., Giovanazzi, S. M., and Peak, M. J. 1985. Superoxide anion is generated from cellular metabolites by solar radiation and its components. *Free Radic. Biol. Med.* 5:381–85.

Denk, W. J., Piston, D. W., and Webb, W. W. 1995. Two-photon molecular excitation in laser-scanning microscopy. In *Handbook of Biological Confocal Microscopy.* James Pawley, Ed. New York: Plenum Press.

Denk, W., Strickler, J., and Webb, W. 1990. Two-photon laser scanning microscopy. *Science* 248:73–76.

Dikstein, S., and Zlotogorski, A. 1994. Measurement of skin pH. *Acta. Dermatol. Venereol. (Stockh)* 185:18–20.

Djaker, N., Lenne, P. F., Marguet, D., Colonna, A., Hadjur, C., and Rigneault, H. 2007. Coherent anti-Stokes Raman scattering microscopy: Instrumentation and applications. *Nucl. Inst. Meth. Phys. Res. A* 571:177–81.

Dong, C. Y., French, T., So, P. T. C., Buehler, C., Berland, K. M., and Gratton, E. 2003. Fluorescence lifetime imaging techniques for microscopy. *Methods in Cell Biology* 72:431–64.

Dong, C. Y., Yu, B., Kaplan, P. D., and So, P. T. C. 2004. Performances of high NA water and oil immersion objective in deep tissue, multiphoton microscopic imaging of excised human skin. *Micr. Res. Tech.* 63:81–86.

Dunn, A. K., Wallace, V. P., Coleno, M., Berns, M. W., and Tromberg, B. J. 2000. Influence of optical properties on two-photon fluorescence imaging in turbid samples. *Appl. Opt.* 39:1194–1205.

Evans, C. L., Potma, E. O., Puoris'haag, M., Cote, D., Lin, C. P., and Xie, X. S. 2005. Chemical imaging of tissue in vivo with video-rate coherent anti-Stokes Raman scattering microscopy. *Proc. Nat. Acad. Sci.* 102:16807–12.

Gadella, T. W. J., Jovin, T. M., and Clegg, R. M. 1993. Fluorescence lifetime imaging microscopy (FLIM): Spatial resolution of microstructures on the nanosecond time scale. *Biophys. Chem.* 48:221–39.

Group, L. 2004. *The burden of skin disease.* American Academy of Dermatology Society of Investigative Dermatology, Cleveland, Ohio, USA.

Guild, J. B., and Webb, W. W. 1995. Line scanning microscopy with two-photon fluorescence excitation. *Biophys. J.* 68:290a.

Hanson, K. M., Behne, M. J., Parry, N. P., Mauro, T. M., Gratton, E., and Clegg, R. M. 2002. Two-photon fluorescence lifetime imaging of the skin stratum corneum pH gradient. *Biophysical Journal* 83:1682–90.

Hanson, K. M., and Clegg, R. M. 2002. Observation and quantification of UV-induced reactive oxygen species in ex vivo human skin. *Photochem. Photobiol.* 76:57–63.

Hanson, K. M., and Clegg, R. M. 2003. Bioconvertible vitamin antioxidants improve sunscreen photoprotection against UV-induced reactive oxygen species. *J. Cosmet. Sci.* 54:589–98.

Hanson, K. M., and Clegg, R. M. 2005. Two-photon fluorescence imaging and reactive oxygen species detection within the epidermis. *Methods Mol. Biol.* 289:413–22.

Hanson, K. M., Gratton, E., and Bardeen, C. J. 2006. Sunscreen enhancement of UV-induced reactive oxygen species in the skin. *Free Radic. Biol. Med.* 41:1205–12.

Haralampus-Grynaviski, N., Ranson, C., Ye, T., Rozanowska, M., Wrona, M., Sarna, T., Simon, J. D. 2002. Photogeneration and quenching of ROS by UVA. *J. Am. Chem. Soc.* 124:3461–68.

Hsia, C. Y., Sun, Y., Chen, W. L., Tung, C. K., Lo, W., Su, J. W., Lin, S. J., Jee, S. H., Jan, G. J., and Dong, C. Y. 2006. Effects of different immersion media in multiphoton imaging of the epithelium and dermis of human skin. *Micros. Res. Tech.* 69:992–97.

Iwai, I., Hatao, M., Nagnauma, M., Kumano, Y., and Ichihasi, M. 1999. UVA induced immune suppression through an oxidative pathway. *J. Invest. Dermatol.* 112:19–24.

Jameson, D. M., Gratton, E., and Hall, R. D. 1984. The measurement and analysis of heterogeneous emissions by multifrequency phase and modulation fluorometery. *Appl. Spectrosc. Rev.* 20:55–106.

Kim, K. H., Buehler, C., and So, P. T. C. 1999. High-speed two-photon scanning microscope. *Appl. Optics* 38:6004–9.

Laiho, L. H., Pelet, S., Hacewicz, T. M., Kaplan, P. D., and So, P. T. C. 2005. Two-photon 3-D mapping of *ex vivo* human skin endogenous fluorescence species based on fluorescence emission spectra. *J. Biomed. Opt.* 10:1–10.

Larson, D. R., Zipfel, W. R., Williams, R. M., Clark, S. W., Bruchez, M. P., Wise, F. W., and Webb, W. W. 2003. Water solutble quantm dots for multiphoton fluorescence imaging *in vivo. Science* 300:1434–36.

Lee, E. H., Faulhaber, D., Hanson, K. M., Ding, W., Peters, S., Kodali, S., and Granstein, R. D. 2004. Dietary lutein reduces ultraviolet radiation-induced inflammation and immunosuppression. *J. Invest. Dermatol.* 122:510–17.

MacKenzie, I. C. 1969. Ordered structure of the stratum corneum of mammalian skin. *Nature* 22:881–82.

Malone, J. C., Hood, A. F., Conley, T., Nurnberger, J., Baldridge, L. A., Clendenon, J.L., Dunn, K. W., and Phillips, C. L. 2002. 3D imaging of human skin and mucosa by two-photon laser scanning microscopy. *J. Cut. Path.* 29:453–58.

Margineanu, A., Hotta, J., Auweraer, M. V. D., Ameloot, M., Stefan, A., Beijonne, D., Engleborghs, Y., et al. 2007. Visualization of membrane rafts using a perylene monoimide derivative and fluorescence lifetime imaging. *Biophys. J.* 93:2877–91.

Masters, B. R., and So, P. T. C. 1999. Multiphoton excitation microscopy and confocal microscopy imaging of in vivo human skin: A comparison. *Microsc. Microanal.* 5:282–89.

Masters, B. R., So, P. T. C., Buehler, C., Barry, N., Sutin, J. D., Mantulin, W. W., and Gratton, E. 2004. Mitigating thermal mechanical damage potential during two-photon dermal imaging. *J. Biomed. Opt.* 9:1265–70.

Masters, B. R., So, P. T. C., and Gratton, E. 1997. Multiphoton excitation fluorescence microscopy and spectroscopy of in vivo human skin. *Biophysical Journal* 72:2405–12.

Mauro, T., Holleran, W. M., Grayson, S., Gao, W. N., Man, M. Q., Kriehuberm E., Behne, M., Feingold, K. R., and Elias, P. M. 1998. Barrier recovery is impeded at neutral pH, independent of ionic effects: Implications for extracellular lipid processing. *Arch. Dermatol. Res.* 290:215–22.

Menon, E. L., Perera, R., Kuhn, R. J., and Morrison, H. 2003. Reactive oxygen species formation by UVA irradiation of UA and the role of trace metals in this chemistry. *Photochem. Photobiol.* 78:567–75.

Mukamel, S. 1995. *Principles of nonlinear optical spectroscopy.* New York: University Press.

Niesner, R., Peker, B., Schlusche, P., Gericke, K. H., Hoffman, C., Hahne, D., and Muller-Goymann, C. 2005. 3D Resolved investigation of the pH gradient in artificial skin constructs by means of FLIM. *Pharm. Res.* 22:1079–87.

Nofsinger, J. B., Liu, Y., and Simon, J. D. 2002. Aggregation of eumelanin mitigates photogeneration of ROS. *Free Rad. Biol. Med.* 32:720–30.

Ohman, H., and Vahlquist, A. 1994. In vivo studies concerning a pH gradient in human stratum corneum and upper epidermis. *Acta Dermatol. Venerol. (Stockh.)* 74:375–79.

Owen, D. M., Auksorius, E., Manning, B. H., Talbot, C. B., Beule, P. A. D., Dunsby, C., Neil, M. A. A., and French, P. M. W. 2007. Excitation-resolved hyperspectral fluorescence lifetime imaging using a UV-extended supercontinuum source. *Opt. Lett.* 32:3408–10.

Pathak, M. A., and Caronare, M. D. 1992. *Biological responses to UVA radiation.* Overland Park, KS: Valdenmar.

Peak, M. J., and Peak, J. G. 1986. *The biological effects of UVA radiation.* New York: Praeger.

Peak, M. J., and Peak, J. G. 1989. Solar-UV-induced damage to DNA. *Photodermatology* 6:1–15.

Pustovavlov, V. K. 1995. Initiation of explosive boiling and optical breakdown as a result of the action of laser pulses on melanomasome in pigmented biotissues. *Kvantovaya Electronika* 22:1091–94.

Ragan, T. R., Huang, H., and So, P. T. C. 2003. In vivo and ex vivo tissue applications of two-photon microscopy. *Methods Enzymol.* 361:481–506.

Shen, Y. R. 2003. *The principles of nonlinear optics.* New York: Wiley.

So, P. T. C., Dong, C. Y., Masters, B. R., and Berland, K. M. 2000. Two-photon excitation fluorescence microscopy. *Annu. Rev. Biomed. Eng.* 2:399–429.

Straub, M., and Hell, S. W. 1998. Fluorescence lifetime three-dimensional microscopy with picosecond precision using a mulifocal multiphoton microscope. *Appl. Phys. Lett.* 73:1769–71.

Szmacinksi, H., and Lakowicz, J. R. 1993. Optical measurements of pH using fluorescence lifetimes and phase-modulation fluorometer. *Anal. Chem.* 65:1668–74.

Tirlapur, U. K., Mulholland, W. J., Bellhouse, B. J., Kendall, M., Cornhill, J. F., and Cui, Z. 2006. Femtosecond two-photon high-resolution 3D imaging spatial volume rendering and microspectral characterization of immunolocalized MHC-II and mLangerin/CD207 antigens in the mouse epidermis. *Micr. Res. Tech.* 69:767–75.

Vile, G. F., and Tyrrell, R. M. 1995. UVA radiation-induced oxidative damage to lipids and proteins in vitro and in human skin fibroblasts is dependent on iron and singlet oxygen. *Free Rad. Biol. Med.* 18:721–22.

Wlaschek, M., Wenk, J., Brenneisen, P., Briviba, K., Schwarz, A., Sies, H., and Scharfetter-Kochanek, K. 1997. Singlet oxygen is an early intermediate in cytokine dependent UVA inducetion of interstitial collagenase in human dermal fibroblasts in vitro. *FEBS Lett.* 413:239–42.

Xu, C., and Webb, W. W. 1997. *Nonlinear and two-photon induced fluorescence,* Vol. 5. New York: Plenum Press.

Yazdanfar, S., Chen, Y. Y., So, P. T. C., and Laiho, L. H. 2007. Multifunctional imaging of endogenous contrast by simultaneous nonlinear and optical coherence microscopy of thick tissues. *Micr. Res. Tech.* 70:628–633.

Yu, B., Kim, K. H., So, P. T. C., Blankschtein, D., and Langer, R. 2002. Topographic heterogeneity in transdermal transport revealed by high-speed two-photon microscopy: Determination of representative skin sample sizes. *J. Inv. Dermatol.* 118:1085–88.

Zipfel, W. R., Williams, R. M., Christie, R., Nikitin, A. Y., Hyman, B. T., and Webb, W. W. 2003. Live tissue intrinsic emission microscopy using multiphoton-excited native fluorescence and SHG. *Proc. Natl. Acad. Sci. USA* 100:7075–80.

3 Ultrafast Fluorescence Microscopes

Tatsuya Fujino and Tahei Tahara

CONTENTS

I. INTRODUCTION

Recent innovation of the ultrafast lasers makes it possible to investigate a variety of dynamics of molecules with very high time resolution. Femtosecond and picosecond time-resolved spectroscopy enables us to investigate ultrafast events including the electronic relaxation, vibrational relaxation, energy transfer, electron transfer, as well as the structure and properties of short-lived transients such as excited states. Although ultrafast spectroscopy has been mainly conducted for homogeneous systems so far, the combination with microscopic techniques is indispensable when we study inhomogeneous systems. Because the interaction between molecules and local environments causes significant changes in dynamical properties in inhomogeneous systems, it is essential to know site-specific physicochemical properties to understand inhomogeneous systems. Ultrafast spectroscopy combined with optical microscopes has been nicely demonstrated by several groups (Itoh et al. 2001; Tamai et al. 1993). For example, Nechay et al. constructed a femtosecond pump-probe scanning near-field optical microscope that realized a temporal resolution of 250 femtoseconds (fs) and a transverse spatial resolution of 150 nm (Nechay et al. 1999). Because ultrafast lasers provide short optical pulses that have high peak power, they also allow us to use nonlinear optical processes in optical microscopy. Actually, we can now readily observe two-photon excitation fluorescence (Konig et al. 1996; Sytsma et al. 1998), second harmonic generation (SHG) (Dombeck et al. 2003; Kobayashi et al. 2002), third harmonic generation (THG) (Canioni et al. 2001; Millard et al. 1999), and coherent anti–Stokes Raman scattering (CARS) (Evans et al. 2007; Hashimoto and Araki 1999; Zumbusch et al. 1999) under optical microscopes.

Fluorescence microscopy, having a high time resolution (time-resolved fluorescence microscope), is a powerful tool to study inhomogeneous systems. For example, with the combination of spatial and temporal resolutions, fluorescence lifetime imaging (FLIM) of biological samples has been performed (Clayton et al. 2002; Gerritsen et al. 2002; Konig et al. 1996; Krishnan et al. 2003; Sytsma et al. 1998). However, compared with other ultrafast spectroscopy combined with optical microscopes, the time resolution of fluorescence microscopes had been limited in the picosecond or nanosecond time regime because the time-resolved measurement was achieved by electronic-basis detection methods such as time-correlated single photon counting or streak camera. For the FLIM measurements, such time resolution is often considered sufficient because the fluorescence lifetime is as long as nanoseconds in usual cases. However, fluorescence dynamics in the femtosecond or picosecond time region contains fruitful information about the molecular dynamics such as solvation, energy transfer, rotation diffusion, as well as various kinds of reactions. Thus, the imaging microscopy has the potential to provide much more fruitful information if we can observe fluorescence with a higher time resolution under the microscope.

In this chapter, we describe two types of ultrafast time-resolved fluorescence microscopes that we recently developed: the femtosecond fluorescence up-conversion microscope (Fujino et al. 2005a; Fujino and Tahara 2003, 2004) and the non-scanning picosecond fluorescence Kerr gate microscope (Fujino et al. 2005b). The femtosecond fluorescence up-conversion microscope achieved femtosecond time resolution in fluorescence microscopy for the first time. Its diffraction limit, space resolution, and femtosecond time resolution enabled us to perform imaging of microscopic samples based on the position-dependent ultrafast fluorescence dynamics. The Kerr gate microscope realized direct detection of time-resolved fluorescence images of the microscopic samples using the optical Kerr gate technique. In the following sections, we first explain the principles of typical time-resolved fluorescence detection methods including fluorescence up-conversion and optical Kerr gate, and then describe the two new ultrafast fluorescence microscopes and their demonstrative applications.

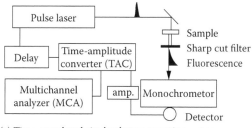

(a) Time-correlated single photon counting system

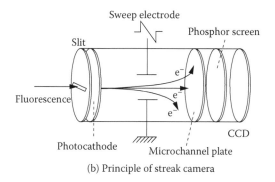

(b) Principle of streak camera

FIGURE 3.1 Schematic diagrams of (a) the time-correlated single photon counting system and (b) principle of streak camera.

II. TECHNIQUES FOR TIME-RESOLVED FLUORESCENCE MEASUREMENTS

A. TIME-CORRELATED SINGLE PHOTON COUNTING AND STREAK CAMERA

Time-correlated single photon counting is the most popular technique in time-resolved fluorescence microscopy (Becker et al. 1999; Duncan et al. 2004; Emiliani et al. 2003; Lassiter et al. 2000; Li et al. 2004; McLoskey et al. 1996; Periasamy et al. 1996; Tinnefeld et al. 2001). A schematic diagram of the time-correlated single photon counting system is depicted in Figure 3.1a. In this technique, fluorescence is detected with a single channel detector that has a fast temporal response such as a photomultiplier, microchannel plate, or avalanche photodiode. The electric signal generated in the detector by one fluorescence photon is amplified, and then the time difference between the resultant electronic pulse and the photoexcitation (trigger from a pulse laser) is converted to the height of the electronic pulse with a time-amplitude converter (TAC). The output of the TAC is analyzed by a multichannel analyzer (MCA), which makes a histogram of photon numbers against the time difference and provides a time-resolved fluorescence trace. The time resolution of the measurement is predominantly determined by the temporal spread of electronic pulses generated in the detector, and the highest time resolution achieved is a few tens of picoseconds.

Another well-known time-resolved fluorescence detection method is the streak camera. The principle of the streak camera is shown in Figure 3.1b. In the streak camera, fluorescence photons passing through a slit are converted into electrons on the photocathode, and the electrons are accelerated in the streak tube. A sweep electrode is set in the streak tube, and a sweeping high voltage is applied to the electrode, which is synchronized with the trigger from the excitation laser pulse. Then, the electrons, which are generated by fluorescence photons that enter the streak camera at different times, are deflected to a different vertical direction and enter different vertical positions of a microchannel plate (MCP). The electron is amplified in MCP and then bombarded against the phosphor screen where the signal is converted to photons again. Finally, the resultant photon signals are detected by a charge-coupled device (CCD). Therefore, the vertical position of the streak image represents the time axis. The time resolution of the streak camera is determined by several factors (e.g., the slit width, jitter, electron repulsion in the streak tube, etc.). Nevertheless, the time resolution is higher than that of the time-correlated single photon counting technique, in general. In a practical sense, the time resolution readily achieved is around 10 picoseconds (ps), although sub-picosecond time resolution is achievable in single-shot measurements.

The time-resolved detection of fluorescence microscopy has been achieved conventionally by these two methods. Therefore, the time resolution of the fluorescence microscope was limited to several tens of picoseconds or nanoseconds (Krishnan et al. 2003; McLoskey et al. 1996).

B. Fluorescence Up-Conversion and Optical Kerr Gate

Contrary to the above-described detection methods, fluorescence up-conversion and optical Kerr gate techniques readily achieve picosecond/femtosecond time resolution (Ippen and Shank 1975; Shah 1988; Takeuchi and Tahara 1998), because they are in the pump-probe measurement, in principle.

The principle of fluorescence up-conversion is depicted in Figure 3.2a. In this method, the sample is excited with femtosecond laser pulses. The fluorescence emitted from the sample is introduced to a nonlinear crystal where the fluorescence is mixed with another femtosecond laser pulse called the "gate pulse" or "monitor pulse." Then, the sum frequency of fluorescence and the gate pulse is generated by the second-order nonlinear process in the crystal. The sum-frequency signal is generated only when the fluorescence and gate pulse are temporally overlapped, and its intensity is proportional to the intensity of the two inputs. Therefore, the intensity of the sum-frequency signal is proportional to the intensity of the portion of the fluorescence that is temporally overlapped with the gate pulse. Consequently, when the intensity of the sum-frequency signal is measured with scanning of the timing of the gate pulse and it is plotted against the delay time of the gate pulse, we obtain a replica of the temporal change of the fluorescence in question. As with other types of pump-probe spectroscopy, the time resolution of the fluorescence up-conversion measurements is determined only by the laser pulses that are used for excitation and gating, and the time resolution is not required

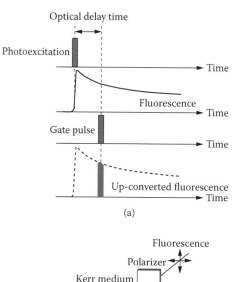

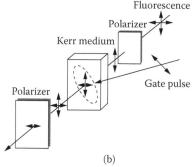

FIGURE 3.2 Principle of (a) fluorescence up-conversion and (b) optical Kerr gate.

for the detector. Therefore, femtosecond time resolution is readily achieved with recent laser light sources based on the Ti:sapphire laser.

The principle of the optical Kerr gate technique is shown in Figure 3.2b. The Kerr gate technique uses the transient birefringence induced in a transparent Kerr medium, such as glass plate or CS_2, by irradiation of high-intensity laser pulses. In this method, fluorescence excited with ultrashort laser pulses is focused into a Kerr medium that is located between a pair of crossed polarizers. While the Kerr medium is isotropic, the fluorescence passing through the first polarizer is blocked by the second polarizer because of its orthogonal polarization. Then, at a certain delay time, an intense ultrashort optical pulse (the gate pulse) is introduced to the Kerr medium to induce transient birefringence. The polarization of the gate pulse is tilted (typically 45°) against the polarizations of the two polarizers, so that the polarization of the fluorescence is changed from linear to ellipse when it passes through the Kerr medium. Consequently, a portion of the fluorescence is transmitted through the second polarizer and reaches the detector, while the transient birefringence is induced in the Kerr media. Because this optical setup works as an ultrafast optical shutter, this method is also often called the "Kerr shutter." The time resolution of this method is determined by the duration of excitation and gate pulses as well as the response time of the Kerr medium.

III. FEMTOSECOND FLUORESCENCE UP-CONVERSION MICROSCOPE

Here we first describe the ultrafast fluorescence microscope, which uses the fluorescence up-conversion method. This microscope simultaneously achieves femtosecond time resolution and submicron space resolution (Fujino and Tahara 2003, 2004).

A. APPARATUS OF THE FEMTOSECOND FLUORESCENCE UP-CONVERSION MICROSCOPE

The schematic diagram of the femtosecond fluorescence up-conversion microscope is depicted in Figure 3.3. A mode-locked Ti:sapphire laser that was pumped by an Nd:YVO$_4$ laser provided femtosecond pulses (800 nm, 9.0 nJ, 75 fs) at a repetition rate of 76 MHz. The output from the Ti:sapphire laser was frequency doubled by a 1-mm LBO crystal (400 nm, 0.6 nJ), and the generated second harmonic pulse was separated from the fundamental by a dichroic mirror. The generated second harmonic pulse was used for excitation of the sample, whereas the residual fundamental pulse (~ 6.4 nJ) was used as the gate pulse for the up-conversion process. This excitation pulse was first introduced into a prism pair to generate negative chirp, in order to compensate for the positive chirp due to the optics in the microscope. The resultant

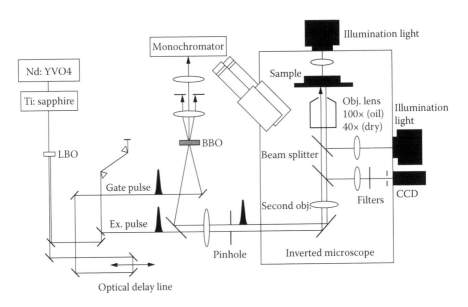

FIGURE 3.3 Schematic diagram of the femtosecond fluorescence up-conversion microscope. (From Fujino, T. and Tahara, T. *J. Phys. Chem. B* 107:5120–5122, 2003. Used with permission)

negatively chirped pulse was focused into a pinhole by a quartz lens ($f = 200$ mm), and then introduced to an inverted optical microscope (Nikon, TE-2000U). Finally, the excitation pulse was guided into a tube lens and was focused on the sample by an objective lens (Nikon, CFI Plan Fluor 100×, N.A. 1.3, oil immersion, or 40×, N.A. 0.75, dry). The pulse energy at the sample point was kept <12 pJ (<1 mW) for ordinary measurements. The fluorescence from the sample was collected by the same objective lens and then guided to the outside of the microscope through the pinhole that passes only the fluorescence emitted from the focal spot. After being separated from the excitation pulse by a dichroic mirror, the fluorescence was focused into a BBO crystal (1-mm thickness) where it was mixed with the gate pulse to be frequency up-converted. The generated sum-frequency signal was collimated and focused on to the slit of a monochromator (Jovin-Yvon, HR-320). An optical band-pass filter was set before the entrance slit to make only up-converted UV signal enter the mono-chromator. The up-converted fluorescence was finally detected by a photomultiplier (Hamamatsu, H6180-01) with a photon counter (Hamamatsu, H8784).

B. Time, Transverse (XY), Axial (Z) Resolution

The time resolution of the femtosecond fluorescence up-conversion microscope was evaluated by measuring the instantaneous rise of the fluorescence of a dye molecule. Figure 3.4a shows the time-resolved fluorescence trace obtained from a thin liquid film of a rhodamine B solution (2×10^{-3} mol dm^{-3} in methanol) in the delay time region from −1 to 1 ps. As clearly seen in the figure, the up-converted fluorescence intensity appeared around the time origin and became almost constant afterwards (dotted circles). This is because the lifetime of the electronically excited singlet (S_1) state of rhodamine B is very long (~1.5 ns) (Smirl et al. 1982), compared with the time range of this measurement. The first derivative of the observed rise-up of the fluores-cence (cross marks) gives the instrumental response of the apparatus. The temporal resolution of the system was evaluated to be 520 fs (FWHM) by the fitting analysis, assuming a Gaussian response function. The same measurement was also carried out with the 40× objective lens, and a time resolution of 460 fs was obtained.

The transverse spatial resolution (XY) of the apparatus was evaluated from the image of the focal spot. The excitation pulses were focused on a thin cover glass with the use of the 100× objective lens, and the image of the focal spot was taken with the CCD camera equipped in the microscope. The obtained image and the spatial profile along a lateral axis are shown in Figure 3.4b. In this measurement, the diameter of the pinhole in the confocal configuration was set at 50 μm, and the pulse energy was sufficiently reduced in order to ensure that the peak intensity of the image was kept within the linearity of the CCD detector. As seen in Figure 3.4b, the spatial profile exhibited weak shoulders beside the main peak. The Fresnel diffraction formula shows that the spatial profile of the focal spot has the form of the Airy disk function. The shoulder in the spatial profile is ascribable to the side peaks of this Airy disk profile. The transverse resolution was evaluated from FWHM of the Lorenz function that was best fitted to the main peak of the observed spatial profile. Since the focal spot on the thin cover glass was not a complete circle, we analyzed the spatial profile along two perpendicular directions, lateral (X) and longitudinal (Y), separately. The

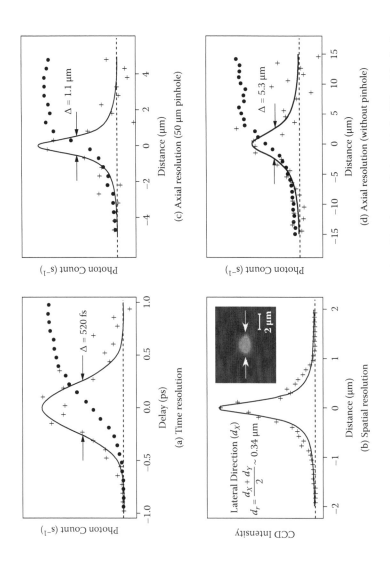

FIGURE 3.4 Performance of the fluorescence up-conversion microscope. (a) Evaluation of the time-resolution with the 100 × objective lens; •, up-converted fluorescence; +, the first derivative. By the fitting analysis, the time-resolution of the microscope was evaluated as 520 fs. (b) Evaluation of the transverse (XY) spatial resolution with the 100 × objective lens. A CCD image of the excitation pulses (inset) and the beam profile along the lateral (X) direction. By the fitting analysis, the transverse resolution was evaluated as 0.34 μm. (c & d) Evaluation of the axial (Z) spatial resolution with the 100 × objective lens; •, up-converted fluorescence; +, the first derivative. By fitting analysis on the first derivative coefficient, the axial resolution was evaluated as 1.1 μm with the 50 μm pinhole (c) and 5.3 μm without pinhole (d). (Rhodamine B, 2 × 10⁻³ mol dm⁻³ in methanol, 600 nm.) (From Fujino, T. and Tahara, T., *Appl. Phys. B* 79:145–151, 2004. Used with permission.)

transverse spatial resolution of the apparatus was finally evaluated to be 0.34 μm as an average of FWHMs of the lateral and longitudinal spatial profiles.

The most advantageous aspect of the confocal microscope is the depth discrimination property, which is also called the optical sectioning property. To evaluate the axial resolution of the apparatus, we also measured fluorescence from a dye solution on a thin cover glass, changing the distance (Z) between the objective lens and the sample (Figure 3.4c). We moved the objective lens from a distance ("negative distance" in the figure) toward the sample and measured the intensity of up-converted fluorescence at each axial position. The dotted circles in Figure 3.4c depict the observed signals from a rhodamine B solution at a fluorescence wavelength of 600 nm. The diameter of the pinhole was set at 50 μm and the 100× objective lens was used in this measurement. Since the sample concentration along the Z axis is considered to be a step function, the first derivative of the observed trace (cross marks) gives the axial resolution of the apparatus. The obtained first derivatives were fitted well with a Lorentz function, and the axial resolution was evaluated to 1.1 μm from FWHM of the best fit. In the case of usual confocal microscopes, the axial spatial resolution totally depends on the pinhole. However, in the fluorescence up-conversion microscope, an axial resolution of 5.3 μm was realized even without a pinhole, as shown in Figure 3.4d. In the up-conversion process, we make a very tight focus of the sample image on the mixing crystal, and the necessity of the spatial overlap with the tight-focused gate pulse practically acts as the second pinhole. This is the reason why the depth discrimination property is realized without a pinhole.

Two-photon excitation can be used for the fluorescence up-conversion microscope, and high axial resolution was achieved without a pinhole in this case. Figure 3.5 shows the up-converted fluorescence from a coumarin 522B solution at a fluorescence wavelength of 520 nm observed in the same manner of Figure 3.4d without pinhole. In this measurement, a fundamental laser pulse at 800 nm was used for excitation. The axial resolution with two-photon excitation was evaluated to be 0.97 μm (FWHM) by fitting for the first derivative of the obtained data. This result indicates

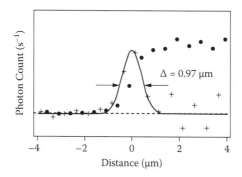

FIGURE 3.5 Evaluation of the axial (Z) spatial resolution with two-photon excitation (objective lens, 100×). By fitting analysis, the axial resolution was evaluated as 0.97 μm.

that the two-photon excitation is effective to improve the axial resolution of the fluorescence up-conversion microscope.

C. APPLICATION 1: FEMTOSECOND TIME-RESOLVED FLUORESCENCE FROM A FLUORESCENT BEAD

In the first example, we describe time-resolved fluorescence measurements of a fluorescent bead (Fujino and Tahara 2004). Figure 3.6a shows the CCD image of a commercial fluorescent bead that has a diameter of ~4.85 μm (Mag Sphere). This bead was laser trapped near the focus point by the excitation pulse. In fact, when the irradiation

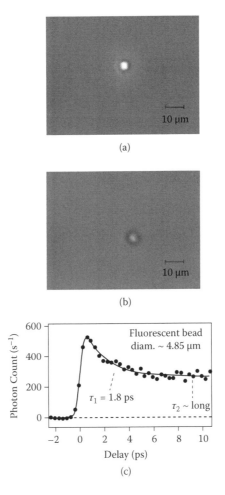

(a)

(b)

(c)

FIGURE 3.6 The CCD image of a fluorescent bead under (a) laser trapping, and (b) without laser trapping. (c) The femtosecond time-resolved fluorescence at 520 nm observed with 100× objective lens. (Form, Fujino, T. and Tahara, T., *Appl. Phys. B* 79:145–151, 2004.)

of the excitation (trapping) pulse is stopped, the bead starts moving by the Brownian motion as shown in Figure 3.6b (Ashkin 1970; Ashkin et al. 1986). The bead exhibits broad fluorescence with the intensity maximum around 520 nm, which is ascribable to the fluorescence of the coumarin dye contained in the bead. The up-converted fluorescence from one bead is depicted in Figure 3.6c for the delay time region from −2 to 10 ps. The polarization of the excitation pulses was set parallel to the gate pulses. Obviously, the observed decay consists of two components. The lifetime of the first component was determined to be $\tau_1 = 1.8$ ps, while the accurate determination of the second time constant (τ_2) was difficult owing to its very long lifetime. It is known that the S_1 state of coumarin dyes have lifetimes as long as nanoseconds (>1 ns), so that the τ_2 component was attributed to the fluorescence of the coumarin dye contained in the bead. We attributed the τ_1 component to the fluorescence of the bead itself. The time-resolved fluorescence measurement of a fluorescence bead is not important in the scientific sense. Nevertheless, this experiment clearly demonstrates a high performance of the fluorescence up-conversion microscope, as well as its capability of trapping small objects in liquid during the measurement.

D. APPLICATION 2: FEMTOSECOND FLUORESCENCE DYNAMICS IMAGING FROM TETRACENE-DOPED ANTHRACENE MICROCRYSTAL

The second example of the application of fluorescence up-conversion microscope is imaging of organic microcrystals based on ultrafast fluorescence dynamics (femtosecond fluorescence dynamics imaging) (Fujino et al. 2005a). In this measurement, the site-specific energy transfer rate in a tetracene-doped anthracene microcrystal was measured, and the crystal was visualized based on the observed local ultrafast dynamics.

The tetracene-doped anthracene microcrystals were prepared from a benzene solution of anthracene and tetracene with the molar ratio of 1 : 0.01 (Figure 3.7a) (Huppert and Rojansky 1985). By photoexcitation with 400-nm light, this microcrystal exhibited intense fluorescence, showing vibrational structures at 500, 530, 570, and 620 nm. This fluorescence feature is very similar to that of tetracene in solution (Sarkar et al. 1999), although the peak wavelengths of the mixed microcrystal are red shifted by ~50 nm. Besides, the fluorescence spectrum shows a very good mirror image of the absorption spectrum of tetracene in solution. Therefore, the intense fluorescence from the mixed microcrystal was assigned to the fluorescence from a tetracene monomer embedded in an anthracene crystal (Yoshikawa et al. 2000). Figure 3.8 depicts the time-resolved fluorescence at 530 nm obtained from the mixed microcrystal. The observed data clearly showed a finite rise of the fluorescence. The rise time of the fluorescence was determined to be 12 ps by the fitting analysis using a single exponential function convoluted with the instrumental response. The time-resolved fluorescence measurements were also carried out at other fluorescence intensity maxima of 500, 570, and 620 nm, and the same temporal change of the up-converted fluorescence was observed. Furthermore, we observed the fluorescence decay having the same time constant at the blue edge of the steady-state fluorescence spectrum (470 nm), where the fluorescence from anthracene is observed. These data clearly demonstrate that the time-resolved fluorescence shown in Figure 3.8

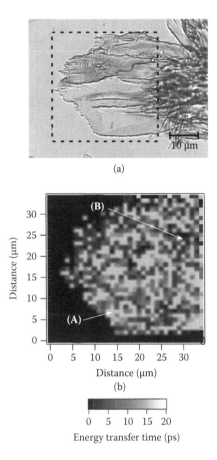

FIGURE 3.7 (a) The CCD camera image of the tetracene-doped anthracene microcrystal used for the femtosecond fluorescence dynamics imaging. (b) The dynamics image obtained from the region indicated by a broken rectangle in (a) (excitation 400 nm; fluorescence 530 nm). (From Fujino, T., Fujima, T., and Tahara, T., *J. Phys. Chem. B* 109:15327–15331, 2005. Used with permission.)

represents the energy transfer dynamics to the tetracene molecule from anthracene that is initially photoexcited by excitation pulses.

The fluorescence dynamics imaging was carried out by monitoring time-resolved fluorescence at 530 nm, where the rise of the tetracene fluorescence is observed, at different positions of the microcrystal. The up-converted fluorescence data were recorded every 1 μm distance. To shorten the measurement time at each point, the up-converted signal was sampled at only four delay time points (−5, 5, 15, 30 ps), and then the energy transfer time, τ, was evaluated by fitting with the following single-exponential function:

$$I(t) - I_{BG} = A\left(1 - \exp\left(-\frac{t}{\tau}\right)\right).$$

(3.1)

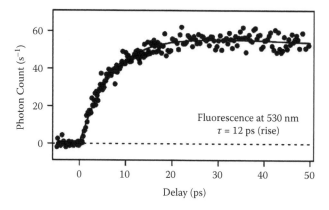

FIGURE 3.8 Femtosecond time-resolved fluorescence observed from a tetracene-doped anthracene microcrystal (excitation 400 nm; fluorescence 530 nm). (From Fujino, T., Fujima, T., and Tahara, T., *J. Phys. Chem. B* 109:15327–15331, 2005. Used with permission.)

The heterogeneity of the sample was visualized in a two-dimensional manner on the basis of the τ value (energy transfer time) evaluated at each sample point.

The measurement was carried out for the region indicated by a broken rectangle in Figure 3.7a, and the image was obtained based on the energy transfer time depicted in Figure 3.7b. As seen in this image, the rise time of 10–15 ps was observed in most parts of the microcrystal (point A in Figure 3.7b, for example). However, there were peculiar points at which the rise time was very short (point B). The mixed microcrystal was prepared by the evaporation of solvent on a cover glass, so that the tetracene molecule is inhomogeneously distributed in the microcrystal. Thus, the difference in the local concentration of tetracene gives rise to the difference in the rise time. In other words, the crystal point (B) is the point where the tetracene (acceptor) concentration is high so that the energy transfer occurs rapidly. The image based on the fluorescence dynamics (fluorescence dynamics image) successfully represents the difference in local concentration of the tetracene molecule in the mixed microcrystal. Although we could not recognize a noticeable difference between the steady-state fluorescence spectra measured at different points, the inhomogeneity in the mixed microcrystal clearly affects the local dynamics of the energy transfer in the crystal.

IV. NON-SCANNING PICOSECOND FLUORESCENCE KERR GATE MICROSCOPE

As described in the previous section, the femtosecond fluorescence up-conversion microscope enabled us to visualize microscopic samples based on position-dependent ultrafast fluorescence dynamics. However, in the imaging measurements using the fluorescence up-conversion microscope, XY scanning was necessary as when using FLIM systems. To achieve non-scanning measurements of time-resolved fluorescence images, we developed another time-resolved fluorescence microscope, which we named "fluorescence Kerr gate microscope" (Fujino et al. 2005b). This non-scanning time-resolved fluorescence microscope enables us to simultaneously measure time-resolved fluorescence from different points in microscopic samples and directly obtain time-resolved fluorescence images.

A. APPARATUS OF THE NON-SCANNING PICOSECOND FLUORESCENCE KERR GATE MICROSCOPE

A schematic diagram of the non-scanning picosecond fluorescence Kerr gate microscope is depicted in Figure 3.9a. A femtosecond Ti:sapphire laser with regenerative amplifier provided femtosecond pulses (800 nm, 1 mJ, 110 fs) at a

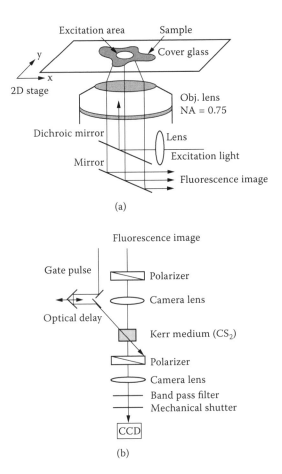

FIGURE 3.9 Schematic diagram of the non-scanning picosecond fluorescence Kerr-gate microscope. (a) The microscope part, and (b) the Kerr gate part. (From Fujino, T., Fujima, T., and Tahara, T., *Appl. Phys. Lett.*, 87:131105–131107, 2005. Used with permission.)

repetition rate of 1 kHz. A part of the output was frequency doubled by an LBO crystal, and the generated second harmonic (400 nm, ~1 µJ) was introduced into an inverted microscope for photoexcitaion. A lens was placed before the (first) objective lens (40×, NA = 0.75, Nikon), so that the excitation light was defocused on the sample and irradiated a large region (~80 µm). The fluorescence from this area was collected by the same objective lens and then guided to the outside of the microscope. Then, the fluorescence was focused by the second objective lens (camera lens), and it was introduced into the Kerr gate medium as shown in Figure 3.9b. The first polarizer of the Kerr gate was set parallel and the second one was perpendicular to the polarization of the excitation light, and the extinction ratio of the crossed polarizer was ~5 × 10^{-4}. The polarization of the gate pulse (800 nm, ~20 µJ) was set at 45° against the excitation polarization by a $\lambda/2$ plate, and it was focused into the Kerr medium (CS_2 in a 1-mm quartz cell). The fluorescence that was temporally overlapped with the gate pulse passed through the second polarizer, and it was detected by a thermoelectrically cooled CCD detector (TEA/CCD-1024-EM/1UV, Princeton Instruments) after wavelength selection by band-pass filters. The CCD was placed on the image plane of the sample, so that it directly measured the image of the fluorescent sample. The Kerr medium was located at the image plane of the first objective lens to avoid the effect of the spatial intensity profile of the gate pulse. Because the fluorescence emitted from each sample point was spread equally on the first objective lens (and also on its image plane), this configuration achieved equal Kerr gate efficiency over a fluorescence image.

B. Time, Transverse (XY) Resolution

The time-resolution of the non-scanning picosecond fluorescence Kerr gate microscope was evaluated by measuring the instantaneous rise of the fluorescence of a dye molecule, as in the case of the fluorescence up-conversion microscope. Figure 3.10a shows a time-resolved fluorescence trace obtained from a thin liquid film of a coumarin 522B solution (1 × 10^{-3} mol dm^{-3} in methanol) in the delay time region from −3 to 3 ps. Instead of the CCD detector, the PMT tube was used for this measurement, and the fluorescence intensity from the whole region of the fluorescence was monitored. As clearly seen in the figure, time-resolved fluorescence intensity appeared around the time origin and became almost constant afterward (dotted circles). By the fitting analysis of the first derivative of the time-resolved fluorescence trace, the temporal resolution of the fluorescence Kerr gate microscope was evaluated as 1.4 ps.

The spatial resolution of this system was evaluated from the steady-state fluorescence image of a commercial fluorescent bead with a diameter of 4.85 µm (Figure 3.10b). This steady-state fluorescence image was measured by setting the two polarizers in the Kerr gate setup parallel to each other, without the gate light irradiation. The fluorescence intensity profile was extracted from this image, and the spatial resolution was evaluated from the first derivative of the intensity profile. It was estimated as ~ 0.98 µm at 520 nm fluorescence (Figure 3.10c). Considering that the pixel size of the CCD detector (26 μm) and that 20 pixels were used for monitoring one bead, the magnification of this optical setup was evaluated as M ~ 20 × 26/4.85 ~ 110.

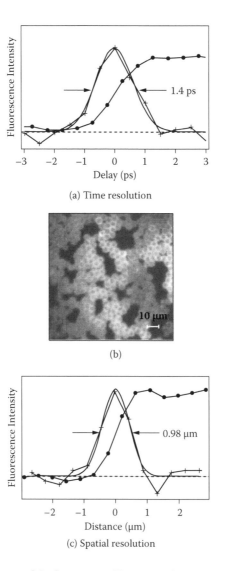

(a) Time resolution

(b)

(c) Spatial resolution

FIGURE 3.10 Performance of the fluorescence Kerr-gate microscope. (a) Evaluation of time-resolution; •, fluorescence intensity; +, the first derivative. CS_2 in a 1-mm quartz cell was used for Kerr medium. By the fitting analysis, the time resolution of the microscope was evaluated as 1.4 ps. (b) steady-state fluorescence image of fluorescent beads, and (c) evaluation of spatial resolution using the image of (b); •, fluorescence intensity; +, the first derivative. By fitting analysis, the transverse spatial resolution was evaluated as 0.98 μm. (Coumarin 522B, 1×10^{-3} mol dm^{-3} in methanol, 600 nm.)

C. APPLICATION: PICOSECOND FLUORESCENCE IMAGING OF α-PERYLENE MICROCRYSTALS

The Kerr gate microscope was used to measure time-resolved fluorescence images of α-perylene microcrystals. Figure 3.11a depicts a steady-state fluorescence image of α-perylene microcrystals that were used for the time-resolved measurements. This steady-state fluorescence image (i.e., the time-integrated fluorescence image) was measured by setting the two polarizers of the Kerr gate setup parallel to each other, without the gate pulse irradiation. As seen in this figure, the excitation light simultaneously photoexcited a large area of the sample. The edge of each crystal strongly emitted fluorescence compared with the center of the crystal, because of the waveguiding property of organic crystals (Aaviksoo and Reinot 1992). Figures 3.11b–l show the time-resolved fluorescence images of the same α-perylene microcrystals. The signal accumulation time to obtain an image was 5 min. The time-resolved fluorescence images at each delay time were obtained by subtracting the fluorescence image measured at a negative delay time (–5 ps) from the detected image, which was necessary to cancel out the signal leaked from the crossed polarizer. These figures

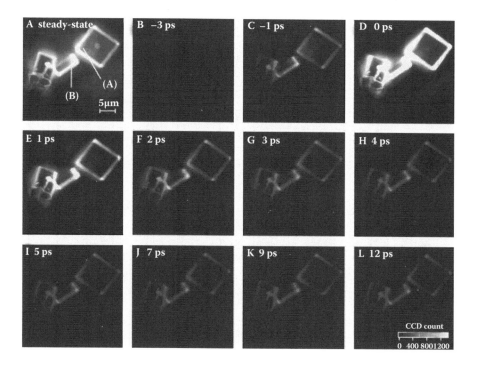

FIGURE 3.11 Fluorescence images of α-perylene microcrystal measured by the fluorescence Kerr-gate microscope. (A) Steady-state fluorescence image of α-perylene microcrystal. (B) – (L) time-resolved fluorescence images (520 nm). Fluorescence image measured at a negative delay time (–5ps) was subtracted from the image taken at each delay time. (From Fujino, T., Fujima, T., and Tahara, T., *Appl. Phys. Lett.* 87:131105–131107, 2005. Used with permission.)

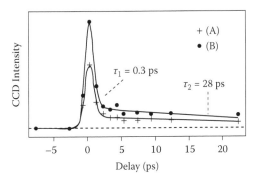

FIGURE 3.12 Time-resolved fluorescence traces measured at the positions (A) and (B) indicated in Fig. 3.11a. Both time-resolved data were fitted by double exponential functions with the time constant of $\tau_1 = 0.3$ ps and $\tau_2 = 28$ ps. (From Fujino, T., Fujima, T., and Tahara, T., *Appl. Phys. Lett.* 87:131105–131107, 2005. Used with permission.)

clearly showed that the fluorescence images of the α-perylene microcrystals were well time resolved: the strong fluorescence of α-perylene microcrystals was observed immediately after photoexcitation, and it decays with increasing the delay time. Also in the time-resolved fluorescence image, crystal edges emitted fluorescence strongly compared with the crystal center. This is because the light-guiding phenomenon in the microcrystal occurs within the time resolution of this microscope.

The temporal intensity profiles measured at two specific sample points (A and B in Figure 3.11a) are depicted in Figure 3.12. In this delay time region, the obtained decays were well fitted by double exponential functions that were convoluted with the instrumental response. The time constant for the rapid decay component was evaluated to be $\tau_1 = 0.3$ ps, and the second long component was $\tau_2 = 28$ ps, for both traces. The rapid decay component observed immediately after photoexcitation was assigned to the decay of the fluorescence of the free exciton, and the second long component was attributed to the decay of the self-trapped exciton (the Y state and E state) (Fujino and Tahara 2003). The observed lifetime of the free exciton was much shorter than that reported in a previous study (~2 ps) (Fujino and Tahara, 2003). This lifetime shortening of the free exciton is due to significant exciton-exciton annihilation under high excitation conditions by amplified laser pulses.

V. CONCLUDING REMARKS

The time-resolved techniques that are usually used for FLIM are based on electronic-basis detection methods such as the time-correlated single photon counting or streak camera. Therefore, the time resolution of the FLIM system has been limited by several tens of picoseconds. However, fluorescence microscopy has the potential to provide much more information if we can observe the fluorescence dynamics in a microscopic region with higher time resolution. Given this background, we developed two types of ultrafast time-resolved fluorescence microscopes, i.e., the femtosecond fluorescence up-conversion microscope and the non-scanning picosecond fluorescence Kerr gate microscope. These new time-resolved fluorescence microscopes enabled us to measure site-specific ultrafast fluorescence dynamics with unprecedented high time resolution. In the picosecond and femtosecond time region, a variety of important phenomena take place, such as solvation, molecular rotation, energy transfer, electron transfer, as well as various types of ultrafast chemical reactions. As indicated by demonstrative measurements described in this review, these new time-resolved fluorescence microscopes have high potential to visualize microscopic samples based on site-specific ultrafast dynamics that directly represent local physicochemical properties of the sample. This "fluorescence dynamics imaging" is a generalization of fluorescence lifetime imaging, and it can be a powerful method to study inhomogeneous systems that have microscopic structures.

REFERENCES

Aaviksoo, J., and Reinot, T. 1992. Ballistic propagation of luminescence pulse in anthracene crystal flakes. *Mol. Cryst. Liq. Cryst.* 217: 147.

Ashkin, A. 1970. Acceleration and trapping of particles by radiation pressure *Phys. Rev. Lett.* 24: 156.

Ashkin, A., Dziedzic, J.M., Bjorkholm, J.E., and Chu, S. 1986. Observation of a single-beam gradient force optical trap for dielectric particles *Opt. Lett.* 11: 288.

Becker, W., Hickl, H., Zander, C., et al. 1999. Time-resolved detection and identification of single analyte molecules in microcapillaries by time-correlated single-photon counting (Tcspc). *Rev. Sci. Instrum.* 70: 1835.

Canioni, L., Rivet, S., Sarger, L., et al. 2001. Imaging of Ca2+ intracellular dynamics with third harmonic generation microscope. *Opt. Lett.* 26: 515.

Clayton, A. H. A., Hanley, Q. S., Arndt-Jovin, D. J., Subramaniam, V., and Jovin, T. M. 2002. Dynamic fluorescence anisotropy imaging microscopy in the frequency domain (Rflim). *Biophys. J.* 83: 1631.

Dombeck, D. A., Kasischke, K. A., Vishwasrao, H. D., et al. 2003. Uniform polarity microtubule assemblies imaged in native brain tissue by second harmonic generation microscopy. *Proc. Natl. Acad. Sci. USA* 100: 7081.

Duncan, R.R., Bergmann, A., Cousin, M.A., Apps, D.K., and Shipston, M.J. et al. 2004. Multidimensional time correllated single photon counting (TCSPC) fluorescence lifetime imaging microscopy (flim) to Detect Fret in Cells. *J. Microsc.* 215: 1.

Emiliani, V., Sanvitto, D., Tramier, M., et al. 2003. Low-intensity two-dimensional imaging of fluorescence lifetimes in living cell. *Appl. Phys. Lett.* 83: 2471.

Evans, C. L., Xu, X., Kesari, S., et al. 2007. Chemically-selective imaging of brain structures with CARS microscopy. *Opt. Express* 15: 12076.

Fujino, T., Fujima, T., and Tahara, T. 2005. Femtosecond fluorescence dynamics imaging using a fluorescence up-conversion microscope. *J. Phys. Chem. B* 109: 15327.

Fujino, T., Fujima, T., and Tahara, T. 2005. Picosecond Time-Resolved Imaging by Nonscanning Fluorescence Kerr Gate Microscope. *Appl. Phys. Lett.* 87: 131105.

Fujino, T., and Tahara, T. 2004. Characterization and performance of femtosecond fluorescence upconversion microscope. *Appl. Phys. B* 79: 145.

Fujino, T., and Tahara, T. 2003. femtosecond fluorescence up-conversion microscopy: Excitation dynamics in α-perylene microcrystal. *J. Phys. Chem. B* 107: 5120.

Gerritsen, H. C., Asselbergs, M.A.H., Agronskaia, A. V., and Sark, W. G. J. H. M. Van. 2002. Fluorescence lifetime imaging in scanning microscope: Acquisition speed, photon economy and lifetime resolution. *J. Microsc.* 206: 218.

Hashimoto, M., and Araki, T. 1999. Coherent anti-Stokes Raman scattering microscope. *Proc. SPIE* 3749: 496.

Huppert, D., and Rojansky, D. 1985. Picosecond study of electronic energy transfer in tetracene-doped anthracene crystal. *Chem. Phys. Lett.* 114: 149.

Ippen, E. P., and Shank, C. V. 1975. Picosecond response of a high-repetition-rate CS_2 optical Kerr gate. *Appl. Phys. Lett.* 26: 92.

Itoh, T., Asahi, T., and Masuhara, H. 2001. Femtosecond light scattering spectroscopy of single gold nanoparticles. *Appl. Phys. Lett.* 79: 1667.

Kobayashi, M., Fujita, K., Kaneko, T., et al. 2002. Second harmonic generation microscope with microlens array scanner. *Opt. Lett.* 27: 1324.

Konig, K., So, P.T.C., Mantulin, W.W., Tromberg, B.J., and Gratton, E. 1996. Two-photon excited lifetime imaging in cell. *J. Microsc.* 183: 197.

Krishnan, R. V., Biener, E., Zhang, J. H., Heckel, R., and Herman, B. 2003. Probing subtle fluorescence dynamics in cellular proteins by streak camara based fluorescence lifetime imaging microscopy. *Appl. Phys. Lett.* 83: 4658.

Lassiter, S.J., Stryjewski, W., Jr., B.L. Legendre, et al. 2000. Time-resolved florescence imaging of slab gels. *Anal. Chem.* 72: 5373.

Li, Q.A., Ruckstuhl, T., and Seeger, S. 2004. Deep-UV laser-based fluorescence lifetime imaging microscopy of single molecules. *J. Phys. Chem. B* 108: 8324.

McLoskey, D., Birch, D. J. S., Sanderson, A., et al. 1996. Multiplex single-photon counting. I. A time-correlated fluorescence lifetime camera. *Rev. Sci. Instrum.* 67: 2228.

Millard, A. C., Wiseman, P. W., Fittinghoff, D. N., et al. 1999. Third harmonic generation microscopy by use of a compact femtosecond fiber laser source. *Appl. Opt.* 38: 7393.

Nechay, B.A., Siegner, U., Achermann, M., et al. 1999. Femtosecond near-field scanning optical microscopy. *J. Micros.* 194: 329.

Periasamy, A., Wodnicki, P., Wang, X.F., et al. 1996. Time-resolved fluorescence lifetime imaging microscopy using a picosecond pulsed tunable dye laser system. *Rev. Sci. Instrum.* 67: 3722.

Sarkar, N., Takeuchi, S., and Tahara, T. 1999. Vibronic relaxation of polyatomic molecule in nonpolar solvent. *J. Phys. Chem. A* 103: 4808.

Shah, J. 1988. Ultrafast luminescence spectroscopy using sum frequency generation. *IEEE J. Quantum Electron.* 24: 276.

Smirl, A.L., Clark, L.B., Stryland, E.W. Van, and Russell, B.R. 1982. Population and rotational kinetics of the rhodamine B. *J. Chem. Phys.* 77: 631.

Sytsma, J., Vroom, J.M., de Grauw, C.J., and Gerritsen, H.C. 1998. Time-gated fluorescence lifetime imaging using two photon excitation. *J. Microsc.* 191: 39-51.

Takeuchi, S., and Tahara, T. 1998. Femtosecond Ultraviolet-Visible Fluorescence Study of the Excited State Proton Transfer Reaction of 7-Azaindole Dimer. *J. Phys. Chem. A* 102: 7740.

Tamai, N., Asahi, T., and Masuhara, H. 1993. Femtosecond transient absorption microscope combined with optical trapping technique. *Rev. Sci. Instrum.* 64: 2496.

Tinnefeld, P., Herten, D.P., and Sauer, M. 2001. Photophysical dynamics of single molecules studied by spectrally-resolved fluorescence lifetime imaging microscopy (SFLIM). *J. Phys. Chem. A* 105: 7989.

Yoshikawa, H., Sasaki, K., and Masuhara, H. 2000. Picosecond near-field microspectroscopic study of a single anthracene microcrystal in evaporated anthracene-tetracene film: Inhomogeneous inner structure and growth mechanism. *J. Phys. Chem. B* 104: 3429.

Zumbusch, A., Holtom, G. R., and Xie, X. S. 1999. Vibrational microscopy using coherent-anti-Stokes Raman scattering. *Phys. Rev. Lett.* 82: 4014.

4 Multicontrast Nonlinear Imaging Microscopy

Richard Cisek, Nicole Prent, Catherine Greenhalgh, Daaf Sandkuijl, Adam Tuer, Arkady Major, and Virginijus Barzda

CONTENTS

I. INTRODUCTION

Nonlinear light-matter interactions have been successfully applied to create new visualization contrast mechanisms for optical microscopy. Nonlinear optical microscopy employs femtosecond and picosecond lasers to achieve a high photon flux density by focusing the beam onto a sample with a high numerical aperture (NA) microscope objective. A focused ultrafast pulsed laser beam provides sufficient peak intensity for the generation of nonlinear optical responses, but concomitantly retains a low average power so that biological specimens can be imaged for extended periods of time without inducing phototoxic effects. So far, the following nonlinear effects have been implemented in an optical microscope: second harmonic generation (SHG) (Hellwart and Christen 1974; Sheppard et al. 1977; Mizutani et al. 2000), sum frequency generation (Florsheimer 1999), third harmonic generation (THG) (Barad et al. 1997; Muller et al. 1998a), multiphoton excitation fluorescence (MPF) (Denk et al. 1990; Barzda et al. 2001; Tirlapur and König 2001), optical Kerr effect (Potma et al. 2001), and coherent anti-Stokes Raman scattering (CARS) (Duncan et al. 1982; Zumbusch et al. 1999; Muller et al. 2000). Several nonlinear responses can be induced simultaneously with the same ultrashort laser pulse; therefore, multicontrast imaging microscopy schemes with parallel detection of MPF, SHG, and THG signals have been successfully implemented (Campagnola et al. 1999; Chu et al. 2001; Barzda et al. 2005; Sun 2005). The parallel images can be compared and correlated, giving rich information about the structural architecture and molecular distribution in the sample (Barzda et al. 2005; Greenhalgh et al. 2005).

Nonlinear microscopy has several advantages when compared with classical microscopy. Poor axial resolution is the main drawback of classical transmission and fluorescence microscopes. This can be improved by introducing confocal detection, or by employing nonlinear optical contrast mechanisms (Pawley 1995; Fung and Theriot 1998; Hepler and Gunning 1998; Feijó and Moreno 2004). Since nonlinear optical effects are proportional to the second or third power of the fundamental light intensity, a focused femtosecond laser beam only yields enough intensity for nonlinear interactions within the focal volume. This focal confinement of nonlinear signal generation determines the excitation confocality and renders inherent optical sectioning. Three-dimensional (3D) images can be obtained by raster scanning the focal volume in the transverse direction and translating the sample or the excitation objective along the optical axis.

Nonlinear imaging requires substantially higher pulse energies than linear fluorescence microscopy. Therefore, the laser excitation wavelengths have to be tuned away from the linear absorption bands of the sample. Usually, infrared (IR) excitation wavelengths are employed that also provide deeper penetration of light into the biological tissue, reaching up to a few hundred microns penetration depth in highly scattering specimens (Chu et al. 2001).

Although MPF microscopy is the most frequently used nonlinear contrast mechanism, the first nonlinear microscope was introduced by employing second harmonic generation (Hellwart and Christen 1974; Sheppard et al. 1977). SHG microscopy has been extensively reviewed in several articles (Yamada and Lee 1998; Gauderon et al. 2001; Millard et al. 2003; Mertz 2004; Sun 2005). The renaissance of nonlinear microscopy started with the implementation of the MPF microscope and its application for biological imaging (Denk et al. 1990). The introduction of stable solid-state Ti:sapphire femtosecond lasers largely facilitated the development of nonlinear microscopy. MPF imaging of biological structures has been reviewed in a number of recent papers (König 2000; So et al. 2000; Williams et al. 2001; Zipfel et al.

2003; Feijó and Moreno 2004). THG microscopy was introduced by Barad et al. (1997), and instrumentation as well as applications were recently reviewed (Squier and Muller 2001; Sun 2005). In recent years CARS microscopy attracted significant attention from the perspective of instrument development as well as biological applications. CARS microscopy was introduced by Duncan et al. (1982). Reviews on CARS microscopy can be found in Cheng and Xie (2004) and Volkmer (2005).

Nonlinear microscopy is a rapidly growing area of research. With the exception of the MPF contrast mechanism, which is firmly established as a valuable approach for biomedical imaging, other nonlinear contrast mechanisms are going through the stage of major technical development. Most of the work on nonlinear microscopy has been published in engineering literature; however, biological investigations are starting to emerge, especially with epi-detected SHG, which is possible to record with commercially available two-photon excitation microscopes. Harmonic generation microscopy applications are partly limited by the lack of commercial instrumentation offered by the major microscope manufacturers. Fortunately, with simple modifications to a confocal or MPF microscope, SHG and THG microscopy can readily be implemented.

This chapter introduces principles of three nonlinear contrast mechanisms—MPF, SHG, and THG—for microscopic imaging. The instrumentation of nonlinear multicontrast microscopy is described, and the methods of multicontrast image analysis are presented. The later parts of the chapter focus on several examples of multicontrast nonlinear microscopy applications in biological imaging.

II. NONLINEAR CONTRAST MECHANISMS IN MICROSCOPIC IMAGING

A. NONLINEAR LIGHT–MATTER INTERACTIONS

Light–matter interactions can be described via an induced polarization, i.e., the induced dipole moment per unit volume. Ultrafast laser pulses, which are used in laser scanning microscopes, have high enough intensity to induce a nonlinear polarization in various materials. For intense optical electric field E, the polarization vector P can be expanded in the power series (Boyd 1992)

$$P(t) = \chi^{(1)}E(t) + \chi^{(2)}E(t)E(t) + \chi^{(3)}E(t)E(t)E(t)\cdots, \tag{4.1}$$

where $\chi^{(1)}$ is a linear susceptibility tensor representing effects, such as linear absorption and refraction; $\chi^{(2)}$ is the second-order nonlinear optical susceptibility; and $\chi^{(3)}$ is the third order nonlinear susceptibility. Second harmonic generation is a second-order process, whereas two-photon excitation fluorescence and third harmonic generation are both third-order processes. Equation (4.1) shows that the same excitation source can induce several nonlinear effects simultaneously.

Nonlinear effects are characterized by new components of the E-field generated from the acceleration of charges as the nonlinear polarization P^{NL} [second, third, and higher terms in Equation (4.1)] drives the electric field. When linear absorption is negligible, waves generated via nonlinear interactions can be described by the inhomogeneous wave equation

$$\nabla^2 \boldsymbol{E} - \frac{n^2}{c^2} \frac{\partial^2 \boldsymbol{E}}{\partial t^2} = \frac{4\pi}{c^2} \frac{\partial^2 \boldsymbol{P}^{NL}}{\partial t^2}, \tag{4.2}$$

where n represents the linear refractive index and c is the speed of light in vacuum.

Harmonic generation differs significantly from fluorescence. Second and third harmonic generation are parametric processes, while fluorescence is nonparametric. Parametric processes are described by a real susceptibility, while nonparametric processes have a complex susceptibility associated with them (Boyd 1992). For parametric processes, the initial and final quantum-mechanical states are the same. In the time between these states, the population can reside momentary in a "virtual" level, which is a superposition of one or more photon radiation fields and an eigenstate of the molecule. Since parametric processes conserve photon energy, no energy is deposited into the system. In nonparametric processes, the initial and final states are different, so there is a transfer of population from one real level to another. Nonparametric interactions lead to photon absorption, making samples more prone to the effects of bleaching and thermal damage. Near the resonant molecular transitions, parametric processes are resonantly enhanced; however, at these wavelengths absorption of laser radiation also increases. A trade-off between the laser intensity and choice of radiation wavelength has to be made for obtaining the strongest nonlinear signals without inducing signal bleaching or photodamage. It is very important to know the nonlinear absorption spectrum for choosing the optimal wavelength for excitation. In the following sections, specific imaging conditions are described for each of the three aforementioned nonlinear processes (MPF, SHG, and THG).

B. MULTIPHOTON EXCITATION FLUORESCENCE

Two-photon excitation fluorescence is currently the most widely used nonlinear contrast mechanism for microscopic investigations. The first experimental demonstration of two-photon excitation fluorescence was provided in 1961 (Kaiser and Garrett 1961), even though the first theoretical description of two-photon excitation fluorescence stems back to 1931 (Göppert-Mayer 1931). Three-photon absorption was demonstrated a few years later by Singh and Bradley (1964). Two-photon absorption is a third-order nonlinear effect, whereas three-photon absorption is a fifth-order nonlinear effect. The transition rate for two-photon absorption, R, depends on the square of the intensity, I, as follows (see Boyd 1992):

$$R = \frac{\sigma^{(2)} I^2}{\hbar \omega}, \tag{4.3}$$

where $\sigma^{(2)}$ is the two-photon absorption cross section and $\hbar\omega$ is energy of the excitation photon. The quadratic dependence of absorption or fluorescence on the excitation intensity can be used to determine whether a sample is excited via two-photon excitation. Following nonlinear absorption, the excitation-relaxation dynamics proceed as if the excited state were populated via linear absorption. If the nonlinear excitation process is not clearly determined, the detected fluorescence signal is commonly denoted as multiphoton excitation fluorescence.

Two-photon absorption occurs when the energy of a molecular transition matches the combined energy of two photons. Quantum mechanically, the absorption probability is proportional to the two-photon transition moment M_{ng} from the ground state, g, to the excited state, n, via intermediate state, m, and can be expressed as follows (Boyd 1992; Abe 2001):

$$M_{ng} = \sum_{m(\neq n, \neq g)} \left[2 \frac{(\mu_{nm} \cdot E)(\mu_{mg} \cdot E)}{\hbar\omega - \hbar\omega_{mg}} \right] - 2 \frac{(\Delta p_{ng} \cdot E)(\mu_{ng} \cdot E)}{\hbar\omega}, \quad (4.4)$$

where μ_{nm}, μ_{mg}, and μ_{ng} are the transition dipole moments between the n and m, m and g, and n and g states, respectively. The term $\Delta p_{ng} = p_{nn} - p_{gg}$ is the change in the static dipole moment between the final n and initial g states. $\hbar\omega_{mg}$ is the energy difference between the states m and g. Equation (4.4) has two distinct terms: (i) the first term describes the two-photon excitation process via one photon excitation to a virtual state, m, and subsequent excitation to the final electronic state, n; (ii) the second term expresses direct two-photon excitation to the final state via change in the static dipole moment between the n and g states. According to the first process, the ground and the final excited states have the same symmetry; therefore, the excitation is one-photon forbidden and two-photon allowed. This situation usually appears for nonpolar molecules. For polar molecules, two-photon absorption can proceed via the second mechanism, if a large change in the dipole moment occurs during excitation. In this case, the two-photon transition dipole moment is proportional to the one-photon transition dipole; thus, two-photon absorption bands will be similar to the linear absorption. The two-photon absorption spectrum of molecules may have similar bands as linear absorption, but new bands may appear and some bands might be missing.

Fluorescence is not a coherent optical response in contrast to SHG or THG; therefore, phase matching and interference effects are not observed in MPF. This simplifies the interpretation of microscopic images where the fluorescence can be directly related to the presence of fluorophores in the sample. However, the direct assignment of fluorescence intensity to the concentration of fluorophores cannot be applied. Some fluorescing molecules might be quenched while others might be highly fluorescent, leading to different fluorescence intensities for the same concentrations of the molecules. Better understanding of fluorescence properties can be achieved with simultaneous measurement of fluorescence intensity and decays by using fluorescence lifetime imaging microscopy (FLIM) (Barzda et al. 2001). Fluorescence lifetime imaging techniques are extensively reviewed by Becker (2005). Fluorescence imaging is advantageous in the specificity it has. Alongside the naturally occurring fluorophores, a wide variety of labels are available that enable a user to image specific organelles and cells within biological samples.

C. SECOND HARMONIC GENERATION

The efficiency of the second harmonic is highly dependent on the nonlinear properties of the molecules and structural organization of the sample. This is reflected in the elements of second order of nonlinear susceptibility tensor [see Equation (4.1)]. Second harmonic can be generated in a media that does not have inversion symmetry. Therefore, a nonzero $\chi^{(2)}$ is determined by asymmetric molecules or asymmetric microcrystalline structures. Isotropic and cubic lattice structures do not produce SHG, although molecules may have a nonvanishing $\chi^{(2)}$. Inversion symmetry can be broken at an interface between two materials (Bloembergen and Pershan 1962). This provides a powerful tool for studying second-order hyperpolarizabilities of molecules adsorbed onto surfaces (Shen 1989). A sufficient strength of SHG can be achieved from a monolayer of molecules at an interface, or molecules asymmetrically arranged in lipid membranes (Moreaux et al. 2000a, 2001). Biological membranes often have different lipid and protein content in the two leaflets. If membranes are stained with nonlinear dye that accumulates only in one leaflet, intense SHG can be observed. When two such asymmetric membranes appear adhered to each other to construct a symmetric arrangement, the SHG vanishes (Moreaux et al. 2000a). Different strategies have been developed to maximize $\chi^{(2)}$ via optimizing asymmetry of the molecules (Marder et al. 1991) or improving the alignment of non-centrosymmetric molecules in a macroscopic structure (Verbiest et al. 1998).

SHG can also be efficiently generated in bulk material. Bulk SHG in biological specimens appears from the microcrystalline structures, such as calcite and starch granules (Mizutani et al. 2000; Baconnier and Lang 2004) or anisotropic bands of myocytes (Greenhalgh et al. 2007). The orientation of the structural axes with respect to the direction of the beam propagation, for example, has been shown to affect the SHG efficiency in myocytes and thin layers of silicon crystal (Chu et al. 2004; Kolthammer et al. 2005). In most cases bulk-generated SHG signals appear to be stronger than surface SHG, due to larger amounts of coherently phased molecules present in a focal volume compared to the excited molecules at an interface.

In addition to the symmetry and structural constraints, phase-matching conditions have to be met for efficient SHG to occur in a bulk media. Due to wavelength dependency of the refractive index, a phase shift or walk off between the fundamental and second harmonic wave may occur (Armstrong et al. 1962). The phase-matching conditions are usually achieved by angle or temperature tuning of nonlinear crystals that have different refractive indices along the ordinary and extraordinary axes. At a certain orientation of the crystal, the fundamental and second harmonic waves propagate at the same speed, leading to the efficient conversion of the nonlinear signal (Boyd 1992). In nonlinear microscopy, phase-matching conditions are usually satisfied for at least some of the incoming rays due to the wide-angle cone of light produced with a high NA microscope objective.

SHG microscopy has been applied for a variety of biological investigations. SHG is very efficiently generated in collagen (Freund et al. 1986; Campagnola et al. 2002; Lin et al. 2005), actin-myosin complexes of myocytes (Campagnola et al. 2002; Chu et al. 2004; Barzda et al. 2005), tubulin (Campagnola et al. 2002), plant starch granules and other formations of polysaccharides (Mizutani et al. 2000; Chu et al. 2001;

Cox et al. 2005), chloroplasts (Chu et al. 2001; Prent et al. 2005a), and aggregates of light harvesting chlorophyll a/b pigment-protein complexes of photosystem II (LHCII) (Prent et al. 2005b). In the last few years, the number of SHG microscopy applications dramatically increased and continue to grow mainly due to the realization that appreciable scattered or backward-generated SHG signals can be recorded in epi-detection mode with standard two-photon excitation microscopes.

D. THIRD HARMONIC GENERATION

Third harmonic generation microscopy is a relatively new imaging modality that was introduced by Barad et al. (1997). In contrast to SHG, third harmonic can be generated in a homogeneous media; THG is a third-order nonlinear process, and therefore it is dipole allowed (Boyd 1992). Already in the early stages of research on harmonic generation in gases, it was realized that third harmonic generation vanishes under tight focusing conditions in a homogeneous media with normal dispersion; however, when focal symmetry is broken by focusing the beam at the entrance window of the gas cell the THG signal can be obtained (Ward and New 1969). This effect constitutes the basis for contrast mechanism of THG microscopy revealing interfaces in microscopic structures (Barad et al. 1997).

The loss of observable THG in the far field with tight focusing of the beam in homogenous normal dispersion media can be described with the paraxial wave equation [Equation (4.2)] assuming slow spatial variation of electric field amplitudes along the beam propagation direction (z direction). The solution of the paraxial wave equation for the amplitude of third harmonic ($A_{3\omega}$) can be written as follows (Boyd 1992):

$$A_{3\omega}(z) = \frac{i6\pi\omega}{nc} \chi^{(3)} A_\omega^3 J_{3\omega},$$

where

$$J_{3\omega}(\Delta k, z_0, z) = \int_{z_0}^{z} \frac{e^{i\Delta k z'}}{\left(1 + \frac{2iz'}{b}\right)^2} dz', \tag{4.5}$$

where A_ω is the amplitude of the fundamental electric field, $J_{3\omega}$ is the phase matching integral, b is the confocal parameter, $\Delta k = 3k_\omega - k_{3\omega}$ is the wave vector mismatch, and z_0 is the z value at the entrance of the nonlinear medium. The phase-matching integral can be solved analytically for the case of a tightly focused beam in a homogeneous medium. The limits of integration can be replaced by the $-\infty$ to ∞, and the solution can be expressed as follows:

$$J_{3\omega}(\Delta k, z_0, z) = \begin{cases} 0, & \Delta k \leq 0 \\ \frac{1}{2}\pi b^2 \Delta k e^{-\frac{b\Delta k}{2}}, & \Delta k > 0 \end{cases}. \tag{4.6}$$

For normally dispersive materials, the integral and therefore the amplitude of the third harmonic equals zero, even in the case of perfect phase matching where $\Delta k = 0$.

The lack of THG in the far field can be further understood by considering the Gouy shift, or phase anomaly, that a beam undergoes when passing through the focus of a microscope objective. Because of the π phase shift that the beam experiences at the focus, the third harmonic wave generated before the focus is converted back into the fundamental beam after the focus, resulting in a loss of THG signal in the far field (Ward and New 1969; Boyd 1992). THG is observed if an interface of two materials with different refractive indexes or third-order nonlinear susceptibilities is introduced in the focal volume. Although THG appears at interfaces, it is primarily a volume effect since THG is generated in the bulk of the media on both sides of the interface (Saeta and Miller 2001). Third harmonic generation from different structures has been modeled by conducting numerical integration of the phase-matching integral [Equation (4.6)] (Naumov et al. 2001; Schins et al. 2002).

Although third harmonic was first generated and observed from a calcite crystal in the early 1960s (Terhune et al. 1962), efficient THG at an air-dielectric interface was generated in the mid 1990s (Tsang 1995). It was shown that multilayer structures can enhance THG intensity provided the right periodicity matching for the excitation wavelength is achieved (Tsang 1995; Kolthammer et al. 2005). It was also demonstrated that THG is sensitive to the orientation of the interface with respect to the principal direction of propagation of the laser beam (Muller et al. 1998a). For biological applications, THG was used to visualize biological membranes (Muller et al. 1998a), cell walls (Squier et al. 1998), and multilayer structures such as grana of chloroplasts (Muller et al. 1998a; Millard et al. 1999; Chu et al. 2001; Prent et al. 2005a), aggregates of photosynthetic pigment-protein complexes (Prent et al. 2005a), and crista of mitochondria (Barzda et al. 2005). THG was also observed in fixed epithelial, neuron, and muscle cells (Yelin et al. 2002), rhizoids from green algae (Squier et al. 1998), erythrocytes (Millard et al. 1999), cultured neurons and yeast cells (Yelin and Silberberg 1999), human glial cells (Barille et al. 2001), *in vivo* muscle cells (Chu et al. 2004; Barzda et al. 2005), *Drosophila* embryos (Supatto et al. 2005), and sea urchin larval spicules (Oron et al. 2003). THG microscopy was recently reviewed by several authors (Squier and Muller 2001; Sun 2005).

E. MULTICONTRAST MICROSCOPY

Although the three contrast mechanisms of nonlinear microscopy have been described above, it is important to return to the fact that the same excitation source can induce several nonlinear effects simultaneously. Therefore, a sample can be imaged with multiple contrast mechanisms by using a single illumination source and several detectors. Multicontrast microscopy appears to be very beneficial when different optical responses reveal different functional structures of the same biological object. The multicontrast imaging approach provides the possibility to investigate the localization and interaction between the visualized functional structures. Dynamic visualization of multiple structures is likely the most beneficial aspect of multicontrast microscopy. The advantages of multicontrast microscopy far outweigh the difficulties that can arise during the development of the setup.

Parallel images, recorded using different contrast mechanisms, can be directly compared on a pixel-by-pixel basis. Although SHG, THG, and MPF images originate from the same structure, their image contrast mechanisms are fundamentally different. The comparison of images obtained with coherent contrast mechanisms, i.e., generated harmonics, and noncoherent contrast, i.e., fluorescence, can be very challenging. For example, homogeneous structures cannot be visualized in SHG or THG, but might be visible in the fluorescence. Visualization of interfaces differs for different contrast mechanisms. While the maximum intensity appears at the central position of the interface for SHG and THG signals, fluorescence intensity reaches only half of the onset at the interface position between fluorescing and nonfluorescing structures. The maximum MPF signal intensity is observed when the full focal volume is immersed in the fluorescing media. Therefore, image comparison and interpretation will always have to be taken with caution.

It is always beneficial to deconvolute the images with the point-spread function of a particular contrast mechanism. The point spread functions are different for the different nonlinear responses. Non-deconvoluted images appear to be blurred due to the finite size of the focal volume that is scanned across the structure. If deconvolution is not performed, comparison may lead to artifacts, where two neighboring structures revealed by different contrast mechanisms may appear to overlap because of the effect of blurring. The comparison algorithms of two and three parallel images have been developed (Greenhalgh et al. 2005) and are discussed in the following chapter.

Simultaneous imaging with MPF, SHG, and THG contrasts has been used to visualize mitochondria and anisotropic bands of sarcomeres in myocytes (Barzda et al. 2005), chloroplasts (Chu et al. 2001; Prent et al. 2005a), and photosynthetic pigment-protein complexes (Prent et al. 2005b). A multicontrast MPF and SHG microscope was used to image labeled lipid vesicles (Moreaux et al. 2000b), labeled neuroblastoma cells (Campagnola et al. 1999), muscle and tubulin structures (Campagnola et al. 2002), and labeled neurons (Moreaux et al. 2001). The THG and MPF simultaneous detection was used for imaging human glial cells (Barille et al. 2001), while THG and SHG contrasts were used to monitor mitosis in a live zebrafish embryo (Chu et al. 2003a).

III. INSTRUMENTATION OF THE MULTICONTRAST NONLINEAR MICROSCOPE

Nonlinear laser scanning microscopes are modified versions of traditional confocal laser scanning microscopes. Instead of continuous wave (CW) laser sources, picosecond and femtosecond lasers are employed to efficiently drive nonlinear light-matter interactions within the focal volume created by a high numerical aperture (NA) microscope objective. In contrast to confocal microscopy, where optical sectioning is achieved by removing out-of-focus excitations via a spatial filter, the confocality in nonlinear microscopes is achieved as a direct result of the nonlinear intensity dependence. Focusing a pulsed laser beam provides enough intensity to achieve nonlinear interactions and provides optical sectioning with high spatial resolution. By scanning the focal spot in the xy plane, where the beam propagation direction is defined to be along the z-axis, 2D images can be created by synchronizing the signal detection with the scanning mirror position. In addition, by scanning the sample at different positions along the z-axis, 3D images can be constructed by rendering the 2D slices with 3D visualization software.

Depending on the desired nonlinear interaction, a nonlinear laser scanning microscope can involve more than one laser beam. For example, second and third harmonic generation imaging can be accomplished with a single laser beam, while CARS and sum frequency generation require two beams with different wavelengths. In principle, the setup can be extended to three laser beams of different wavelengths for nondegenerate four-wave mixing microscopy. In addition, the application of multiple beams of the same wavelength can be used to produce multiple foci, which would decrease the image acquisition times and provide simultaneous imaging at different depths (Amir et al. 2007). This technique is advantageous for imaging dynamic systems because the time delay between 2D images at different optical depths is eliminated. For the highest spatial resolution, a phase delay between the pulses of the multiple beams is introduced to avoid interference between the foci. Single beam setups are always easier to implement and are typically more robust when used in clinical or biology lab environments. This chapter focuses only on single laser beam systems.

The functionality of nonlinear microscopes can be enhanced by implementing a multicontrast detection system. Traditionally, three parallel detection channels have been used in confocal and two-photon excitation fluorescence microscopes, where the emission signal is usually divided into different spectral ranges by dichroic mirrors and optical filters, or by separating the signal with a dispersive optical element. Similarly, spectral separation can be applied for detecting different nonlinear optical responses. Parallel images acquired simultaneously based on different nonlinear contrast mechanisms carry complementary information about the same sample structure. Multicontrast detection methods provide the possibility of making a direct correlation of the parallel images on a pixel-by-pixel basis. In addition, a simultaneous detection scheme eliminates the problem of artifacts from signal bleaching or movement of the sample occurring during imaging.

Since multiphoton excitation fluorescence is emitted isotropically, it can be collected in either forward or backward direction. Conveniently, backward fluorescence detection, epi-fluorescence, uses the excitation objective for collecting the emitted fluorescence photons. Many research groups implement nonlinear excitation fluorescence imaging by coupling femtosecond lasers into a confocal microscope and using a non-descanned port for efficient signal detection (see discussion by Zipfel et al. 2003). Harmonic signals are generated more efficiently in the forward direction; therefore, they are usually detected in transmission mode. Larger modifications of a confocal microscope are required for implementing transmission mode detection. The easiest way of building a harmonic microscope is by using a high NA condenser and transmission detector that exists on some models of laser scanning confocal microscopes (see, for example, Millard et al. 2003; Cox et al. 2005). In addition, three-channel detection requires extensive modifications to a commercial microscope (Sun 2005) or making a home-built microscope setup (Barzda et al. 2005). The instrumentation of nonlinear microscopes was extensively reviewed by Squier and Muller (2001), Sun (2005), and Barzda (2008). Nonlinear microscopes have three main functional parts: the optical setup of the microscope, the synchronized detection/scanning system, and the femtosecond laser source. In the following sections, a detailed description of each of these functional units is provided.

A. THE MICROSCOPE

This section describes, as an example, the three-channel multicontrast nonlinear microscope built in our laboratory. Figure 4.1 depicts a typical scheme for the microscope setup. Two mirrors, schematically shown by M1, direct the laser into the microscope box. Inside the microscope, the beam is expanded with a telescope (lenses L1 and L2) to fill the clearance aperture of the scanning mirrors. The same telescope also spatially filters the beam with an appropriate pinhole (PH) located at the focus between the two lenses. After the first telescope, the beam is coupled to two galvanometric mirrors that raster scan the beam in both the lateral x and y directions. A second telescope, consisting of an achromatic lens (L3) and a tube lens (L4), is used to expand the beam to match the entrance aperture of the excitation objective (EO). The tube lens is designed to correct for aberrations of the objective. It is advisable to use a tube lens that matches the objective and is produced by the same manufacturer. After the second telescope, the collimated beam is transmitted through a dichroic mirror (DM1) and coupled into the excitation objective (EO). Almost all nonlinear microscopes are constructed using commercially available refractive objectives.

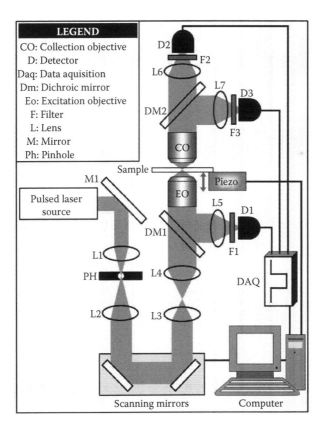

FIGURE 4.1 The setup of the nonlinear multicontrast microscope. Microscope components are defined in the figure legend. See text for detailed description of the setup.

Since most objectives are designed for the visible spectral range, there are only a few objectives that work optimally in the infrared region. Users are advised to contact objective manufacturers for specifications on the use of their objectives with infrared femtosecond lasers. For achieving the highest resolution specified for an objective, a high uniformity of the excitation beam across the entrance aperture of the objective is required. Overfilling the entrance aperture often helps to achieve good uniformity and the specified NA of the objective. It is recommended to test the alignment of the microscope on a regular basis by recording the point spread function (PSF) of the microscope. Nonlinear signals generated in the focal volume can be detected in the forward and backward directions. Multiphoton excitation fluorescence is usually collected in the backward direction with the same objective (EO) used for excitation. A dichroic mirror (DM1) that is specifically chosen for the given fundamental and fluorescence emission wavelengths reflects the fluorescence photons to a lens (L5) so they can be focused into the photomultiplier tube (D1). Interference or band-pass filters (F1) are used in front of the detector for filtering scattered fundamental and harmonic light. Since excitations in nonlinear microscopes are confined to the focal volume of the objective, the use of a pinhole in front of the detector is not required for achieving optical sectioning. Harmonic signals are usually detected in the forward direction with a high NA objective (CO). The collected and collimated signal beam is passed through a dichroic mirror (DM2) that separates the second and third harmonic signals. The second harmonic is focused onto the detector (D2) by the lens (L6), and filtered by the interference filter (F2). Similarly, the third harmonic is focused onto the detector (D3) with the lens (L7), and filtered with the interference filter (F3). A photon-counting data acquisition card (DAQ) is used to receive the pulses in parallel from all three detectors. To obtain optical sections at different depth, the sample is translated along the optical axis with a piezo stage. A computer interface synchronizes the photon counting signal with the x and y positions from the scanning mirrors and the z position from the piezo stage. This renders three simultaneously acquired 3D images with different nonlinear contrast mechanisms.

B. SCANNING AND DETECTION SYSTEMS

For obtaining a three-dimensional image, the focal volume of the high NA objective has to be scanned across a 3D region of interest. It is possible to translate the sample with respect to the laser beam, or scan with the beam while keeping the sample fixed. Sample translation is usually accomplished by a piezo stage, with scanning speeds of several frames per second. Faster scanning rates can be achieved by raster scanning the beam in the lateral directions. Galvanometric mirrors are usually employed for lateral scanning. For video rate scanning, resonance scanners or rotating polygon mirrors can be used. Axial scanning is most commonly performed by translating the sample or by translating the excitation objective. Axial scanning can also be performed by changing the divergence of the beam with adaptive optics (Amir et al. 2007). Translation of the microscope objective gives satisfactory results in epi-detection mode. For detecting the signals in transmission mode the focal point of the excitation and collection objectives must overlap; therefore, axial scanning by sample translation is preferred over the objective translation.

Detection is a very important component of the microscope. Photomultiplier tubes, avalanche photodiodes, and charge-coupled device (CCD) cameras can be used for detection. CCD cameras can be easily interfaced with the microscope; however, if the CCD camera is not synchronized with the scanners, multiple scans are necessary to acquire one image. Scattering samples require point or line scanning synchronized with the detection. Photomultipliers are usually employed for the non-descanned mode due to the large area of the photocathode. In the descanned detection mode, small area avalanche photodiodes or spectrometers can be used. The descanned mode usually has higher losses due to the collection beam passing through more optical elements compared to the non-descanned configuration.

Nonlinear optical responses can be detected using integration, photon counting, or lock-in detection methods. Commercial manufacturers implement mostly the signal integration approach. However, linear excitation fluorescence and nonlinear signals that are emitted from the microscopic samples typically generate less than one photon per excitation pulse. Therefore, it is more appropriate to use the photon counting detection method. The photon counting method has a low signal saturation threshold. This has to be taken with caution but usually does not present a big problem, because excitation laser power can be reduced.

The excited fluorescence radiates uniformly in all directions and is usually detected in the backward direction. Since a confocal pinhole is not used in nonlinear microscopes, the fluorescence can be collected without descanning. In addition to fluorescence, scattered second and third harmonic signals can be recorded in the epi-detection mode. However, strongest second and third harmonic signals are generated in the forward direction due to the phase-matching conditions of coherent radiation. The harmonic signals in the forward direction are collected with an objective (Barzda et al. 2005) or a high NA condenser (Millard et al. 2003). Either one detector with appropriate filters or several detectors detecting different signals separated by dichroic mirrors can be used. Alternatively, the nonlinear response signal can be coupled to a monochromator, and a whole spectrum can be recorded at each pixel (Sun 2005). In this case, the harmonic and fluorescence images are constructed by obtaining the pixel intensities from a selected range of the spectrum. Nonlinear signals such as MPF, SHG, and THG are simultaneously generated; therefore, if simultaneously recorded, the images can be directly compared and statistical image analysis methods can be applied (Barzda et al. 2005).

C. Laser Sources

Each molecule has its characteristic linear and nonlinear absorption spectrum. Therefore, it is very important to have a widely tunable wavelength femtosecond laser source for proper choice of sample excitation. For nonlinear imaging, the laser should be tuned away from a linear absorption of the sample. Infrared excitation usually works well with most biological samples. The tuning window for nonlinear excitation lies in the approximate range between 800 nm and 1500 nm. The use of shorter wavelengths is limited by the linear absorption of endogenous biomolecules and the longer wavelengths are limited by water absorption. Another technical limitation arises from THG signals falling in the ultraviolet (UV) wavelength range.

Most widely used femtosecond Ti:sapphire lasers emitting at 800 nm produce third harmonic signals at 266 nm. For THG detection, UV transmitting coverslips and objectives as well as other UV optics must be used along the third harmonic detection optical path. In addition, thick biological samples may attenuate THG signal due to high absorption in the UV. Therefore, excitation sources such as Yb:KGW or Cr:forsterite femtosecond lasers emitting at 1030 nm and 1250 nm, respectively, are much better choices than Ti:sapphire lasers. Current advances in turnkey Yb-doped pico- and femtosecond fiber lasers may boost the nonlinear biological imaging applications, especially for the third harmonic contrast mechanism.

The pulse duration is an important parameter for efficient generation of SHG and THG signals. The shortest possible pulses are desired for most efficient nonlinear excitation. However, there is a limited pulse peak power that can be applied to a biological specimen, above which dielectric breakdown takes place, manifesting itself in the occurrence of cavitations and visible structural changes (Muller et al. 1998a). Therefore, a balance between the highest peak power and the average laser power has to be achieved. In practical terms, pulses of a few hundred femtoseconds work well. In addition, very short pulses (<20 fs) are significantly broadened due to group velocity dispersion when passing through the glass material of optical components. Microscope objectives, in particular, contain many optical elements that can lead to considerable pulse broadening. If desired, dispersion precompensation can be used for obtaining very short optical pulses at the focal point of the objective (Muller et al. 1998b).

In addition to the pulse duration, the repetition rate of the laser pulses provides another parameter for optimization of nonlinear excitations. As a standard, a repetition rate around 80 MHz is used in most commercial Ti:sapphire lasers. Two trends of decreasing or increasing the repetition rate has been suggested (Barzda et al. 2001; Chu et al. 2003). If nonlinear excitation appears in the two or three photon absorption range, it is desirable to keep the duration time between pulses long enough to permit full relaxation of the excited states. This reduces exciton-exciton annihilation effects and diminishes the generation of triplet states, which lead to the production of singlet oxygen (Barzda et al. 2001) and bleaching of the chromophores. If the excitation falls outside the fundamental and two- or three-photon absorption bands, excitation to the virtual levels does not depopulate the ground state and does not deposit energy into the sample. Therefore, higher pulse repetition rates can be used, and since the peak intensities of the pulses must be kept below the critical level, increasing the repetition rate linearly increases the signal intensity (Chu et al. 2003).

The polarization of the laser also plays a significant role in nonlinear microscopic imaging. The SHG intensity distribution in the image can be different for linear and circular polarized excitation. For example, a linearly polarized laser renders an image of a starch granule with a characteristic dark line across it, while circular polarization produces a uniform image of the granule except for the dark central region (see the section on imaging of starch). However, circularly polarized excitation does not generate the third harmonic signal (Fittinghoff et al. 2005). Therefore, it is very important to have linearly polarized light for efficient imaging with THG. Optimization of linear or circular polarization at the focal spot can be conveniently achieved, respectively, by maximizing or minimizing the THG signal intensity at the air-glass interface. Dichroic mirrors can introduce significant ellipticity to the laser beam; therefore, it is advisable to carefully characterize and optimize the excitation polarization in the microscope.

IV. IMAGE ANALYSIS METHODS FOR MULTICONTRAST MICROSCOPY

A. STRUCTURAL CROSS-CORRELATION IMAGE ANALYSIS FOR TWO IMAGES

Systematic comparison of 2D images is necessary for an adept interpretation of signals originating from different imaging modalities. A pixel-by-pixel comparison algorithm was adopted for this purpose, and a logical comparison of the mathematical criterion is used to separate image regions into areas of overlap (correlation) and nonoverlap (anticorrelation) between the two signals (Greenhalgh et al. 2005). The algorithm, termed *structural crosscorrelation image analysis* (SCIA), assumes that images are spatially aligned, which occurs automatically with the multicontrast imaging microscopy method. The product $Z_{i,k}$ of normalized corresponding pixels in two source images, $A_{i,k}$ and $B_{i,k}$, provides good co-localization results:

$$Z_{i,k} = A_{i,k} \cdot B_{i,k}, \text{ where } A_{i,k} = \frac{a_{i,k}}{a_{max}} \text{ and } B_{i,k} = \frac{b_{i,k}}{b_{max}}, \qquad (4.7)$$

where i and k are the pixel coordinates in the two dimensional images, and $a_{i,k}$ and $b_{i,k}$ stand for the intensity values in the original images.

First, thresholds are applied to images by inspecting the noise levels that differ for each channel. A correlated image A¿B is constructed from the two original images A and B using equation (4.7), where ¿ represents the logical intersection. Pixels with a correlation value of $Z = 0$ (i.e., signal that was generated in one image and not the other) are said to be uncorrelated with each other, and are separated into two images, uncorrelated A (A − A¿B) and uncorrelated B (B − A¿B), where the minus represents the logical "and not." The three images, A¿B, A − A¿B, and B − A¿B are independent (i.e., they do not share any pixels) and can therefore be combined into a three-color image where pixels in each resultant image are colored independently of their intensity. In addition, each data set comprised of the correlated and two uncorrelated 2D images can be reconstructed into a 3D image. Each of the three 3D images can be rendered separately or combined together in a multicolor 3D volume and rendered afterward for analysis of structural relations and spatial correlations in the investigated structure.

B. STRUCTURAL CROSS-CORRELATION IMAGE ANALYSIS FOR THREE IMAGES

SCIA can be applied for correlation of three images obtained with multicontrast nonlinear microscopy. The two-image SCIA can be applied to correlate separately each two of the three images. However, direct correlation of all three images with SCIA algorithm renders quicker and more transparent results. To illustrate three-channel SCIA, an example is shown in Figure 4.2 where three channels (A, B, and C)

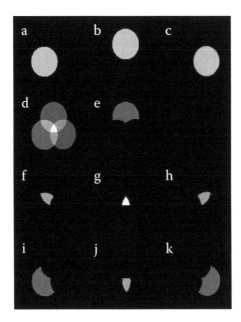

FIGURE 4.2 Example of three channel structural crosscorrelation image analysis (SCIA) showing initial images (a), (b), (c) and the correlation overlay image (d). The SCIA image (d) is composed of 7 images each colored with a different shade or color. The SCIA images are also presented separately: (e) uncorrelated channel B, (f) correlation between channels A and B but not C, (g) correlation between all three channels, (h) correlation between channels B and C but not A, (i) uncorrelated channel A, (j) correlation between channels A and C but not B, and (k) uncorrelated channel C.

are circles arranged in the Venn style with partial overlaps (Figure 4.2a,b,c) and the resulting correlation image is shown in Figure 4.2d. The correlation overlay image is composed of seven mutually exclusive images shown in different shades. Adopting identical notation as in the previous section, one three-channel correlation image is produced, which shows the correlated and uncorrelated areas where all three original images correlate, $A \cup B \cup C$, shown in white, and three two-correlating images are produced, $A \cup B - C$ in yellow, $A \cup C - B$ in magenta, and $B \cup C - A$ in cyan. In addition, three fully uncorrelated images are produced, A-B-C in red, B-A-C in green, and C-A-B in blue.

The advantage of the SCIA technique as compared to conventional image overlap analysis is that each correlation can be analyzed separately, and accordingly switched on and off to focus only on the desired correlations between the images. An imaging example of the three-image SCIA is presented in Figure 4.3. Respectively, Figure 3a, b, and c show the MPF, SHG, and THG images of a rhodamine B–labeled potato starch granule imaged with multicontrast nonlinear microscopy. Panel (d) of Figure 4.3 presents the SCIA correlation. The dye was added to illustrate the SCIA principle with three channels and is not required for second or third harmonic generation in starch granules.

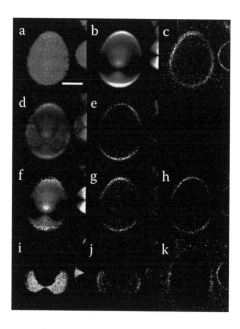

FIGURE 4.3 Three channel SCIA applied to images of a potato starch granule dyed with rhodamine B stain and imaged with vertically aligned linearly polarized excitation simultaneously with (a) two photon excitation fluorescence, (b) second harmonic generation, and (c) third harmonic generation. The image (d) presents the combined three channel correlation image highlighting correlations with different shades. The 7 correlation images are also presented separately: (e) uncorrelated channel B, (f) correlation between channels A and B but not C, (g) correlation between all three channels, (h) correlation between channels B and C but not A, (i) uncorrelated channel A, (j) correlation between channels A and C but not B, and (k) uncorrelated channel C. The scale bar represents 12 μm.

The example in Figure 4.3 shows that the THG does not correlate with the SHG at the top and bottom edges of the starch granule due to the larger SHG signal and the smaller point spread function of THG, which makes the THG seem to be more inside the starch as shown in white and yellow in panel (d). The correlation shows that THG at the left and right edges of the starch is not correlated with MPF (shown in blue). This indicates that the dye is not concentrated at the edge of the starch granule. The example in Figure 4.3 shows an elegant way of discriminating structures uniquely expressed by the different contrast mechanisms. SCIA is used to discriminate different features in complex biological structures and provides a better understanding of the spatial co-localization of the structures revealed by the different nonlinear contrast mechanisms.

V. NONLINEAR MICROSCOPY OF STARCH GRANULES

Starch granules from different plants including algae *Chara coralline* (Mizutani et al. 2000), flowering plant *Lantana camara* (Cox et al. 2005), corn (Chu et al. 2001), potato, rice, and tapioca can be considered as nonlinear optical materials, which produce strong second harmonic generation. The starch granule is organized into a concentric multilayer structure possessing central symmetry with the hilum, the nucleus of the starch granule, located in the center. Starch granules show strong birefringence and when observed with polarization microscopy, exhibit the characteristic "Maltese cross." Starch granules are primarily composed of two glucose polymers: amylose, which is a long linear polymer of glucose, and amylopectin, which consists of shorter linear glucose polymers branched in such a way as to allow the chains to produce double helical arrangements around one another (Glaring et al. 2006). Starch granules consist of alternating layers of hard crystalline and soft semicrystalline material (Gallant et al. 1997; Buleon et al. 1998), which give rise to the characteristic "growth ring" pattern when observed in cross section by optical microscopy. Atomic force microscopy (AFM) has shown blocklets of up to 400 nm in diameter located in the crystalline layers, and 20 nm blocklets embedded in the semicrystalline layers (Gallant et al. 1997). The blocklets are made of alternating crystalline and amorphous lamella (Gallant et al. 1997). Although amylose content is known to affect the crystallinity of the amylopectin, the structure of semicrystalline layers is still not well understood. The inherent sensitivity of SHG signal to non-centrosymmetric macrostructural order renders itself as a highly valuable tool for measuring the crystallinity and structural organization in the starch granule.

SHG imaging of starch was first performed by Mizutani et al. (2000), who showed that starch granules in plant cells generate intense SHG when imaged with 1064 nm laser. The SHG signal in starch originates from well-ordered non-cetrosymmetric nanocrystallites (Chu et al. 2001; Mizutani and Sano 2003; Cox and Feijo 2004; Cox, Moreno, and Feijo 2005). The starch granule shows a linear polarization dependent SHG intensity distribution across the granule (Chu et al. 2001; Cox and Feijo 2004; Cox et al. 2005). The polarization dependency of SHG appears to be due to the radial arrangement of chiral nanostructures in the starch granule (Chu et al. 2001; Cox and Feijo 2004; Cox et al. 2005). Starch granules have been employed as an efficient nonlinear optical material for laser pulse characterization with the frequency-resolved optical gating (FROG) technique (Amat-Roldan et al. 2004), as well as for backscattered SHG autocorrelation measurements (Anisha et al. 2007). In addition, starch has been imaged with third harmonic generation microscopy (Figure 4.3c). The THG image reveals the interface between the starch granule and the surrounding media, which is similar to the THG structural pattern observed from polystyrene beads (Clay et al. 2006).

Figure 4.4 shows high-resolution SHG images of starch granules from potato. The images were recorded with vertical (Figure 4.4a) and horizontal (Figure 4.4b) linear polarization, and circular polarization (Figure 4.4c) of the laser beam. Standard potato starch was immersed in water and immobilized by gelling in a very thin layer of polyacrylamide gel sandwiched between microscope coverslips. The SHG image reveals several characteristic features of the starch granule, which are common to many types of starch including corn, potato, tapioca, rice, and pea, although shape of the starch from various plants tends to be quite different. The observed ring structure in Figure 4.4 is due to a difference in the generation of SHG between the hard crystalline and soft semicrystalline layers. The SHG pattern produced by varying the orientation of linear polarization (Figures 4.4a and b) is due to the radial organization

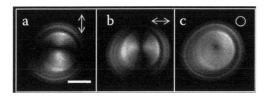

FIGURE 4.4 SHG images of a potato starch granule at different polarizations of the laser, where the direction of linear polarization is indicated on the top left corner, in (a) and (b), while the polarization is circular in (c) as indicated by the circle on the top left corner. The scale bar is 8 μm.

of SHG emitters within the starch (Chu et al. 2001; Cox and Feijo 2004; Cox et al. 2005). Here we show an SHG image obtained with circularly polarized light, which confirms the radial arrangement of crystallites showing the entire starch structure homogeneously highlighted independently of the radial orientation of the granule (Figure 4.4c). Interestingly, however, a small round "hole" of the microscope point spread function (PSF) size is found at the center of nucleation (hilum) of all centro-symmetrically shaped starch granules. The absence of SHG in the center of starch granules may appear to be due to a centrosymmetric or isotropic arrangement of the hilum. In some starch granules a larger dark central region than PSF is observed, indicating that the hilum region has cracks and/or a randomly arranged structure.

The SHG produced from starch granules is very intense and conveniently shows the direction of linear polarization or circular polarization in the focal plane of the high NA microscope objective. The starch granule can be conveniently used for microscope alignment and optimization of the laser beam polarization in the micro-scope. Furthermore, SHG images and relative signal intensity from starch granules can be used as a measure of microscopic starch crystallinity, which is beneficial for agricultural industries and food quality control.

VI. MULTICONTRAST NONLINEAR IMAGING OF PHOTOSYNTHETIC SYSTEMS

Green plants perform light harvesting and photosynthetic energy conversion by using a highly organized structure of pigment-protein complexes embedded in the thyla-koid membranes of chloroplasts (see, for example, Mustardy and Garab 2003). The thylakoid membrane is a dynamic structure that is highly responsive to illumination conditions resulting in a correlation between photosynthetic activity and structural changes in the chloroplast (Barzda et al. 1996). Light-harvesting chlorophyll a/b pig-ment-protein complex of photosystem II (LHCII) is the most abundant photosynthetic complex in chloroplasts. The LHCII contains non-covalently bound chlorophyll and carotenoid pigments that have large transition dipole moments as well as high first- and second-order hyperpolarizabilities, and are therefore ideally suited as inherent labels for fluorescence imaging as well as for harmonic generation (for overview see Barzda 2008). The second and third harmonic generation intensity strongly depends on the spatial organization of the photosynthetic pigments. Depending on the spatial distribution of these pigments in thylakoids, harmonic radiation from each pigment molecule will constructively or destructively interfere, resulting in a strong or weak far field harmonic signal (Boyd 1992; Mertz 2001). While chloroplast photosynthetic activity can be monitored with chlorophyll fluorescence, the structure can be studied with harmonic generation microscopy. Multicontrast nonlinear microscopy opens a new opportunity for studying photosynthetic mechanisms on a single chloroplast level by simultaneously investigating the functional activity with multiphoton excitation fluorescence and imaging the structural dynamics with second and third harmonic generation contrast mechanisms (Barzda 2008). In the following, we present the multicontrast nonlinear microscopy studies of several photosynthetic structures.

A. IMAGING AGGREGATES OF THE GREEN PLANT LIGHT-HARVESTING PIGMENT-PROTEIN COMPLEX LHCII

The pigment-protein complexes in thylakoid membranes of chloroplasts in green plants are organized in highly ordered arrays (Dekker and Boekema 2005). When extracted out of the membrane with a detergent, under a high ion concentration, LHCII tends to assemble into multilamellar, semicrystalline, or random aggregates depending on the lipid content isolated together with the LHCII (Simidjiev et al. 1997). The multilamellar aggregates of LHCII serve as a good model system for studying the photophysical properties of photosynthetic membranes.

Figure 4.5 shows MPF, SHG, and THG images of aggregated LHCII from *Pisum savitum* (pea). The fluorescence image reveals a heterogeneous structure of the LHCII aggregate. The fluorescence intensity cannot be directly assigned to the concentration of LHCII due to the quenching properties of these complexes. However, the fluorescence signal can be used as a natural indicator for determining the location of the pigment-protein complexes.

The SHG image of the LHCII aggregate is shown in Figure 4.5b. Heterogeneity of the SHG signal from the aggregate can be observed. The SHG intensity depends on the macroscopic arrangement of LHCII within the aggregate. Randomly aggregated pigment-protein complexes do not generate SHG, while heterogeneities such as non-centrosymmetric lamellar formations, aggregate edges, or non-centrosymmetric semicrystalline structures give rise to SHG. The domains of higher SHG intensity can be assigned to more ordered structures, which give constructively interfering signals

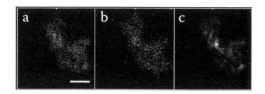

FIGURE 4.5 Aggregates of light-harvesting pigment-protein complex (LHCII) imaged with (a) MPF, (b) SHG, and (c) THG contrasts. The scale bar is 3 μm.

from the SHG radiating dipole arrays. Comparison of the fluorescence and SHG images shows a high correlation of both signals. This can be arguably explained by a higher concentration of pigments in highly ordered domains giving rise to stronger SHG signals. Lower pigment concentration containing disordered structures show both lower SHG and fluorescence. Interestingly, aggregation induces significant fluorescence quenching in LHCII (Horton et al. 1991); therefore, correlation between SHG and MPF is not straightforward and a variation of the macrostructural arrangement of pigments and their quenching properties largely determine the distribution of nonlinear properties in the aggregate.

The THG image of LHCII aggregates (Figure 4.5c) reveals several structural domains. The THG structural domains are more pronounced and do not always correlate with SHG and MPF domains. The overall signal strength of THG appears to be higher than SHG or MPF. THG is not dependent on the symmetry of the aggregate, but it requires the structure to have heterogeneities, such as interfaces between materials with different refractive indices or third-order nonlinear susceptibilities (Muller et al. 1998a; Barad et al. 1997). Areas with low THG intensity as well as low SHG intensity are randomly organized. The LHCII domains that show high THG intensity and have low SHG are probably organized in a multilamellar arrangement with the lamellae planes oriented perpendicular to the laser beam. The multilamellar arrangement enhances THG signal, while the low SHG could be due to the destructive interference of SHG resulting from an antiparallel arrangement of pigments in alternating layers of the lamella, similar to the SHG cancellation observed in the adjacent region of two vesicles with asymmetrically labeled lipid membranes (Moreaux et al. 2000b). The areas highlighted by SHG but not THG have non-centrosymmetric domains arranged in a continuous three-dimensional structure without interfaces. Alternatively, the multilayer membranes with planes parallel to the propagation of the beam may not produce THG, while giving strong SHG and fluorescence signal. The structures exhibiting high SHG, THG, and fluorescence possess interfaces oriented perpendicular to the beam propagation as well as non-centrosymmetric organization in the domains. The results obtained from imaging LHCII aggregates are very helpful in understanding the structural organization of chloroplasts.

B. HIGH-RESOLUTION IMAGING OF CHLOROPLASTS

Nonlinear microscopy can be used to image chloroplasts *in vivo* as well as *in situ* (Muller et al. 1998a; Chu et al. 2001; Tirlapur and König 2001; Prent et al. 2005a). LHCII is the most abundant pigment-protein complex in chloroplasts; therefore, results obtained from imaging LHCII aggregates can be used for understanding the multicontrast microscopic images of chloroplasts. Chloroplasts are comprised of a complex network of photosynthetic grana and stroma membranes. Grana, the multilamellar stacks of photosynthetic thylakoid membranes, are interconnected via stroma membranes into one large membranous network extended over the whole space of the chloroplast (Mustardy and Garab 2003).

Figure 4.6 shows a dark adapted, freshly isolated chloroplast imaged with a multicontrast nonlinear microscope coupled to Yb:KGW laser radiating at 1030 nm (Major et al. 2006). The Yb:KGW laser preferentially excites carotenoids via the

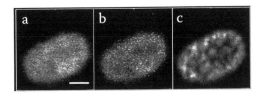

FIGURE 4.6 Isolated dark adapted chloroplast imaged with (a) MPF, (b) SHG, and (c) THG. The chloroplast is edge aligned with thylakoid membranes oriented perpendicular to the image plane. The scale bar is 3 μm.

two-photon excitation process (Walla et al. 2002). The excitation from carotenoids is efficiently transferred to chlorophylls that emit fluorescence at wavelengths peaking around 680 nm. The fluorescence image (Figure 4.6a) shows a relatively homogeneous distribution of intensity over the entire volume of the chloroplast. The image was recorded by collecting red fluorescence between 650 and 730 nm. At these wavelengths, the fluorescence from photosystem II residing in the grana and photosystem I found in the stroma were not resolved. Although there is a higher concentration of pigment molecules in the grana, the quenching properties of the grana apparently hinder the extra fluorescence intensity.

The SHG images of chloroplasts often show a relatively homogeneous intensity over the whole area except for high intensity spots attributed to starch granules (Mizutani et al. 2000) (not shown). In high light conditions, chloroplasts synthesize starch, which is deposited in starch granules (Li et al. 2006). The SHG from starch granules is very intense compared to SHG from the rest of the chloroplast. Therefore, in the presented investigation (Figure 4.6) the plant was dark adapted for 12 hr prior to chloroplast isolation in order for the plant to consume its starch so that SHG signal originating only from thylakoids could be obtained. Although stacks of thylakoid membranes contain large amounts of LHCII arranged in a chiral organization, giving anomalously strong circular dichroism signals (Garab et al. 1988), the SHG is apparently not enhanced in the grana as compared to the stroma region. Destructive interference of SHG from each thylakoid membrane appears to be due to the alternating arrangement of the membranes, both in the stack of the thylakoids in the grana as well as in the two antiparallel stroma membranes. Therefore, the SHG signal from the grana and stroma membranes appears to be weak and has a homogeneous intensity distribution.

Pronounced structural details inside chloroplasts can be observed with THG microscopy (Millard et al. 1999; Prent et al. 2005a; Barzda 2008). Since multilayer structures enhance THG, as compared to monolayer interfaces (Tsang 1995), the multilamellar structure of the grana in chloroplasts, similar to the LHCII aggregates, enhance the intensity of THG. Thylakoid membranes oriented perpendicular to the excitation beam propagation (face oriented) direction render the largest THG signal. For the edge-aligned chloroplasts, where thylakoid membranes in the grana are oriented parallel to the beam propagation direction, appreciable THG signals are also observed (Figure 4.6c). The THG signal is also enhanced by the second hyperpolarizability of the carotenoid molecules. Due to the highly ordered pigment-protein complexes in thylakoid membranes, the orientation-dependent THG signals can help to elucidate the ultrastructure of the chloroplast.

The structure of *in vivo* chloroplasts can be thoroughly investigated using multi-contrast imaging microscopy. Without the addition of extrinsic dyes, the structural organization of chloroplast can be elucidated from the relative signal strengths of different nonlinear contrast mechanisms. Microscopic imaging with nonlinear signals from chloroplasts at different orientations yields information about the three-dimensional organization of the grana. A deduction of granum ultrastructure below the diffraction limit is possible by analyzing and comparing intensities of the nonlinear signals. Moreover, the intensity variation of different nonlinear responses over time provides information about the dynamic changes of organization in pho-tosynthetic structures. The naturally pigmented photosynthetic organelles produce intense nonlinear signals. Therefore, multicontrast imaging microscopy has great potential for studying *in vivo* and *in situ* photosynthetic structures.

VII. MULTICONTRAST NONLINEAR MICROSCOPY OF MUSCLE CELLS

The inherent anisotropy found in the anisotropic (A-) bands of muscle cells makes them ideal candidates for biological investigations using second harmonic genera-tion microscopy. In addition, SHG microscopy can be performed on living myo-cytes, which is advantageous over electron microscopy techniques that require fixing and dehydration of the sample. Tomography studies first showed that the second harmonic could be generated from muscle (Guo et al. 1997), but it took five years before the first muscle cells were imaged with SHG microscopy (Campagnola et al. 2002).

Mouse skeletal muscles imaged with SHG microscopy revealed bright SHG bands, with a periodicity on the same order of a sarcomere length, which is the fundamental unit in a muscle cell. Several studies have been dedicated to elucidate the origin of SHG signal from the striated muscles. It has been shown that SHG originates from the A-band region, where the semicrystalline arrangement of myosin molecules in the myofibrils is responsible for the generation of SHG (Campagnola et al. 2002). The ori-entation and symmetry of the myofibrils affects the efficiency of the second harmonic response (Chu et al. 2004; Greenhalgh et al. 2007), and it has been suggested that this dependency could be valuable in understanding muscular dynamics and intracellular regulatory processes, especially when combined with other nonlinear contrast mecha-nisms, such as MPF and THG (Both et al. 2004; Barzda et al. 2005).

Muscle structures can also be investigated with third harmonic generation microscopy. In multicontrast investigations, THG signal reveals that the isotro-pic (I-) bands are interlaced with the A-bands, which are visualized with SHG (Chu et al. 2004). Furthermore, it has also been shown that the mitochondria sur-rounding cardiomyocytes produce efficient THG (Barzda et al. 2005). Figure 4.7a shows the simultaneously detected THG (dark gray) image, which reveals mostly sarcolemal mitochondria. Contrarily, Figure 4.7b shows the cardiomyocyte image in SHG (light gray), where the characteristic banded structure of myofibrils is clearly depicted. Merging of the THG and SHG images shows that the signals are originating

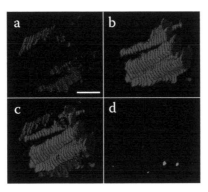

FIGURE 4.7 Rendered three dimentional images of cardiomyocytes from mouse. An uncorrelated THG image is presented in (a), an uncorrelated SHG in (b), the combined image is shown in (c), and the correlation between SHG and THG is shown in (d). The scale bar is 10 μm.

from separate structures (Figure 4.7c). Figure 4.7d depicts the SCIA crosscorrelation analysis image, which has very little positive correlation supporting the fact that SHG and THG signals are generated from different structures (please refer to correlation section). This highlights the advantages of the multicontrast imaging abilities of nonlinear microscopy. In addition, it should be noted that no labels were required to observe SHG from the A-bands of sarcomeres or THG from mitochondria, and since the excitation wavelength can be moved to a nonabsorbing region, muscle cells can be imaged for extended periods of time without any damage.

The SHG signal generated in the A-bands of the myocytes is extremely strong; therefore, it can be used to image muscle contraction at sub-video rates without damage to the cellular structures (Greenhalgh et al. 2006). Noninvasive dynamic visualization of a contracting muscle cell with nonlinear microscopy opens new possibilities for contractility research. Recently it has been shown that intensity fluctuations in the SHG signal can be directly correlated with the contraction state of the muscle (Greenhalgh et al. 2007). This phenomenon is a direct result of the change in semicrystalline arrangement of myosins and the antiparallel arrangement of myosin molecules in the central region of the A-band. Typically, the A-band exhibits a double-peaked profile, where there is a reduction in SHG intensity at the center. However, forced stretching experiments revealed that the SHG intensity at the center of the A-band increases, and consequently the double-peaked profile converts to a single-peaked profile, with increased tension force applied on the muscle (Prent et al. 2008).

Second harmonic radiation from oppositely oriented molecules interfere destructively; however, the magnitude of the destructive interference depends on the separation distance of the SHG radiators. Figure 4.8a shows a 2D SHG image of a *Drosophila melanogaster* larva muscle at the resting phase of a contraction cycle. The line profile along a myofibril was extracted in time to show the contraction dynamics (Figure 4.8b top). By analyzing the average SHG intensity along the

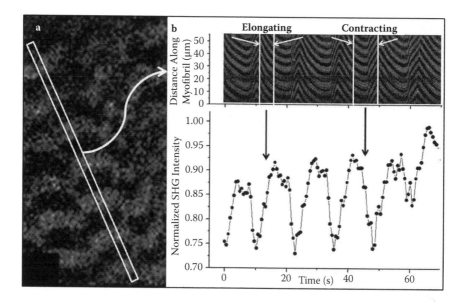

FIGURE 4.8 A rhythmically contracting myocyte from living *Drosophila melanogaster* larva imaged with SHG microscopy for a period of 70 s. A SHG intensity image during the resting phase of the contraction cycle of the larva is shown in (a). The rectangle in (a) shows the 52 μm area along the myofibril that is selected for analysis in (b). The image at the top of (b) shows the selected myofibril profile in time. A contracting and an elongating time segment have been highlighted for clarity. A graph of the average SHG intensity evolution for the selected myofibril profile during the 70 s is shown at the bottom of (b). A clear correlation between muscle contractility state and SHG intensity is observed.

myofibril, a direct relationship between the contraction dynamics and the SHG intensity is observed (Figure 4.8b bottom). Therefore, the SHG intensity can be used as an internal sensor for measuring nano-displacements and tension forces in the myofilaments at the single sarcomere level. The new dynamic local sensing of tension forces in each sarcomere facilitates fundamental studies of muscle contraction dynamics at the subcellular level (Prent et al. 2008).

VIII. MULTICONTRAST NONLINEAR MICROSCOPY OF H&E STAINED HISTOLOGICAL SECTIONS

Histological investigations of biological tissue benefited tremendously from the staining of different cellular structures with various organic dyes. Due to lack of strong structural contrast in biological samples imaged with bright field microscopy, stains are used to label specific structures and increase image contrast. In nonlinear microscopy, contrast is typically provided by the biological structure itself. As outlined in the previous sections, SHG occurs in the media with inherent non-centrosymmetric arrangements of molecules, and THG can be generated at interfaces

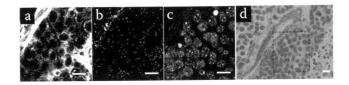

FIGURE 4.9 High resolution multicontrast microscopy images of mouse prostate tumor tissue stained with hematoxylin and eosin (H&E) dye. The panels present (a) multiphoton excitation fluorescence, (b) second harmonic generation, and (c) third harmonic generation images as well as (d) white light image. The scale bar is 10 μm.

between the media with different refractive indices. Despite the advantages of using harmonic generation microscopy for label-free imaging, a higher structural specificity is often required for biomedical imaging and cancer diagnostics. This brings about the need for labels that highlight specific biological structures in harmonic generation microscopy. In analogy with the fluorescence labels used for traditional microscopy, dyes that have high second- or third-order nonlinearity can be used for specific labeling and enhancement of harmonic generation signals. Labels that enhance harmonic generation have been termed *harmonophores*.

In this section, we demonstrate that hematoxylin, the standard histological stain used in hematoxylin and eosin (H&E) staining, enhances the THG microscopic signal, while eosin provides the MPF labeling. H&E staining is the standard method used to highlight specific areas in thin tissue sections. Hematoxylin has an affinity for nucleic acids, ribosomes, and mitochondria, and, consequently, stains these structures a dark purplish-blue color. Eosin, a pinkish-red dye, accumulates in the intracellular and extracellular protein material, such as collagen, connective tissue, borders of the cell membrane, and red blood cells. H&E staining currently serves as a gold standard for pathological investigations.

The potential use of harmonic generation microscopy for imaging of H&E stained pathological sections was investigated by imaging transgenic adenocarcinoma mouse prostate (TRAMP) tissue. TRAMP investigations help in the understanding of the molecular development of a tumor, which is a potentially important factor in the clinical and therapeutic treatment of patients. THG and SHG images of cancerous and normal tissue revealed pronounced morphological differences indicating the potential use of harmonophores and harmonic generation microscopy in pathological investigations and cancer diagnostics (Tuer et al. 2008).

Figure 4.9 shows the images obtained from mouse prostate cancer tissue using the three-channel multicontrast microscope with 1030-nm-wavelength laser excitation. The white-light image of the tissue is also presented in Figure 4.9d, where the boxed area indicates the scanned region. Figures 4.9a–c show the three images recorded in parallel with the multicontrast microscope. Figure 4.9a presents the MPF recorded in the spectral range between 550 nm and 600 nm; Figure 4.9b shows SHG obtained at 515 nm, and Figure 4.9c shows the THG image detected at 343 nm. The MPF signal originates mostly from the extracellular matrix, and the darker areas inside the tissue appear at the places where the cells reside. In the SHG image, second harmonic also appears in the extracellular matrix, where it primarily originates from collagen structures (Williams et al. 2005). Also, due to the proximity of the MPF and SHG detection wavelength regions, there is a small leakage of MPF signal into the SHG channel. However, the leakage is much smaller than the signal originating from the collagen fibrils. The THG image appears to be the most informative as it clearly highlights the cells in the prostate tissue and even reveals details about the intracellular structure.

Clearly, the anticorrelation between the fluorescence and THG signals indicates that the fluorescing dye does not cause strong THG, while the non-fluorescing dye provides strong THG signal (Tuer et al. 2008; Yu et al. 2008). It is known that hematoxylin has low fluorescence yield (Tuer et al. 2008; Yu et al. 2008), while eosin is highly fluorescing (De et al. 2005). Therefore, the THG signal that does not correlate with the fluorescence can be assigned to hematoxylin, while fluorescing areas are highlighted by eosin. The presented example demonstrates that harmonophores can be beneficially used for specific visualization of cellular structures. It is also apparent that multicontrast nonlinear microscopy adds invaluable information about cellular structures to the widely used bright field investigations of H&E stained histological sections, and can be efficiently used for morphological studies as well as cancer diagnostics.

REFERENCES

Abe, S. 2001. Two-photon probe of forbidden exciton states in symmetric aggregates of asymmetric molecules. *Chem. Phys.* 264:355–63.

Amat-Roldan, I., Cormack, I. G., Loza-Alvarez, P., and Artigas, D. 2004. Starch-based second-harmonic-generated collinear frequency-resolved optical gating pulse characterization at the focal plane of a high-numerical-aperture lens. *Opt. Lett.* 29:2282–84.

Amir, W., Carriles, R., Hoover, E. E., Planchon, T. A., Durfee, C. G., and Squier, J. A. 2007. Simultaneous imaging of multiple focal planes using a two-photon scanning microscope. *Opt. Lett.* 32:1731–33.

Anisha, T. K. N., Gualda, E. J., Cormack, L. G., Soria, S., and Loza-Alvarez, P. 2007. Backward second harmonic generation from starch for in situ, real time pulse characterization in multiphoton microscopy. *Multiphoton Microsc. Biomed. Sci. VII* 6442:64421S1–8.

Armstrong, J. A., Bloembergen, N., Ducuing, J. , and Pershan, P. S. 1962. Interactions between light waves in a nonlinear dielectric. *Phys. Rev.* 127:1918–39.

Baconnier, S., and Lang, S. B. 2004. Calcite microcrystals in the pineal gland of the human brain: Second harmonic generators and possible piezoelectric transducers. *IEEE Trans. Dielectrics and Electrical Insulation* 11:203–9.

Barad, Y., Eisenberg, H., Horowitz, M., and Silberberg, Y. 1997. Nonlinear scanning laser microscopy by third harmonic generation. *Appl. Phys. Lett.* 70:922–24.

Barille, R., Canioni, L., Rivet, S., Sarger, L., Vacher, P., and Ducret, T. 2001. Visualization of intracellular Ca^{2+} dynamics with simultaneous two-photon-excited fluorescence and third-harmonic generation microscopes. *Appl. Phys. Lett.* 79:4045–47.

Barzda, V. 2008. Non-linear contrast mechanisms for optical microscopy. In *Biophysical Techniques in Photosynthesis.* T. J. Aartsma and J. Matysik, Eds. 35–54. Urbana, IL, USA: Springer.

Barzda, V., de Grauw, C. J., Vroom, J., Kleima, F. J., van Grondelle, R., van Amerongen, H., and Gerritsen, H. C. 2001. Fluorescence lifetime heterogeneity in aggregates of LHCII revealed by time-resolved microscopy. *Biophys. J.* 81:538–46.

Barzda, V., Greenhalgh, C., Aus der Au, J., Elmore, S., Van Beek, J. H. G. M., and Squier, J. 2005. Visualization of mitochondria in cardiomyocytes by simultaneous harmonic generation and fluorescence microscopy. *Opt. Exp.* 13:8263–76.

Barzda, V., Istokovics, A., Simidjiev, I., and Garab, G. 1996. Structural flexibility of chiral macroaggregates of light-harvesting chlorophyll a/b pigment-protein complexes. Light-induced reversible structural changes associated with energy dissipation. *Biochemistry* 35:8981–85.

Becker, W. 2005. Advanced time-correlated single photon counting techniques. In *Chemical physics*. A. W. Castleman, J. P. Toennies, and W. Zinth, Eds. Berlin: Springer.

Bloembergen, N., and Pershan, P. S. 1962. Light waves at boundary of nonlinear media. *Phys. Rev.* 128:606–22.

Both, M., Vogel, M., Friedrich, O., von Wegner, F., Kunsting, T., Fink, R. H. A., and Uttenweiler, D. 2004. Second harmonic imaging of intrinsic signals in muscle fibers in situ. *J. Biomed. Opt.* 9:882–92.

Boyd, R. W. 1992. *Nonlinear optics*, 2nd ed. Amsterdam: Academic Press.

Buleon, A., Colonna, P., Planchot, V., and Ball, S. 1998. Starch granules: Structure and biosynthesis. *Int. J. Biol. Macromol.* 23:85–112.

Campagnola, P. J., Millard, A. C., Terasaki, M., Hoppe, P. E., Malone, C. J., and Mohler, W. A. 2002. Three-dimensional high-resolution second-harmonic generation imaging of endogenous structural proteins in biological tissues. *Biophys. J.* 82:493–508.

Campagnola, P. J., Wei, M. D., Lewis, A., and Loew, L. M. 1999. High-resolution nonlinear optical imaging of live cells by second harmonic generation. *Biophys. J.* 77:3341–49.

Cheng, J. X., and Xie, X. S. 2004. Coherent anti-Stokes Raman scattering microscopy: Instrumentation, theory, and applications. *J. Phys. Chem. B* 108:827–40.

Chu, S. W., Chen, I. H., Liu, T. M., Chen, P. C., Sun, C. K., and Lin, B. L. 2001. Multimodal nonlinear spectral microscopy based on a femtosecond Cr : forsterite laser. *Opt. Lett.* 26:1909–11.

Chu, S. W., Chen, S. Y., Chern, G. W., Tsai, T. H., Chen, Y. C.,. Lin, B. L., and Sun, C. K. 2004. Studies of $\chi((2))/\chi((3))$ tensors in submicron-scaled bio-tissues by polarization harmonics optical microscopy. *Biophys. J.* 86:3914–22.

Chu, S. W., Chen, S. Y., Tsai, T. H., Liu, T. M., Lin, C. Y., Tsai, H. J., and Sun, C. K. 2003a. In vivo developmental biology study using noninvasive multi-harmonic generation microscopy. *Opt. Exp.* 11:3093–99.

Chu, S. W., Liu, T. M., and Sun, C. K. 2003b. Real-time second-harmonic-generation microscopy based on a 2-GHz repetition rate Ti : sapphire laser. *Opt. Exp.* 11:933–38.

Clay, G. O., Millard, A. C., Schaffer, C. B., Aus-Der-Au, J., Tsai, P. S., Squier, J. A., and Kleinfeld, D. 2006. Spectroscopy of third-harmonic generation: Evidence for resonances in model compounds and ligated hemoglobin. *J. Opt. Soc. Am. B-Opt. Phys.* 23:932–50.

Cox, G., and Feijo, J. 2004. Second harmonic imaging of plant polysaccharides. *Proc. SPIE* 5323: 335–342. *Microscopy in the biomedical sciences IV.*

Cox, G., Moreno, N., and Feijo, J. 2005. Second-harmonic imaging of plant polysaccharides. *J. Biomed. Opt.* 10:0240131–6.

De, S., Das, S., and Girigoswami, A. 2005. Environmental effects on the aggregation of some xanthene dyes used in lasers. *Spectrochim. Acta. A-Mol. Biomol. Spectrosc.* 61:1821–33.

Dekker, J. P., and Boekema, E. J. 2005. Supramolecular organization of thylakoid membrane proteins in green plants. *Biochim. Biophys. Acta.-Bioenerg.* 1706:12–39.

Denk, W., Strickler, J. H., and Webb, W. W. 1990. Two-photon laser scanning fluorescence microscopy. *Science* 248:73–76.

Duncan, M. D., Reintjes, J., and Manuccia, T. J.. 1982. Scanning coherent anti-Stokes Raman microscope. *Opt. Lett.* 7:350–52.

Feijó, J. A., and Moreno, N. 2004. Imaging plant cells by two-photon excitation. *Protoplasma* 223:1–32.

Fittinghoff, D. N., der Au, J. A., and Squier, J. 2005. Spatial and temporal characterizations of femtosecond pulses at high-numerical aperture using collinear, background-free, third-harmonic autocorrelation. *Opt. Comm.* 247:405–26.

Florsheimer, M. 1999. Second-harmonic microscopy: A new tool for the remote sensing of interfaces. *Phys. Stat. Sol. A-Appl. Res.* 173:15–27.

Freund, I., Deutsch, M., and Sprecher, A. 1986. Connective-tissue polarity: Optical second-harmonic microscopy, crossed-beam summation, and small-angle scattering in rat-tail tendon. *Biophys. J.* 50:693–712.

Fung, D. C., and Theriot, J. A. 1998. Imaging techniques in microbiology. *Curr. Opin. Microbiol.* 1:346–51.

Gallant, D. J., Bouchet, B., and Baldwin, P. M. 1997. Microscopy of starch: Evidence of a new level of granule organization. *Carbohyd. Polymers* 32:177–91.

Garab, G., Faludidaniel, A., Sutherland, J. C., and Hind, G. 1988. Macroorganization of chlorophyll a/b light-harvesting complex in thylakoids and aggregates: Information from circular differential scattering. *Biochemistry* 27:2425–30.

Gauderon, R., Lukins, P. B., and Sheppard, C. J. R. 2001. Optimization of second-harmonic generation microscopy. *Micron* 32:691–700.

Glaring, M. A., Koch, C. B., and Blennow, A. 2006. Genotype-specific spatial distribution of starch molecules in the starch granule: A combined CLSM and SEM approach. *Biomacromolecules* 7:2310–20.

Göppert-Mayer, M. 1931. Uber elementarakte mit zwei quantensprungen. *Ann. Phys. (Leipzig)* 5:273–94.

Greenhalgh, C., Cisek, R., Prent, N., Major, A., Aus der Au, J., Squier, J., and Barzda, V. 2005. Time and structural image analysis of microscopic volumes, simultaneously recorded with second harmonic generation, third harmonic generation, and multiphoton excitation fluorescence microscopy. *Proc. SPIE* 5969:59692F1–F8.

Greenhalgh, C., Prent, N., Green, C., Cisek, R., Major, A., Stewart, B., and Barzda, V. 2007. Influence of semicrystalline order on the second-harmonic generation efficiency in the anisotropic bands of myocytes. *Appl. Opt.* 46:1852–59.

Greenhalgh, C., Stewart, B., Cisek, R., Prent, N., Major, A., and Barzda, V. 2006. Dynamic investigation of *Drosophila* myocytes with second harmonic generations microscopy. *Proc. SPIE* 6343:6343081–8.

Guo, Y. C., Ho, P. P., Savage, H., Harris, D., Sacks, P., Schantz, S., Liu, F., Zhadin, N., and Alfano, R. R. 1997. Second-harmonic tomography of tissues. *Opt. Lett.* 22:1323–25.

Hellwart, R., and Christen, P. 1974. Nonlinear optical microscopic examination of structure in polycrystalline ZnSe. *Opt. Comm.* 12:318–22.

Hepler, P. K., and Gunning, B. E. S. 1998. Confocal fluorescence microscopy of plant cells. *Protoplasma* 201:121–57.

Horton, P., Ruban, A. V., Rees, D., Pascal, A. A., Noctor, G., and Young, A. J. 1991. Control of the light-harvesting function of chloroplast membranes by aggregation of the LHCII chlorophyll protein complex. *Febs Lett.* 292:1–4.

Kaiser, W., and Garrett, C. G. B. 1961. 2-Photon excitation in Caf2–Eu2+. *Phys. Rev. Lett.* 7:229–31.

Kolthammer, W. S., Barnard, D., Carlson, N., Edens, A. D., Miller, N. A., and Saeta, P. N. 2005. Harmonic generation in thin films and multilayers. *Phys. Rev. B* 72:04544601–15.

König, K. 2000. Multiphoton microscopy in life sciences. *J. Microsc.* 200:83–104.

Li, J. H., Guiltinan, M. J., and Thompson, D. B. 2006. The use of laser differential interference contrast microscopy for the characterization of starch granule ring structure. *Starch-Starke* 58:1–5.

Lin, S. J., Hsiao, C. Y., Sun, Y., Lo, W., Lin, W. C., Jan, G. J., Jee, S. H., and Dong, C. Y. 2005. Monitoring the thermally induced structural transitions of collagen by use of second-harmonic generation microscopy. *Opt. Lett.* 30:622–24.

Major, A., Cisek, R., and Barzda, V. 2006. Femtosecond Yb : KGd(WO4)(2) laser oscillator pumped by a high power fiber-coupled diode laser module. *Opt. Exp.* 14:12163–68.

Marder, S. R., Beratan, D. N., and Cheng, L. T. 1991. Approaches for optimizing the first electronic hyperpolarizability of conjugated organic-molecules. *Science* 252:103–6.

Mertz, J. 2001. Nonlinear microscopy. *C. R. Acad. Sci. IV* 2:1153–60.

Mertz, J. 2004. Nonlinear microscopy: New techniques and applications. Review of overview of authors work in multiphoton microscopy. *Curr. Opin. Neurobiol.* 14:610–16.

Millard, A. C., Campagnola, P. J., Mohler, W., Lewis, A., and Loew, L. M. 2003. Second harmonic imaging microscopy. *Biophotonics B* 361:47–69.

Millard, A. C., Wiseman, P. W., Fittinghoff, D. N., Wilson, K. R., Squier, J. A., and Muller, M. 1999. Third-harmonic generation microscopy by use of a compact, femtosecond fiber laser source. *Appl. Opt.* 38:7393–97.

Mizutani, G., and Sano, H. 2003. Starch image in living water plants observed by optical second harmonic microscopy. In *Science, technology and education of microscopy: An overview*, ed. A. Mendez-Vilas, 499–504. Badajoz, Spain: Formatex.

Mizutani, G., Sonoda, Y., Sano, H., Sakamoto, M., Takahashi, T., and Ushioda, S. 2000. Detection of starch granules in a living plant by optical second harmonic microscopy. *J. Luminesc.* 87–89:824–26.

Moreaux, L., Sandre, O., Blanchard-Desce, M., and Mertz, J. 2000a. Membrane imaging by simultaneous second-harmonic generation and two-photon microscopy. *Opt. Lett.* 25:320–22.

Moreaux, L., Sandre, O., Charpak, S., Blanchard-Desce, M., and Mertz, J. 2001. Coherent scattering in multi-harmonic light microscopy. *Biophys. J.* 80:1568–74.

Moreaux, L., Sandre, O., and Mertz, J. 2000b. Membrane imaging by second-harmonic generation microscopy. *J. Opt. Soc. Am. B-Opt. Phys.* 17:1685–94.

Muller, M., Squier, J., De Lange, C. A., and Brakenhoff, G. J. 2000. CARS microscopy with folded BoxCARS phasematching. *J. Microsc.* 197:150–58.

Muller, M., Squier, J., Wilson, K. R., and Brakenhoff, G. J. 1998a. 3D microscopy of transparent objects using third-harmonic generation. *J. Microsc.* 191:266–74.

Muller, M., Squier, J., Wolleschensky, R., Simon, U., and Brakenhoff, G. J. 1998b. Dispersion pre-compensation of 15 femtosecond optical pulses for high-numerical-aperture objectives. *J. Microsc.* 191:141–50.

Mustardy, L., and Garab, G. 2003. Granum revisited. A three-dimensional model—Where things fall into place. *Trends Plant Sci.* 8:117–22.

Naumov, A. N., Sidorov-Biryukov, D. A., Fedotov, A. B., and Zheltikov, A. M. 2001. Third-harmonic generation in focused beams as a method of 3D microscopy of a laser-produced plasma. *Opt. Spectrosc.* 90:778–83.

Oron, D., Tal, E., and Silberberg, Y. 2003. Depth-resolved multiphoton polarization microscopy by third-harmonic generation. *Opt. Lett.* 28:2315–17.

Pawley, J. B. 1995. *Handbook of biological confocal microscopy*, 2nd ed., ed. J. B. Pawley. New York: Plenum Publishing.

Potma, E. O., de Boeij, W. P., and Wiersma, D. A. 2001. Femtosecond dynamics of intracellular water probed with nonlinear optical Kerr effect microspectroscopy. *Biophys. J.* 80:3019–24.

Prent, N., Cisek, R., Greenhalgh, C., Aus der Au, J., Squier, J., and Barzda, V. 2005a. Imaging individual chloroplasts simultaneously with third- and second-harmonic generation and multiphoton excitation fluorescence microscopy. In *Photosynthesis: Fundamental Aspects to Global Perspectives*. A. Van der Est and D. Bruce, Eds. 1037–39. Lawrence, KS: Allen Press.

Prent, N., Cisek, R., Greenhalgh, C., Sparrow, R., Rohitlall, N., Milkereit, M. S., Green, C., and Barzda, V. 2005b. Application of nonlinear microscopy for studying the structure and dynamics in biological systems. *Proc. SPIE* 5971:5971061–68.

Prent, N., Green, C., Greenhalgh, C., Cisek, R., Major, A., Stewart, B., and Barzda, V. 2008. Inter-myofilament dynamics of myocytes revealed by second harmonic generation microscopy. *J. Biomed. Opt.* 13:0413181–7.

Saeta, P. N., and Miller, N. A. 2001. Distinguishing surface and bulk contributions to third-harmonic generation in silicon. *Appl. Phys. Lett.* 79:2704–06.

Schins, J. M., Schrama, T. , Squier, J., Brakenhoff, G. J., and Muller, M. 2002. Determination of material properties by use of third-harmonic generation microscopy. *J. Opt. Soc. Am. B-Opt. Phys.* 19:1627–34.

Shen, Y. R. 1989. Surface-properties probed by second-harmonic and sum-frequency generation. *Nature* 337:519–25.

Sheppard, C. J. R., Gannaway, J. N., Kompfner, R., and Walsh, D. 1977. Scanning harmonic optical microscope. *IEEE J. Quantum Electron.* 13:D1.

Simidjiev, I., Barzda, V., Mustardy, L., and Garab, G. 1997. Isolation of lamellar aggregates of the light-harvesting chlorophyll a/b protein complex of photosystem II with long-range chiral order and structural flexibility. *Anal. Biochem.* 250:169–75.

Singh, S., and Bradley, L. T. 1964. Three-photon absorption in napthalene crystals by laser excitation. *Phys. Rev. Lett.* 12:612–14.

So, P. T. C., Dong, C. Y., Masters, B. R., and Berland, K. M. 2000. Two-photon excitation fluorescence microscopy. *Ann. Rev. Biomed. Eng.* 2:399–429.

Squier, J. A., and Muller, M. 2001. High resolution nonlinear microscopy: A review of sources and methods for achieving optimal imaging. *Rev. Sci. Instrum.* 72:2855–67.

Squier, J. A., Muller, M., Brakenhoff, G. J., and Wilson, K. R. 1998. Third harmonic generation microscopy. Review of dynamic imaging with THG in living organisms was first demonstrated. The point scanning source was used for THG imaging. Excitation at 1.2 μm, 250 kHz. Interface orientation dependency in respect to the laser beam was shown in glass beats. *Opt. Exp.* 3:315–24.

Sun, C. K. 2005. Higher harmonic generation microscopy. *Microsc. Tech.* 95:17–56.

Supatto, W., Debarre, D., Moulia, B., Brouzes, E., Martin, J. L., Farge, E., and Beaurepaire, E. 2005. In vivo modulation of morphogenetic movements in *Drosophila* embryos with femtosecond laser pulses. *Proc. Nat. Acad. Sci. USA* 102:1047–52.

Terhune, R. W., Maker, P. D., and Savage, C. M. 1962. Optical harmonic generation in calcite. *Phy. Rev. Lett.* 8:404–6.

Tirlapur, U. K., and König, K. 2001. Femtosecond near-infrared lasers as a novel tool for non-invasive real-time high-resolution time-lapse imaging of chloroplast division in living bundle sheath cells of *Arabidopsis*. *Planta* 214:1–10.

Tsang, T. Y. F. 1995. Optical third-harmonic generation at interfaces. *Phys. Rev. A* 52:4116–25.

Tuer, A., Bakueva, L., Cisek, R., Alami, J., Dumont, D. J., Rowlands, J., and Barzda, V. 2008. Enhancement of third harmonic contrast with harmonophores in multimodal non-linear microscopy of histological sections. *Proc. of SPIE* 6860:6860051–56.

Verbiest, T., Van Elshocht, S., Kauranen, M., Hellemans, L., Snauwaert, J., Nuckolls, C., Katz, T. J., and Persoons, A. 1998. Strong enhancement of nonlinear optical properties through supramolecular chirality. *Science* 282:913–15.

Volkmer, A. 2005. Vibrational imaging and microspectroscopies based on coherent anti-Stokes Raman scattering microscopy. *J. Phys. D-Appl. Phys.* 38 (5):R59–R81.

Walla, P. J., Linden, P. A., Ohta, K., and Fleming, G. R. 2002. Excited-state kinetics of the carotenoid S-1 state in LHC II and two-photon excitation spectra of lutein and beta-carotene in solution: Efficient car S-1 -> Chl electronic energy transfer via hot S-1 states? *J. Phys. Chem. A* 106:1909–16.

Ward, J. F., and New, G. H. C. 1969. Optical third harmonic generation in gases by a focused laser beam. *Phys. Rev.* 185:57–72.

Williams, R. M., Zipfel, W. R., and Webb, W. W. 2001. Multiphoton microscopy in biological research. *Curr. Opin. Chem. Biol.* 5:603–8.

Williams, R. M., Zipfel, W. R., and Webb, W. W. 2005. Interpreting second-harmonic generation images of collagen I fibrils. *Biophys. J.* 88:1377–86.

Yamada, S., and Lee, I. Y. S. 1998. Recent progress in analytical SHG spectroscopy. *Anal. Sci.* 14:1045–51.

Yelin, D., Oron, D., Korkotian, E., Segal, M., and Silbergerg, Y. 2002. Third-harmonic microscopy with a titanium-sapphire laser. *Appl. Phys. B-Lasers Opt.* 74:S97–S101.

Yelin, D., and Silberberg, Y. 1999. Laser scanning third-harmonic-generation microscopy in biology. *Opt. Exp.* 5:169–75.

Yu, C. H., Tai, S. P., Kung, C. T., Lee, W. J., Chan, Y. F., Liu, H. L., Lyu, J. Y., and Sun, C. K. 2008. Molecular third-harmonic-generation microscopy through resonance enhancement with absorbing dye. *Opt. Lett.* 33:387–89.

Zipfel, W. R., Williams, R. M., and Webb, W. W. 2003. Nonlinear magic: Multiphoton microscopy in the biosciences. *Nat. Biotechnol.* 21:1368–76.

Zumbusch, A., Holtom, G. R., and Xie, X. S. 1999. Three-dimensional vibrational imaging by coherent anti-Stokes Raman scattering. *Phys. Rev. Lett.* 82:4142–45.

5 Broadband Laser Source and Sensitive Detection Solutions for Coherent Anti-Stokes Raman Scattering Microscopy

Feruz Ganikhanov

CONTENTS

I. INTRODUCTION

Nonlinear optical phenomena are at the basis of powerful imaging techniques with wide applications in biology and medicine (Zipfel et al. 2003; Campagnola 2003; Oron et al. 2004; Cheng and Xie 2004). Coherent anti-Stokes Raman scattering (CARS) microscopy (Cheng and Xie 2004) is a high sensitivity nonlinear technique that is noninvasive and has chemical selectivity. In the CARS process, light is coherently scattered from Raman active vibrational resonances of a sample, resulting in a strong signal that allows for label-free, high-speed imaging. CARS is a third-order nonlinear optical process that requires at least two laser beams, called "pump" and "Stokes," with optical frequencies λ_p and λ_s, respectively. When the frequency difference $\omega_p - \omega_s$ is tuned to the frequency of a Raman-active molecular vibration (Ω), the anti-Stokes signal is generated at frequency $\omega_{as} = 2\omega_p - \omega_s$. In recent years, CARS microscopy has matured as a biomedical imaging modality.

Developments in laser sources have been essential for the advances in CARS microscopy (Cheng and Xie 2004). Recent CARS microscopy studies have relied on solid-state ultrafast lasers, and have evolved from low-repetition-rate femtosecond amplified systems (Hashimoto et al. 2000) to electronically synchronized high-rep-etition-rate picosecond sources (Potma et al. 2002) to the current state-of-the-art synchronously pumped optical parametric oscillator (OPO) systems (Evans et al. 2005). Broadband laser systems for CARS spectroscopy, though less successful in imaging, have also been used (Paulsen et al. 2003; Kee and Cicerone 2004; Petrov and Yakovlev 2005). An ideal CARS microscopy source would be a compact and turnkey laser system that has broad tunability and optimizes the spatial resolution, penetration depth, and nonlinear photodamage.

II. LASER SOURCE FOR CARS MICROSCOPY

A simple laser source solution for CARS microscopy that meets the above require-ments is described below. The approach exploits a broadly tunable picosecond OPO based on a periodically poled KT:OPO$_4$ (PP-KTP) crystal synchronously pumped by the second harmonic (532 nm) output of a mode-locked Nd:YVO$_4$ laser. The OPO is continuously temperature tunable from degeneracy (1064 nm) to 890 nm for its signal beam and to 1325 nm for its idler beam. By using the signal and idler beams as the pump and Stokes beams for CARS, respectively, it is possible to cover the entire chemically important vibrational frequency range of 100–3700 cm^{-1}. With this new laser system, CARS imaging with sub-wavelength resolution *in vitro* and in biological systems is demonstrated in cells and tissue where higher tissue penetration depths are feasible with the use of the longer wavelength output of this OPO.

The experimental setup is shown in Figure 5.1. Six picosecond (ps)-long pulses at 532 nm and 80 MHz repetition rate were delivered by a frequency-doubled, passively mode-locked Nd:YVO$_4$ laser (Hi-Q Laser Production, Austria). The maximum avail-able average power of the laser was reduced by an external variable attenuator to about a few hundred milliwatts. The OPO gain material is a flux-grown KTiOPO$_4$ crystal,

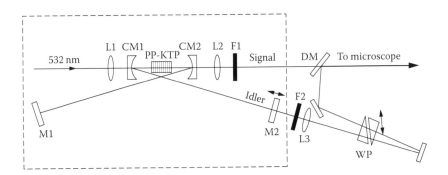

FIGURE 5.1 OPO resonator and external layout used for CARS microscopy. The OPO is singly resonant within the idler beam wavelength range. CM1, CM2, OPO cavity concave ($r = 100$ mm) mirrors with high reflection ($R > 99\%$) in the idler beam wavelength range. CM2 has 70%–80% transmission at the signal beam wavelengths. M1, high-reflecting mirror; M2, 10% output coupler for the idler beam; F1, F2, long-pass filters, $\lambda = 850$ nm; L1, L2, L3, lenses with focal lengths of 50, 150, and 1000 mm, respectively; DM, dichroic mirror; WP, wedged plates for delay control.

$0.5 \times 2.8 \times 10.8$ mm^3 (T × H × L), cut in XY plane and poled with a single grating period of $\Lambda = 8.99$ μm, satisfying the first-order phase-matching condition for second harmonic generation at ~41°C (Emanueli and Arie 2003). The PP-KTP crystal was antireflection coated at a target wavelength of 1064 nm ($R < 0.2\%$) with a residual reflection of up to 1% at ~1300 nm. The length of the crystal is slightly above the temporal walk-off interaction length between the 6 ps pump and idler pulses. The OPO resonator and mirror component details are also shown in Figure 5.1. The distances between the pump focusing lens, concave mirrors, and the crystal were close to the calculated values obtained, assuming optimal focusing conditions for parametric waves considered in detail by Boyd and Kleinman (1970) and taking into account the pump beam spatial parameters. With this cavity design, extremely low pumping thresholds of less than 1 mW near degeneracy for doubly resonant operation and ~40 mW for singly resonant cavity for 924 nm/1254 nm signal/idler wavelength combination were achieved in good agreement with the results of Boyd and Kleinman (1970).

The theoretical tuning curve based on the KTP material refractive index data (Emanueli and Arie 2003), the measured values, and the corresponding Raman shifts as a function of temperature are shown in Figure 5.2a. At the maximum set crystal

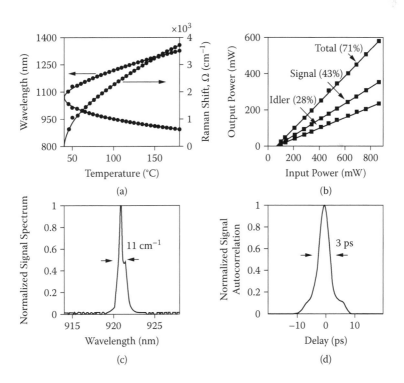

FIGURE 5.2 (a) Experimental (filled circles) wavelength tuning curve and accessible Raman frequencies as a function of the crystal temperature. The solid curves are a result of the calculations. (b) OPO output power versus pump power at the crystal facet; (c) and (d) show the typical signal pulse spectrum and autocorrelation trace at the OPO cavity detuning of minus 36 μm, respectively.

temperature of 180°C, the corresponding OPO signal (pump) and idler (Stokes) pair of wavelengths are 889.4 nm and 1323.9 nm, respectively, corresponding to a Raman shift of 3690 cm^{-1}.

The signal, idler, and total powers are plotted versus the input pump power in Figure 5.2b when the OPO crystal is maintained at temperature 132°C to deliver signal and idler beams at wavelengths of 921 nm and 1260 nm, respectively. Perfectly linear slope efficiencies for the signal, idler, and total powers of 43%, 28%, and 71% can be observed, respectively, up to the maximum pump power level. Output powers for both beams were fairly constant with a ~10%–15% margin throughout the tuning range at fixed pump power.

At maximum OPO output power, the OPO delivers pulses with a typical power spectrum of ~100 cm^{-1} (FWHM) and pulse widths of ~7.5 ps. As shown theoretically (Akhmanov et al. 1968) and experimentally (Khaydarov et al. 1994), when in the appropriate regime of group velocity mismatch and at certain parametric gain values, a variety of intensity profiles and spectra can be observed. For example, at an input pump power level of 715 mW and under a negative single-pass cavity detuning of 36 μm, spectral narrowing down to ~11 cm^{-1} (Figure 5.2c) occurs, leading to a less chirped pulse with tolerable spectral bandwidth (Cheng et al. 2001c) for resonant high contrast CARS imaging. In this case, an approximately 2.5-times shorter signal pulse (Figure 5.2d) was observed. The negative detuning resulted in 20% power drop from the maximum level. Therefore, nearly optimal performance for CARS imaging purposes and reliable long-term day-to-day operation at different crystal temperatures were possible without any passive spectral filtering element in the cavity.

The signal and idler beams were outcoupled from the cavity according to the optical layout shown in Figure 5.1. The idler pulse was delayed and spatially combined with the signal on a dichroic mirror. The two beams were coupled into a scanning microscope (Olympus, FV300/IX70) with a 1.2 numerical aperture water immersion objective lens (× 60, IR UPlanApo, Olympus). Long-pass filters were used in order to block the OPO outputs at shorter wavelengths. The two lenses in the output beam paths were used to achieve optimal spot sizes at the objective lens entrance pupil. The total average power in the image plane of the microscope was attenuated to less than 30 mW.

In order to determine the spatial resolution of the system, various sized polystyrene beads were imaged at a Raman shift of 2850 cm^{-1}. This experimental condition was achieved by choosing a signal-idler pair at wavelengths of 924 nm and 1254 nm. The characteristic lateral (xy) and longitudinal (z) resolutions were found to be diffraction limited to approximately 420 nm and ~1.1 μm (FWHM), respectively.

To demonstrate the versatility and robustness of the laser source, a series of CARS images for different Raman shifts are presented. Figure 5.3a shows an image of polystyrene beads at the C = C stretching frequency of ~1600 cm^{-1}. Figure 5.3b shows the image of a living mouse fibroblast cell (NIH-3T3-L1) cultured with deuterium-labeled oleic acid imaged at the CD_2 symmetric stretch (2100 cm^{-1}). The strong signals arise from high-density lipid structures known as lipid droplets. The same cell type cultured in deuterium-free media is shown in Figure 5.3c with the Raman shift tuned to the 2850 cm^{-1} symmetric CH_2 stretching frequency. To image with the OH stretching vibration at 3375 cm^{-1}, we prepared POPS [(1-palmitoyl-2-oleoyl-sn-glycero-3-phospho-L-serine)] multilamellar vesicles of lipids and water

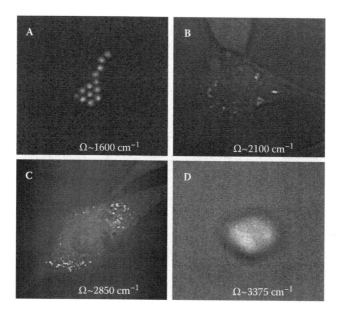

FIGURE 5.3 CARS images at different Raman shifts. (A) 1.5-μm-diameter polystyrene beads imaged at the C = C stretching vibrational frequency, 1600 cm^{-1}. (B) NIH 3T3-L1 cell cultured with deuterium-labeled oleic acid imaged at the CD$_2$ symmetric stretching vibration, 2100 cm^{-1}. Image size, 47 μm × 47 μm. (C) NIH 3T3-L1 cell imaged at the CH$_2$ symmetric stretching vibration, 2850 cm^{-1}. Image size, 78 μm × 78 μm. (D) 5-μm size POPS multilamellar vesicle imaged at the OH stretch, 3375 cm^{-1}.

(Cheng et al. 2001b). The corresponding image of the 5-μm structure containing water is shown in Figure 5.3d.

The described CARS source is optimized to image highly heterogeneous tissue samples. It is well known that the Rayleigh scattering cross section for a media with sub-wavelength size features is inversely proportional to the fourth power of the wavelength. Therefore, longer pump and Stokes wavelengths should increase the depth of penetration into tissue. With our system operating in the 900–1300 nm range, water absorption due to direct IR transitions is negligible. It should be noted that higher penetration depths come with a cost of reduced spatial resolution.

An example of high-contrast resonant imaging of tissue structures with this source is shown in Figure 5.4a. Here, the white adipose tissue of a mouse *omentum majus* is imaged at a depth of ~10 μm from the surface at a Raman shift of 2850 cm^{-1} (λ_{pump} = 924 nm; λ_{stokes} = 1254 nm). In contrast, for the same Raman shift, two synchronized Ti:sapphire lasers (Potma et al. 2002) would typically have pump and Stokes wavelengths of ~710 nm and ~890 nm, respectively, and are much more strongly scattered in turbid tissue.

To prove this point, the CARS imaging penetration depth of this OPO to that of a synchronized Ti:sapphire system have been compared. Figure 5.4b shows an image obtained at ~130 μm depth from the skin surface of a mouse ear by tuning into the CH stretching vibrational band using the OPO laser. At this depth, resonant lipid-rich

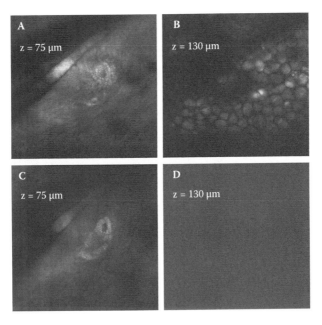

FIGURE 5.4 Forward CARS images of mouse ear tissue at the aliphatic CH2 stretching frequency, 2850 cm^{-1}, at two different depths of $z = 75$ μm and $z = 130$ μm, taken with two different sources: A and B with the OPO, and C and D with the two picosecond-Ti:sapphire-lasers-based system.

adipocytes in the subcutaneous layer are clearly visible. When the Ti:sapphire system was used at this depth, no contrast was observed. The longer wavelength OPO allowed for penetration depths almost 70 microns deeper than the Ti:sapphire system.

Multiphoton processes caused by the high peak power of ultrafast pulses significantly contribute to sample photodamage (Hopt and Neher 2001; Nan et al. 2006); these processes are reduced at longer excitation wavelengths. No sample photodamage was observed at excitation powers of 20–30 mW for the wavelengths provided by this OPO.

The described source's stable operation, broad tunablity with a single nonlinear crystal, and improved penetration depth make it an optimal one for CARS imaging in chemical and biomedical research.

III. IMPROVING DETECTION SENSITIVITY OF CARS MICROSCOPY

Detection sensitivity is one of the key issues in CARS microscopy. This is an especially acute problem in applications where chemical selectivity of CARS perfectly suits the tracking of small changes in cells related to specific protein and DNA distributions, external drug delivery/distribution, etc. There is, however, a component in CARS signal that is not associated with a particular vibration resonance and therefore does not carry chemically specific information. Unfortunately, in many cases, it can distort and even overwhelm the resonant signal of interest. In modeled approach, the CARS response originates from the third-order nonlinear susceptibility, which is the sum of a resonant contribution, $\chi_R^{(3)}(\Omega)$, and a nonresonant electronic component, $\chi_{NR}^{(3)}$. The total detected CARS signal is given by

$$I_{\text{CARS}}(\Omega) \propto \left| \chi_R^{(3)}(\Omega) + \chi_{NR}^{(3)} \right|^2 = \left| \chi_R^{(3)}(\Omega) \right|^2 + \left(\chi_{NR}^{(3)} \right)^2 + 2\,\text{Re}\left\{ \chi_R^{(3)}(\Omega) \right\} \chi_{NR}^{(3)}. \quad (5.1)$$

The frequency dependence of the three terms in Equation (5.1) is shown in Figure 5.5a. The nonresonant term can often obscure the resonant CARS signal of interest, making it difficult to identify the chemically selective contributions to an image. This is especially true when imaging biological materials as the aqueous environment gives rise to a substantial nonresonant response that often overwhelms the resonant signal.

Several approaches have been developed in an attempt to suppress this nonresonant background. The signal-to-background ratio was improved by matching the bandwidths of the laser sources to the Raman linewidths (Cheng et al. 2001c). Epi-sensitive (Volkmer et al. 2001), polarization-sensitive detection (Cheng et al. 2001a), and time-resolved (Volkmer et al. 2002) CARS, while effective in eliminating the nonresonant background, have the disadvantage of severely attenuating the resonant signal. In order to increase the sensitivity of CARS microscopy, a method of suppressing the nonresonant background while preserving the full strength of the resonant contribution is needed.

An interferometric CARS microscopy (Potma et al. 2006) approach has been reported recently. This is an extension of the CARS technique that offers much greater resonant sensitivity while simultaneously suppressing the nonresonant background. A drawback of the technique is the increased complexity due to the need for interferometric precision. The local oscillator beam, which provides the signal enhancement, also serves as a source of noise that can affect the ultimate S/N ratio for the detected CARS signal. In addition, image artifacts can arise from inhomogeneity of the index of refraction across the sample, limiting the ultimate detection sensitivity.

A new approach to enhance the vibrational contrast and sensitivity of CARS imaging is offered and described in this chapter. Consider an isolated resonance centered at vibrational frequency Ω_R with FWHM linewidth Γ (Figure 5.5a) probed with a narrow-band source. If the source is rapidly switched between two frequencies, ω_1 and ω_2, with a frequency difference $\delta = \omega_1 - \omega_2$, the frequency modulation results in an amplitude modulation of the CARS signal, $\Delta I(\delta) = I(\omega_1) - I(\omega_2)$, that can be extracted using phase-sensitive detection (Figure 5.5b). In this approach, a resonant spectral feature effectively becomes a frequency modulation (FM) to amplitude modulation (AM) converter. The nonresonant contribution, which is essentially spectrally flat, does not contribute to the detected modulated signal, and therefore is efficiently suppressed.

When the concentration of resonant species in a sample is high under this suppression condition, the quadratic term $|\chi_R^{(3)}(\Omega)|^2$ in Equation (5.1) is the greatest contribution to the detected signal. At low concentrations, however, the linear term $2\,\text{Re}\{\chi_R^{(3)}(\Omega)\}\chi_{NR}^{(3)}$ in equation (1) becomes the dominant factor. It is advantageous to detect this term, as it is proportional to the resonant specie concentration, enabling immediate concentration measurements though CARS microscopy. It is important

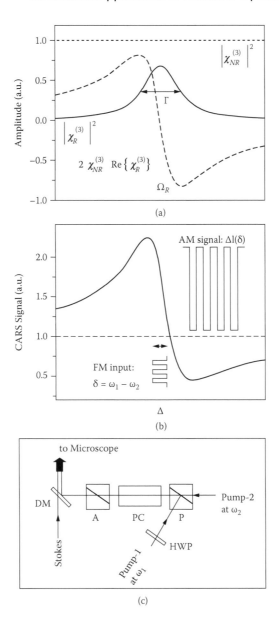

FIGURE 5.5 (a) Components of the CARS signal. The last three terms in Equation 5.1 are plotted versus detuning $\Delta = \omega_p - \omega_s - \Omega_R$. Ω_R is the center frequency of a homogeneously broadened Raman line with linewidth Γ. The curves are calculated with an assumption that $\chi^{(3)}_{NR} = 1.2\,\chi^{(3)}_{NR}(\Delta = 0)$. (b) Schematic of the FM CARS process. Solid curve, sum of the contributions from (a). Dashed curve, nonresonant background. The resonance acts as an FM-to-AM converter, resulting in an amplitude-modulated signal that can be detected by using a lock-in amplifier. (c) Schematic of the experimental setup. PC, Pockels cell; P, two-port Glan–Taylor prism; A, Glan–Thompson prism; DM, dichroic mirror, HWP, $\lambda/2$ plate. A circled cross and an arrow indicate pump-1 and pump-2 polarizations, respectively.

to note that this heterodyne term contains a factor of $\chi_{NR}^{(3)}$, which implies that the detected signal is amplified by the nonresonant response of the solvent. The above approach is implemented by modulating the optical frequency of the pump beam, ω_p, at a high enough rate (>500 kHz) to separate the modulated signal from the lower frequency laser noise.

The experimental setup (Figure 5.5c) consists of three pulsed lasers coupled into a modified laser-scanning microscope (Olympus FV300). The Stokes beam is ~10% of the output from a passively mode-locked, fixed-frequency Nd:YVO$_4$ laser (High-Q, Austria, 7 ps, 1064 nm, 76 MHz rep. rate). The 90% of the output of the Nd:YVO$_4$ source is used to synchronously pump an intracavity doubled optical parametric oscillator (OPO) (Levante, APE-Berlin), producing tunable 5 ps near-IR radiation for use as a pump beam (pump-1). The second pump beam (pump-2) is provided by a mode-locked Ti:Al$_2$O$_3$ oscillator (Coherent, Mira 900S) delivering tunable 3 ps pulses that are electronically synchronized (Potma et al. 2002) to the Nd:YVO$_4$ source. A half-wave plate inserted into the pump-1 beam path is used to rotate the polarization so that pump-1 and pump-2 are perpendicularly polarized. The two pump beams are then combined in a two-port Glan-Taylor prism and sent collinearly into a Pockels cell (ConOptics, Model 350-160). Square waveforms with a 50% duty cycle, derived from a pulse delay generator synchronized to the laser pulse train, supply a modulation frequency of ~500 kHz to the Pockels cell. When the waveform is in the "low" state, pump 1 is allowed to pass through the exit analyzer. When the waveform is in the "high" state, the polarization of both beams is rotated by $\pi/2$, such that pump 2 now passes unattenuated though the analyzer while pump-1 is blocked. The modulated pump beams are spatially combined with the Stokes beam on a dichroic mirror and the combined beams are directed into the scanning microscope. The CARS signal from the sample is detected by a PMT (Hamamatsu, R3896) and fed into a lock-in amplifier (Stanford Research Systems, Model 844). The lock-in reference is provided by the external signal supplied from the pulse generator driving the Pockels cell. The half-wave plate introduced into the pump-1 beam path can be used in conjunction with the Glan-Taylor prism to balance the intensity of the two beams for maximum nonresonant background suppression.

FM-CARS is most advantageous when the resonant signal is comparable to or smaller than the nonresonant background. In this situation, images acquired with FM-CARS have substantially better contrast than those acquired through normal CARS microscopy. Figure 5.6a shows normal forward-CARS and FM-CARS images of weakly scattering 0.36-μm polystyrene beads on a glass surface. Pump-1 is tuned to the peak of the vinyl CH stretching band at 3050 cm^{-1}, while pump 2 targets the spectrally flat region at 3000 cm^{-1}. Figure 5.6a is the normal CARS image at 3050 cm^{-1}, showing a resonant signal only slightly stronger than that of the nonresonant background. The same image acquired using the FM-CARS technique demonstrates considerable suppression of the nonresonant background (Figure 5.6b). The improvement in signal-to-background ratio is shown in the intensity profiles through the beads (Figures 5.6c and d). The nonresonant signal suppression of FM-CARS is immediately applicable to in vivo imaging, where nonresonant CARS signals can be as strong as the resonant signal of interest. Figures 5.6e and 5.6f are normal

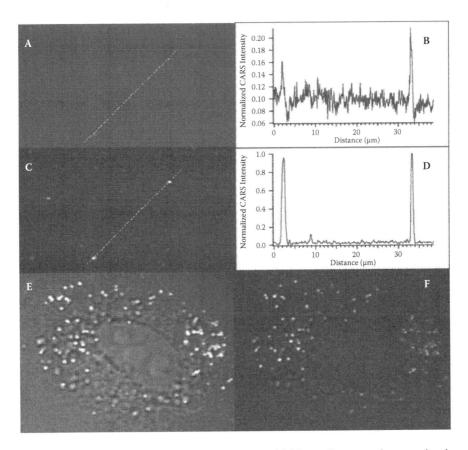

FIGURE 5.6 Forward CARS microscopy image (A) of 0.36-μm-diameter polystyrene beads taken at $\Omega_R = 3050$ cm^{-1}. (C), FM CARS image of the same area ($\omega_{p1} - \omega_s = 3050$ cm^{-1}; $\omega_{p2} - \omega_s = 3000$ cm^{-1}). (B), (D), Corresponding cross-sectional profiles of the images along the indicated line show the magnitude of nonresonant background suppression. (E), Forward CARS image of a fixed A549 human lung cancer cell cultured with deuterium-labeled oleic acids taken at $\Omega_R = 2100$ cm^{-1}. (F), FM CARS image obtained when toggling between 2060 and 2100 cm^{-1}. Nonresonant background components have been significantly reduced by the FM CARS method.

forward CARS and FM-CARS images of a fixed human lung cancer cell (A549), respectively. The cell was cultured with deuterated oleic acid before fixation and imaged at the CD_2 stretching frequency (pump-1, 2100 cm^{-1}) and at an off-resonance frequency (pump-2, 2060 cm^{-1}). The normal forward CARS image (Figure 5.6e), taken at a Raman shift of 2100 cm^{-1}, exhibits many nonresonant cellular features that make it unclear which cellular components contain the deuterated compound. When FM-CARS is used to image the cell (Figure 5.6f) the nonresonant signals vanish, revealing only the resonant signals of the deuterated lipid droplets.

In addition to truly resonant imaging, FM-CARS also allows for increased detection sensitivity over normal CARS microscopy. To quantify the increased sensitivity, solutions of methanol dissolved in water were used. Methanol is an ideal test compound as it is well characterized by Raman spectroscopy (Schwartz et al. 1980) and contains only a single CH_3 moiety that gives rise to two relatively narrow ($\Gamma_{FWHM} \sim 25$ cm^{-1}), well-spaced, Lorentzian-like peaks in the CH stretching region. For this experiment, pump 1 was tuned to target 2928 cm^{-1}, which corresponds to the symmetric CH_3 stretch of methanol, while pump 2 was tuned to target 3048 cm^{-1}, where there is no vibrational resonance. As considered earlier, the FM-CARS intensity in terms of detuning ($\Delta_{1,2} = \omega_{p1,p2} - \omega_s - \Omega$) is equal to $\Delta I(\delta) = I(\Delta_1) - I(\Delta_2)$. At relatively low concentrations, $I(\delta)$ can be expressed in terms of the fraction of maximum solute concentration, n, by the following equation:

$$I_{CARS}(\Delta, n) = I_{CARS}^{H_2O} \left[\frac{\left(\frac{\Gamma}{2}\right)^2}{\Delta^2 + \left(\frac{\Gamma}{2}\right)^2} \right] \left(Rn^2 - 2\sqrt{R}\left(\frac{2\Delta}{\Gamma}\right)n \right) \qquad (5.2)$$

where $I_{CARS}^{H_2O}$ is the nonresonant CARS intensity from pure water, and R is the ratio of peak CARS signal from pure methanol to $I_{CARS}^{H_2O}$. The FM-CARS signal is maximized at $\Delta_{1,2} = \pm \frac{\Gamma}{2}$. The R parameter can be readily measured experimentally at the resonance maximum, which is $R = 24$ for this experiment.

Figure 5.7 presents the experimentally measured values of the FM-CARS microscopy signal versus concentration for two distinctly different lock-in detection

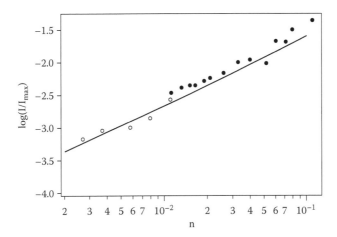

FIGURE 5.7 FM CARS signal versus dilution factor, *n*. *I* is the FM CARS signal intensity from methanol dissolved in water, while I_{max} is the CARS signal intensity from a pure methanol sample. Filled circles represent the experimental data points taken at a detector bandwidth of 25 kHz. Open circles correspond to data taken when the detector bandwidth was set to 1.6 Hz. The solid line is a plot of Equation 5.2.

bandwidths. Since FM-CARS makes use of a lock-in amplifier, the noise floor for the detected resonant signal can be reduced by narrowing the detector bandwidth to achieve better resonant signal detection sensitivity, with the ultimate sensitivity reached at a infinitely narrow bandwidth. While a detection bandwidth of $\Delta f_1 = 25$ kHz (filled circles) achieves significantly better sensitivity than seen in normal CARS, the ultimate sensitivity of our configuration is reached at a bandwidth of $\Delta f_2 = 1.6$ Hz (open circles). Equation (5.2) (solid line) provides a direct fit to the data with no adjustable parameters. The efficient removal of nonresonant background originating from water with a narrow detection bandwidth (Δf_2) allowed us to detect resonant signal from methanol with only 5×10^5 oscillators in the probed volume of 100 atto-liters, as opposed to approximately 4×10^8 oscillators achieved with normal forward CARS in the same experiment. The minimum detectable signal for both bandwidths differs from the expected value since they should scale linearly with the value of Δf. This suggests that the CARS signal noise spectrum has significant components in the sub-Hz region, which most likely originate from beam-pointing instability as well as laser intensity and spatial mode fluctuations.

Improvements to this technique, including dual-wavelength laser sources (Barros and Becker 1993; Leitenstorfer et al. 1995), OPOs with fast electro-optic tuning (Ewbank et al. 1997), and acousto-optic tunable filters for rapid wavelength modulation will very likely improve the detection limit by eliminating sources of noise. The technique can also be configured to detect small changes in a Raman spectrum through appropriate choice of the modulated wavelengths. The presented approach vastly enhances one's ability to distinguish resonant features from the nonresonant background, providing resonant images with an improvement of nearly three orders of magnitude in sensitivity for chemical species at low concentrations.

IV. CARS ENDOSCOPE

For many applications in biomedicine, it is desirable to perform tissue imaging in situ. Endoscopy, in particular, has been widely used in medicine for applications ranging from surgical interventions to disease diagnosis. Recent advances in endoscopic techniques have used contrast based on optical coherence tomography (OCT) (Tearney et al. 1997), one- and two-photon fluorescence (Flusberg et al. 2005; Myaing et al. 2006; Jung and Schnitzer 2003), and second harmonic generation (Fu et al. 2006), leading to immediate applications in clinical environments. While OCT has been very successful for imaging tissue morphology, its contrast is solely based on refractive index differences. Endogenous fluorescence endoscopy is only sensitive to intrinsic biological fluorophores, which limits its applications. Second harmonic endoscopy is extremely sensitive to non-centrosymmetric structures, but few biological materials have this optical property. An endoscope based on the vibrational contrast of CARS would offer label-free chemically selective imaging in situ, making it an ideal tool for biomedical applications. A prototype CARS endoscope capable of imaging specimens with submicron lateral resolution and ~12-µm axial resolution is presented.

Figure 5.8 shows a schematic of the proof-of-principle CARS endoscope system. The laser source consisted of a passively mode-locked 10 W Nd:YVO$_4$ laser (High-Q Laser GmbH) operating at 1064 nm, which delivered transform-limited 7 ps pulses

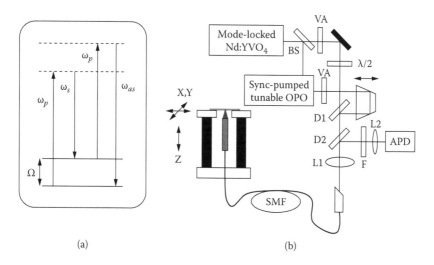

(a)   (b)

FIGURE 5.8 (a) CARS energy diagram. (b) Experimental setup: BS, 15% beam splitter; VA, variable attenuator; $\lambda/2$, half-waveplate; D1, 950 nm longpass dichroic mirror; D2, 750 nm longpass dichroic mirror; F, three 670 nm bandpass filters; L1, aspheric lens; L2, 10 cm concave lens.

at a repetition rate of 76 MHz. The majority of the Nd:YVO$_4$ output (9 W) was used to synchronously pump an OPO (APE-Berlin). The OPO generates a 5 ps pulse train with up to 2 W of average output power and wavelength tuning range of 780–920 nm. A portion of the 1064 nm output (1.5 W) was used as a Stokes beam, while the tunable OPO output provided the pump beam. The pump and Stokes beams were combined on a dichroic mirror (Chroma, 950DCXR) and directed to coupling optics at the proximal fiber end. The combined pump and Stokes beam was coupled by an aspheric lens into a 1-m-long step index, nonpolarization-maintaining silica fiber designed for single mode propagation at 830 nm (0.12 numerical aperture, 5.6-μm mode field diameter). The fiber bandwidth allowed for single mode propagation up to 200 nm above the designed wavelength, enabling efficient propagation of the pump and Stokes frequencies. Since the fiber was able to efficiently propagate a wide range of pump wavelengths, the endoscope could easily be used to probe any biologically relevant Raman active vibrational resonance. The size and divergence of the pump and Stokes beams were independently adjusted before the dichroic mirror to maximize the coupling efficiency. The typical coupling efficiency of the beams into the fiber was better than 40%. A 4-mm-diameter focusing unit using an aspheric lens (2.4 mm diameter, 0.43 numerical aperture, 1.1-mm working distance) was attached to the distal fiber tip. Samples were raster scanned by a three-dimensional piezo stage (PI, P547.3CL) with respect to the fixed fiber assembly. The backward-propagating (epi) CARS photons generated in focus were collected by the focusing unit and separated from the excitation beams at the proximal fiber end with a dichroic mirror (Chroma, 750DCXR). The CARS signal was filtered with three 670 nm band-pass filters (Ealing 42-7393) and focused with a 100-mm concave lens onto an avalanche photodiode (PerkinElmer, SPCM-AQR-14).

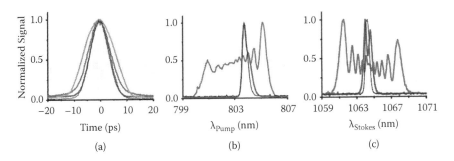

FIGURE 5.9 (a) Autocorrelation traces of pump (dark gray) and Stokes (gray) pulses after propagation in 1m long fiber. Cross-correlation trace after 1m of fiber propagation (light gray). The faintly seen black trace is the autocorrelation of the pump before fiber propagation. Measured pump (b) and Stokes (c) power spectra after the OPO (gray), 1-m (dark gray) and 5-m (gray) long fiber propagation at 400 mW and 700 mW of pump and Stokes fiber output, respectively . The fiber input powers were 1 W and 1.3 W, respectively.

The temporal characteristics for both pulses were precisely estimated using the material dispersion, modal refractive index, and nonlinear optical parameters for a fused silica fiber. Temporal broadening for the 5-7 to-ps transform-limited pulses in the 1-m length fiber was negligible and estimated to be less than 0.1%, using the well-known formula for short optical pulse temporal broadening in dispersive media (Yariv 1997). Auto- and cross-correlation traces for the two pulses (Figure 5.9a) confirmed this prediction. Pulse spectral broadening due to fiber propagation can potentially have a large effect on the spectral resolution and image contrast of the CARS endoscope (Cheng et al. 2001c). Spectral broadening due to self-phase modulation begins to be noticeable when the nonlinear phase shift becomes close to 2π. The per-pulse energy corresponding to this condition is found to be (Boyd 2003)

$$\varepsilon_{2\pi} = \frac{\pi\lambda t_p D_{\text{eff}}^2}{4n_2 L} \tag{5.3}$$

where t_p is the pulsewidth (~6 ps), D_{eff} is the mode field diameter (5.6 μm), n_2 is the nonlinear refractive index of silica (2.6×10^{-16} cm^2/W), and L is the fiber length. For a fiber length of 1 m and a pulse wavelength centered at 800 nm, the 2π-shift occurs at pulse energies of nearly 4 nJ, which is equivalent to approximately 300 mW average power at a repetition rate of 76 MHz. Measured power spectra of the pump and Stokes pulse trains before and after propagation through different fiber lengths are presented in Figures 5.9b and c. The 2π nonlinear phase shift was observed at average powers of 400 mW and 700 mW for 803.5 nm and 1064 nm, respectively (Figures 5.9b and c, blue traces). When collecting endoscopic images, we typically used 80 mW of power for both laser beams. These average powers are so far below the threshold for self-phase modulation that the pump and Stokes spectra do not undergo detectable broadening. Therefore, at the powers used, fiber propagation does not affect the CARS spectral selectivity (Légaré et al. 2005). At the abovementioned

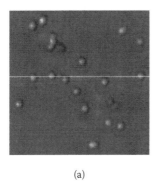 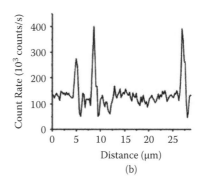

(a)   (b)

FIGURE 5.10 (a) Prototype CARS endoscope image of 0.75-μm polystyrene beads embedded in agarose gel spin-coated on a coverslip ($\Delta\omega = 2845$ cm^{-1}). The image dimension is 29 μm × 29 μm (128 × 128 pixels). The pump and Stokes powers at the sample were 80 mW each, with a pixel dwell time of 1 ms. (b) CARS intensity profile along the white line in (a). The CARS contrast decreased when the system was tuned off the resonance maximum.

power levels spectral broadening on the order of Raman linewidth (>20 cm^{-1}) was observed for fiber length 5 m and more (Figures 5.9b and c, red traces). The performance of the prototype CARS endoscope by imaging size-calibrated polystyrene beads with the pump and Stokes difference frequency set to the aliphatic symmetric CH$_2$ stretch ($\Delta\omega = 2845$ cm^{-1}) has been characterized. Figure 5.10a shows an XY image of 0.75 μm polystyrene beads spin-coated on a coverslip. A Gaussian fit to the bead profile gives a FWHM of 0.76 μm. Smaller beads (0.5 μm) were found to have the same FWHM profile, indicating an upper bound on the lateral resolution of approximately 0.8 μm. The axial resolution upper bound was measured to be ~12 μm using the same procedure. This spatial resolution was expected based on the numerical aperture of the focusing lens and is sufficient for imaging tissue morphology.

The prototype endoscope collects epi-CARS signals, which can come from three different mechanisms (Evans et al. 2005). In the first mechanism, epi-signal is generated from objects that are comparable in size to the excitation wavelengths. In the second mechanism, epi-CARS is generated at the interface of two media with different third-order nonlinear susceptibilities. For these two mechanisms, the epi-CARS signal arises from incomplete destructive interference in the epi-direction (Zumbusch et al. 1999; Cheng et al. 2002). A third mechanism can occur in thick samples (>100 μm); multiple scattering events due to the turbidity of the specimen can redirect a fraction of the intense forward-propagating CARS signal in the epi-direction. This mechanism has been shown to generate the dominant epi-CARS signal in tissues such as skin (Evans et al. 2005).

Thus, for small objects such as the 0.75-μm diameter beads in Figure 5.10a, the epi-CARS signal arose solely from the first mechanism. For larger objects such as the 5-μm beads in Figure 5.11, however, the second epi-CARS mechanism dominated signal generation. Epi-signal generated at the bead surface caused the larger beads to have a ring-shaped appearance. The center of the bead did not contribute to the image since it can be considered as bulk material in which complete destructive

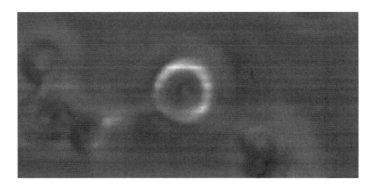

FIGURE 5.11 Prototype CARS endoscope image of 5-μm polystyrene beads embedded in agarose spin-coated on a coverslip ($\Delta\omega = 2845$ cm^{-1}). The ring-like structure is due to signal arising at the edge of the bead from incomplete destructive interference in the epi-direction. The image dimension is 31 μm × 15 μm (139 × 66 pixels). The pixel dwell time was 1 ms.

interference occurred in the epi-direction. The lower-intensity ring features in Figure 5.11 arose from out-of-focus beads.

To mimic imaging thick tissue, an attempt is made to detect the backscattered forward CARS due to the third epi-CARS mechanism from intralipid tissue phantoms (Evans et al. 2005). Although strong forward-CARS signal was redirected backward, the signal could not be detected with the endoscope because the multiple scattering events occurred outside the focal volume. The single-mode fiber, acting as a confocal pinhole, rejected photons scattered in the epi-direction from outside the focal plane. Thus, our current prototype can only collect epi-CARS resulting from the first two mechanisms. For CARS endoscopy to be most useful for tissue imaging, future endoscope designs will need to collect the backscattered CARS signal. Since the backscattered signal is typically orders of magnitude more intense than epi-CARS arising from incomplete destructive interference alone, such endoscope designs will likely require lower excitation powers and could operate at greater acquisition rates.

In this report, the sample was scanned relative to a fixed focus in order to demonstrate delivery and detection through the same fiber. This successful demonstration paves the way for an endoscope that incorporates a distal fiber scanning mechanism. Our proof-of-principle endoscope coupled with such a scanning mechanism would be extremely useful for imaging samples for which the first and second epi-CARS mechanisms dominate, such as nerve bundles.

In order to minimize the size of the focusing unit, our CARS endoscope prototype used a low numerical aperture focusing lens. This differs from CARS microscopes, which normally use achromatic high numerical aperture objectives that give higher spatial resolution, signal generation, and collection efficiency. For epi-signal generated through the first two mechanisms, the performance of the endoscope is expected to be the same as that of a CARS microscope equipped with a similar numerical aperture focusing unit. Since the aspheric lens in our prototype has chromatic aberrations, the spatial overlap of the pump and Stokes can be further optimized for efficient signal generation. The collection efficiency of the anti-Stokes signal is also affected by the chromaticity. An achromatic focusing unit is expected to significantly improve endoscope performance.

It is worth noting that the spectral broadening associated with long fiber length propagation could be used for endoscopic multiplex CARS micro-spectroscopy (Petrov and Yakovlev 2005). This could be realized, for example, by propagating the Stokes laser pulse train at higher average powers through the fiber while reducing the pump power. Future developments, such as backscattered photon collection and distal fiber scanning, will be a step closer to a CARS endoscope capable of label-free in situ biomedical imaging.

V. COMBINING CARS AND THG IMAGING

There is certain interest to develop a characterization tool that would combine nonlinear techniques that potentially provide highly complementary information about biological samples. In this chapter, first results on application of combined CARS and third harmonic generation (THG) microsopy are presented for the case of myelinated axon. Understanding of the normal physiological in vivo as well as pathological processes of the axon and Schwann cell interaction and fine-tuned interplay resulting in balanced myelin production and axon function is of primary importance and the goal.

The THG and CARS processes have a common origin with regard to the material response function, the nonlinear polarization that drives generation of the corresponding signals that are both related to the third-order $\chi^{(3)}$ susceptibility. However, in imaging applications, the space structural– and optical frequency–dependent image properties can be quite different due to a number of specific reasons. Resonant CARS can highlight media constituents through the enhanced part of the corresponding $\chi^{(3)}$. The enhancement is due to characteristic Raman active molecular vibrations of substructures in the media. At the same time, the CARS process generates nonresonant signal from the surrounding environment that is not chemically specific and often can mask the response of a structure of interest. On the other hand, the third harmonic generation does not provide chemical selectivity in the mentioned sense. However, under tight focusing conditions, it is sensitive to interfaces within the focal volume (Tsang 1995; Cheng et al. 2002). A discontinuity in $\chi^{(3)}$ or refractive index, or both, are the contrast mechanisms for THG imaging while signal from homogenous bulk material is absent.

The experiments have been performed on a setup that used the ps-OPO–based CARS system described above and a femtosecond Ti:sapphire laser in conjunction with a commercial laser scanning microscope (Carl Zeiss, model LSM-510). The peripheral nerve samples were gained from C57/B6 wild-type mice. After removing the skin from the lower extremities from freshly sacrificed mice, the saphenous nerve is exposed as it runs very conveniently for excision along the saphenous vein, without too much additional fatty tissue and a favorable tissue thickness of less than 20 μm. A 500-μm long piece is excised and freed from additional fatty tissue as well as the collagenous nerve sheath. The myelinated nerve tissue is fixed for 3–5 hr in 4% PFA or 10% formalin and mounted on 100-μm thick coverslips that are treated with 3-aminopropyltriethoxysilane or a chromium potassium sulfate solution. After

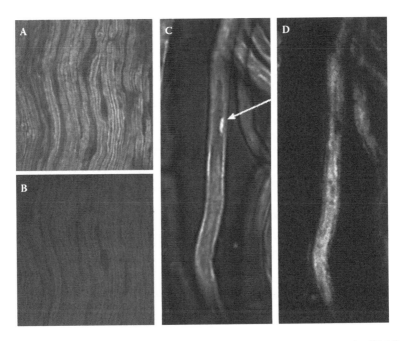

FIGURE 5.12 Images of the peripheral nerve tissue from mouse obtained in CARS and THG microscopy. (a) Resonant CARS image (108 × 108 μm²) obtained near characteristic CH$_2$-type stretching vibration frequency (2850 cm^{-1}); (b) primarily nonresonant CARS image of the same part of the tissue obtained by tuning away from the vibration mode; (c) resonant CARS image (37 × 82 μm²) of a single axon from the peripheral nerve tissue; (d) THG image of the same size and on the same axon as in the previous case.

fixation and mounting, the sample was kept moist with 1 × phosphate buffered saline (PBS) solution to prevent drying artifacts.

Figure 5.12a shows a CARS image obtained on excised and formalin-fixed peripheral nerve when the OPO wavelength was tuned to resonantly enhance the signal via CH$_2$-type stretching vibration. The dominant features on the image correspond to myelinated nerve sheaths. These myelin sheaths are made from Schwann cells that wrap around the axon multiple times. The cell sheaths around the axon are multiply-stacked lipid bilayers that contain 80% dense lipid structures of various lipid compositions with glycolipid as the main component. The axon's core itself contains CH-rich structures, such as its own cell membrane as well as lipid bilayer parts of its organelles that can be seen as less-signal-intense features on the image. The walls are rather homogenous in density along and perpendicular to the axon axis. Since extra care was taken to filter out signal at the anti-Stokes wavelength, the background is primarily due to nonresonant CARS process. That reflects fairly strong contribution to $\chi^{(3)}(\omega_{as},\omega_{OPO},\omega_{OPO},-\omega_L)$ of the medium within and outside of the axon that does not possess a vibration frequency in the vicinity of CH-type Raman active transitions. The $\chi^{(3)}(\omega_{as},\omega_{OPO},\omega_{OPO},-\omega_L)$ values far from characteristic vibration resonances will also primarily depend on physical density of the structures. When the OPO frequency was tuned out sufficiently from resonant peaks in the CARS spectrum, the corresponding image (Figure 5.12b) still revealed the myelin sheaths due to the high-density lipid fat structures that constitute it. Though the resonant enhancement of CARS signal represents the main contrast mechanism, it is important to keep in mind that the CARS image contrast is in fact a complex function of some other parameters. For example, minimum detectable size of a "resonant" substructure in a surrounding "nonresonant" environment will obviously depend on the ratio of the structure's volume to the focal volume, as well as the ratio of the corresponding nonlinearities. The latter can be considered in terms of molecular density and Raman cross section for the characteristic vibration. For the axon core area, faintly seen substructures can be observed. Those are primarily due to density discontinuities formed by the substructures in the surrounding solvent media. The Schwann cell nucleus, surrounded by its own dense lipid bilayer, can clearly be seen in Figure 5.12c (arrow). Obviously, features that are smaller in size and/or less dense than those mentioned cannot be detected due to the high level of the nonresonant CARS signal from the surrounding aqueous medium. A clear demonstration of the fact that for some important applications the CARS technique alone provides rather limited information on tissue morphology as seen in comparison with the THG image is presented in Figure 5.12d. Comparing the image from Figure 5.12d with the one obtained by CARS (Figure 5.12c) for the same axon, the optically inhomogeneous core of the axon across and along the nerve that is not clearly detectable by CARS is revealed in much greater detail in the THG image. Some other observations that can be immediately made from the data are that, unlike for the CARS image, only faint myelin wall contours (due to the wall solvent interface) can be seen, at best, on the THG image, giving us crucial information that the nerve sheath is an optically homogeneous material within the scale of the probed focal volume. The same seems to be true for the nuclear membrane of the Schwann cell nucleus. It is straightforward to show that at the flat interface (i.e., discontinuity in $\chi_h^{(3)}$) of length δ the generated third harmonic intensity $I_{3\omega}$ can be expressed as follows:

$$I_{3\omega} \sim \left| \chi_\delta^{(3)}(3\omega) - \chi_h^{(3)}(3\omega) \right|^2 \delta^2 I_0^3 \tag{5.4}$$

In the above equation the third-order nonlinearity of the host media is designated as $\chi_h^{(3)}(3\omega)$. Based on the discussion above and our data we can conclude that, unlike its walls, the axon core is densely packed with optically inhomogeneous structures in terms of $\chi^{(3)}$ and that there are changes in uniformity along the axon's axis toward the Ranvier nodes. The inhomogeneities are due to heterogeneous aggregates within the axon cytoplasm such as the extensive microtubule network and organelle distribution along the axon axis.

   In regard to Equation (5.4), we have to note that without the above mentioned assumptions the nonlinearities will contain weighting factors that are proportional to the corresponding wave-vector mismatch and inversely proportional to refractive index, thus suggesting that the THG signal is sensitive to the refractive index interface(s) as well. In order to differentiate between contrast mechanisms in THG imaging of soft tissue materials it would be important to know the relationships of the corresponding linear and nonlinear optical parameters. A nonlinear optical susceptibility model has been shown to fit well to characterize wide bandgap materials such as glasses (Boling et al. 1978). The model suggests that third-order nonlinearity is a strongly dependent function (up to eighths power) of the corresponding refractive index. This straightforwardly means that there will be no third harmonic signal for interfaces where change in refractive index is absent, and any change in $n(\omega)$ will be amplified by the power dependence of $\chi^{(3)}$ (3ω) on it. However, this may not be an adequate model to apply to the case of liquids and soft matter due to key factors related to differences in electronics state and structure between the two classes of materials. Experimental techniques and methods for measuring third-order nonlinearities related to THG process in liquids (Kajzar and Messier 1985) have been addressed in the past. An analytical study that would suggest a phenomenological model for $\chi^{(3)}$ (3ω) for structures that constitute tissue-liquid interfaces and $\chi^{(3)}$ dependence on the refractive index and other important macroscopic parameters is required though.

VI. CONCLUSIONS

Overall, the combined approach shows strong promise to reveal important information on structure and morphology of the structurally complex tissue. For the case of myelinated axon, while myelin is routinely detected in CARS via dense CH_2-bond-rich glycolipids, the strong signal masks other, more subtle features and does not allow effective tracking of the change in densities and change in material composition reflected in changes of refractive index. Simultaneous application of CARS and THG microscopy allowed us to detect the rather optically homogeneous myelin sheath around axons on the background of highly inhomogeneous axon cores. These observations and underlying important information would not be possible to obtain by applying only one of the techniques. The combination of these techniques will aid in better understanding the normal physiological in vivo as well as pathological processes of the axon and Schwann cell interaction and fine-tuned interplay resulting in balanced myelin production and axon function severely disrupted in peripheral neuropathies and neurodegenerative diseases such as amyotrophic lateral sclerosis.

REFERENCES

Akhmanov, S. A., Chirkin, A. S., Drabovich, K. N., Kovrigin, A. I., Khokhlov, R. V., and Sukhorukov, A. P. 1968. Nonstationary nonlinear optical effects and ultrashort light pulse formation. *IEEE J. Quantum Electron.* QE-4:598–605.

Barros, M. R. X., and Becker, P. C. 1993. Two-color synchronously mode-locked femtosecond Ti:sapphire laser. *Opt. Lett.* 18:631–33.

Boling, N. L., Glass, A. J., and Owyoung, A. 1978. Empirical relationships for predicting nonlinear refractive index changes in optical solids. *IEEE J. Quant. Electron.* 14:601–8.

Boyd, G. D., and Kleinman, D. A. 1970. Parametric interaction of focused Gaussian light beams. *J. Appl. Phys.* 39:3597–3639.

Boyd, R. W. 2003. *Nonlinear optics.* London: Academic Press.

Campagnola, P. J. 2003. Second-harmonic imaging microscopy for visualizing biomolecular arrays in cells, tissues and organisms. *Nature Biotechnology* 21:1356–60.

Cheng, J.-X., Book, L., and Xie, X. S. 2001a. Polarization coherent anti-Stokes Raman scattering microscopy. *Opt. Lett.* 26:1341–43.

Cheng, J.-X., Pautot, S., Weitz, D. A., and Xie, X. S. 2001b. Ordering of water molecules between phospholipids bilayers visualized by coherent anti-Stokes Raman scattering microscopy. *Proc. Nat. Acad. Sci.* 100:9826–30.

Cheng, J.-X., Volkmer, A., Book, L. D., and Xie, X. S. 2001c. An epi-detected coherent anti-Stokes Raman scattering (E-CARS) microscope with high spectral resolution and high sensitivity. *J. Phys. Chem.* 105:1277–80.

Cheng, J.-X., Volkmer, A., and Xie, X. S. 2002. Theoretical and experimental characterization of coherent anti-Stokes Raman scattering microscopy. *J. Opt. Soc. Am. B* 19: 1363–75.

Cheng, J.-X., and Xie, X. S. 2004. Coherent anti-Stokes Raman scattering microscopy: Instrumentation, theory, and applications. *J. Phys. Chem. B* 108:827–40.

Emanueli, S., and Arie, A. 2003. Temperature-dependent dispersion equations for KTiOPO4 and KTiOAsO$_4$. *Appl. Opt.* 42:6661–65.

Evans, C. L., Potma, E. O., Puoris'haag, M., Cote, D., Lin, C. P., and Xie, X. S. 2005. Chemical imaging of tissue *in vivo* with video-rate coherent anti-Stokes Raman scattering microscopy. *Proc. Nat. Acad. Sci. USA* 102:16807–12.

Ewbank, M. D., Rosker, M. J., and Bennett, G. L. 1997. Frequency tuning a mid-infrared optical parametric oscillator by the electro-optic effect. *J. Opt. Soc. Am. B* 14:666–71.

Flusberg, B. A., Cocker, E. D., Piyawattanametha, W., Jung, J. C., Cheung, E. L. M., and Schnitzer, M. J. 2005. Fiber-optic fluorescence imaging. *Nat. Methods* 12: 941–50.

Fu, L., Jain, A., Xie, H., Cranfield, C., and Gu, M. 2006. Nonlinear optical endoscopy based on a double-clad photonic crystal fiber and a MEMS mirror. *Opt. Express* 14:1027–32.

Hashimoto, M., Araki T., and Kawata, S. 2000. Molecular vibration imaging in the fingerprint region by use of coherent anti-Stokes Raman scattering microscopy with a collinear configuration. *Opt. Lett.* 25: 1768–70.

Hopt, A., and Neher, E. 2001. Highly nonlinear photodamage in two-photon fluorescence microscopy. *Biophys. J.* 80:2029–36.

Jung, J. C., and Schnitzer, M. J. 2003. Multiphoton endoscopy. *Opt. Lett.* 28:902–4.

Kajzar, F., and Messier, J. 1985. Third harmonic generation in liquids. *Phys. Rev. A* 32: 2352–63.

Kee, T. W., and Cicerone, M. T. 2004. Simple approach to one-laser, broadband coherent anti-Stokes Raman scattering microscopy. *Opt. Lett.* 29: 2101–03.

Khaydarov, J. D., Andrews, J. H., and Singer, K. D. 1994. Pulse compression in a synchronously pumped optical parametric oscillator from group-velocity mismatch. *Opt. Lett.* 19:831–33.

Légaré, F., Ganikhanov, F., and Xie, X. S. 2005. Towards an integrated coherent anti-Stokes Raman scattering (CARS) microscope system. *Proc. SPIE* 5971:35–40.

Leitenstorfer, A., Furst, C., and Laubereau, A. 1995. Widely tunable 2-color mode-locked Ti-sapphire laser with pulse jitter of less-than -2-fs. *Opt. Lett.* 20:916–18.

Myaing, M. T., MacDonald, D. J., and Li, X. 2006. Fiber-optic scanning two-photon fluorescence endoscope. *Opt. Lett.* 31:1076–78.

Nan, X., Potma, E. O., and Xie, X. S. 2006. Nonperturbative chemical imaging of organelle transport in living cells with coherent anti-Stokes Raman scattering microscopy. *Biophys. J.* 91:728–35.

Oron, D., Yelin, D., Tal, E., Raz, S., Fachima, R., and Silberberg, Y. 2004. Depth-resolved structural imaging by third harmonic generation microscopy. *J. Struct. Biol.* 147:3–11.

Paulsen, H. N., Hilligse, K. M., Thgersen, J., Keiding, S. R., and Larsen, J. J. 2003. Coherent anti-Stokes Raman scattering microscopy with a photonic crystal fiber based light source. *Opt. Lett.* 28:1123–25.

Petrov, G. I., and Yakovlev, V. V. 2005. Enhancing red-shifted white-light continuum generation in optical fibers for applications in nonlinear Raman microscopy. *Opt. Express* 13:1299–1306.

Potma, E. O., Evans, C. L., and Xie, X. S. 2006. Heterodyne coherent anti-Stokes Raman scattering (CARS) imaging. *Opt. Lett.* 31:241–43.

Potma, E. O., Jones, D. J., Cheng, J.-X., Xie, X. S., and Ye, J. 2002. High-sensitivity coherent anti-Stokes Raman scattering microscopy with two tightly synchronized picosecond lasers. *Opt. Lett.* 27:1168–70.

Schwartz, M., Moradi-Araghi, A., and Koehler, W. H. 1980. Fermi resonance in aqueous methanol. *J. Mol. Struct.* 63:279–85.

Tearney, G. J., Brezinski, M. E., Bouma, B. E., Boppart, S. A., Pitris, C., Southern, J. F., and Fujimoto, J. G. 1997. In vivo endoscopic optical biopsy with optical coherence tomography. *Science* 276:2037–39.

Tsang, T. Y. F. 1995. Optical third-harmonic generation at interfaces. *Phys. Rev. A* 52:4116–25.

Volkmer, A., Book, L., and Xie, X. S. 2002. Time-resolved coherent anti-Stokes Raman scattering microscopy: Imaging based on Raman free induction decay. *Appl. Phys. Lett.* 80:1505–7.

Volkmer, A., Cheng, J.-X., and Xie, X. S. 2001. Vibrational imaging with high sensitivity via epi-detected coherent anti-Stokes Raman scattering microscopy. *Phys. Rev. Lett.* 87:23901–4.

Yariv, A. 1997. *Optical electronics in modern communications.* New York: Oxford University Press.

Zipfel, W. R., Williams R. M., and Webb W. W. 2003. Nonlinear magic: Multiphoton microscopy in the biosciences. *Nature Biotechnol.* 21:1369–75.

Zumbusch, A., Holtom, G., and Xie, S. X. 1999. Vibrational microscopy using coherent anti-Stokes Raman scattering. *Phys. Rev. Lett.* 82:4142–45.

6 Nonlinear Optical Microspectroscopy of Biochemical Interactions in Microfluidic Devices

Vladislav V. Yakovlev, Georgi I. Petrov,
Vladislav Shcheslavskiy, and Rajan Arora

CONTENTS

I. INTRODUCTION

The ability to manipulate fluids in channels with dimensions of tens of micrometers and less has revolutionized biology and chemistry in the same way the invention of the integrated circuit by Kilby and Noyce in the 1950s spawned the progress of microelectronics. This fast developing area of science and technology, called microfluidics, finds more and more applications ranging from chemical analysis to biological and chemical microreactors, replacing bulky devices and sparking new applications (Hansen and Quake 2003; Whitesides 2006). However, the progress of science depends on the development of new tools and instruments capable of real-time molecular-level imaging. Our ability to monitor in situ the transitional changes of chemical and structural composition becomes an important issue for the continuing advancement and successful implementation of this new technology, directing its manufacturing and exploring the new areas of its applications (Viskari and Landers 2006). This chapter discusses the development of an innovative photonics toolbox, which aims at addressing both the structural and chemical analysis in situ of liquids and living cells in microfluidic devices. These new photonic tools include optical harmonics generation and nonlinear Raman microscopy, which perfectly match the space and time scales of the processes in microfluidic channels. Due to their intrinsic noninvasiveness, strong signal level, and unsurpassing chemical and structural sensitivity, they have great potential to become indispensable methods for detecting chemical and structural variations across and along the microfluidic channel, for in vivo analysis of living cells development, for noninvasive chemical recognition, and for in situ monitoring of chemical reactions. These tools can be combined with more traditional light absorption and scattering measurements.

The very first applications of microfluidics were in chemical analysis, for which they offer a number of useful properties: the ability to use very small quantities of reagents, low cost, short times for analysis, and small footprints for the analytical devices (Manz et al. 1992). In part, microfluidics originated from microanalytical methods—gas phase chromatography, high-pressure liquid chromatography, and capillary electrophoresis, which, being enclosed in a capillary tube, revolutionized chemical analysis. However, the recent developments of microfluidic technology, which aimed at exploring the full advantages of laminar liquid flow on a microscale, have shown that there is a huge untapped potential for microscale research and development. So-called "lab-on-a-chip" devices are targeting many potentially important areas of application, such as protein crystallization (Chen and Ismagilov 2006), drug discovery (Dittrich and Manz 2006), cell biology (Lucchertta et al. 2005), systems biology (Breslauer et al. 2006), and chemical microreactor technology (Lowe and Ehrfeld 1999). This daily growing list of promising applications can be continued, but there is an urgent need to develop a set of monitoring tools that could be capable of noninvasive in situ interrogation of structural and chemical properties of the liquid flow. This way, a truly integrated microscopic technology will allow fully automated control of ongoing high-throughput, multiparallel processes (Hong and Quake 2003).

It is hard to find better monitoring tools than optical spectroscopy methods when high spatial and temporal resolution is required in addition to noninvasiveness. Traditionally, absorption, light scattering, chemiluminescence, and fluorescence measurements are used for this purpose (Janasek et al. 2006). Those are well-established techniques that establish a simple way of obtaining useful information. However, absorption and scattering measurements provide very little information about the chemical composition. Fluorescence spectroscopy provides more information but is invasive and can induce undesirable photochemical reactions and be toxic to cells. The signal is also transient and can undergo photobleaching. Vibrational spectroscopy based on Raman scattering provides a great deal of chemical and structural information (Peticolas et al. 1996; Petrich 2001; Navratil et al. 2006), can be noninvasive, and can be used in vivo, but there are several obstacles to its widespread use for diagnostic purposes and in scientific research. Fourier transform infrared spectroscopy (FTIR) and infrared spectroscopy cannot provide sufficient spatial resolution to match the dimensions of microfluidic devices, but Raman microspectroscopy has 200-nm lateral spatial resolution (the diffraction limit of light microscopy) and can be used in live-cell imaging (Huang et al. 2006; Notingher and Hench 2006; Uzunbajakava et al. 2003).

The first problem with Raman spectroscopy is the fluorescence background, which dominates any Raman spectrum taken from a cell or protein solution (Barth and Zscherp 2002). One way to reduce it is to use a longer excitation wavelength (>750 nm) (Shim and Wilson 1997), but this results in a huge reduction of signal (Schrotter and Klockner 1979) and reduced spatial resolution, while keeping the background signal. Most spectroscopists use digital background subtraction, which works to a certain extent (Lieber and Mahadevan-Jansen 2003), since the acquisition of a huge number of useless photons dramatically decreases the signal-to-noise ratio (SNR).

The second problem associated with Raman microspectroscopy is a relatively weak signal, which limits its use as a diagnostic tool. Surface-enhanced Raman spectroscopy (SERS) has achieved a significant step forward in increasing the efficiency of the Raman signal up to 10 orders of magnitude; however, it requires that a metal surface, which forms a plasmonic resonance for signal enhancement, be within 10 nm of the object of interest (Kneipp et al. 2006). This is not possible for noninvasive live-cell in vivo imaging and is hardly useful for the majority of other applications of microfluidic devices, since the injection of metal nanoparticles into a liquid solution is only possible in the final stage of the process (Park et al. 2005). Nonlinear Raman spectroscopy based on coherent anti-Stokes Raman scattering (CARS) has long been considered a promising alternative to Raman spectroscopy; however, both the technical difficulties and the presence of strong, nonresonant background from surrounding water molecules are the major obstacles to implementing this imaging modality in biomedical studies (Hudson 1977; Akhmanov and Koroteev 1981).

Less studied are methods based on harmonics generation. One may think that the technique of second harmonic generation (SHG) is not even applicable to study the flow of liquids since a typical liquid is a medium, which is fully isotropic and does not allow any even order process to be efficiently generated. However, if one takes into account significant concentration gradients occurring on a microscopic scale, the symmetry is lost, giving rise to a very strong second-order nonlinearity, first observed by our group in the late 1980s (Govorkov et al. 1989, 1990). Thus, the method of SHG can be used to study the gradients of liquid flow and their distribution across and along the microfluidic channel. The technique of the third harmonic generation (THG) is gaining significant popularity in the microscopic community, since it allows high-contrast noninvasive imaging of biological interfaces (Debarre et al. 2007). However, recent results obtained by our group (Shcheslavskiy et al. 2003, 2004, 2005a, 2006a,b) and others (Schaller et al. 2000; Debarre et al. 2004, 2005;

Clay et al. 2006) have demonstrated the potential of structural and chemical sensitivity of third harmonic (TH) microscopy.

The outline of this chapter is as follows. First, we discuss the methods of THG microscopy and CARS microspectroscopy and outline the major developments over the past years, emphasizing the application aspect of this work. Then, we discuss the application of these spectroscopy tools for several microfluidic problems, such as live-cell imaging and protein crystallization.

II. THIRD HARMONIC GENERATION MICROSCOPY

A. THEORY OF THIRD HARMONIC GENERATION

The process of THG is driven by third-order optical nonlinearity, $\chi^{(3)}$, which defines the nonlinear optical polarizability at the frequency 3ω as

$$P_i(3\omega) = \chi^{(3)}_{ijkl} E_j(\omega) E_k(\omega) E_l(\omega), \tag{6.1}$$

where ω is the incident laser frequency, and $E_j(\omega)$ is the $j-th$ electric field component of the incident electric field. Unlike tensor, $\chi^{(2)}$, which governs the process of second harmonic generation and which vanishes in the medium with inversion symmetry, $\chi^{(3)}$ is non-vanishing in all nonlinear optical media. However, for an isotropic media at frequencies far from any resonance frequencies, there is only one independent component of this tensor,

$$\chi^{(3)}_{1122} = \chi^{(3)}_{1212} = \frac{1}{3}\chi^{(3)}_{1111}.$$

Considering the amplitude of the incident beam propagating along the z-direction in the form of

$$E_\omega(r,z) = \frac{E_{\omega 0}}{1+i2z/b} e^{-\frac{r^2}{w_0^2(1+i2z/b)}}, \tag{6.2}$$

where
$b = \frac{2\pi n_\omega w_0^2}{\lambda}$ is the confocal parameter of the focused beam,
w_0 is the spot size of the beam waist,
λ is the incident wavelength,
n_ω is the refractive index at the incident wavelength,
$E_{\omega 0}$ is the electric field amplitude of the incident wave.

Solving Maxwell's equations using Equations (6.1) and (6.2) in the slowly varying envelope approximation, one can calculate the amplitude of the electric field of the TH to be (Ward and New 1969; Bjorklund 1975):

$$E_{3\omega}(z) = i\frac{3\pi}{2}\frac{\omega}{cn_{3\omega}} E_{\omega 0}^3 \int_{z_1}^{z_2} \chi^{(3)}_{1111} \frac{e^{-i\Delta kz}}{(1+i2z/b)^2} dz, \tag{6.3}$$

where $n_{3\omega}$ is the refractive index at the wavelength of the third harmonic, and Δk is the wave-vector mismatch between the incident and generated waves. The power of the generated TH signal can be calculated by integrating over the beam's cross section

$$P_{3\omega} = \frac{cn_{3\omega}}{8\pi} \int\limits_0^\infty 2\pi r \left| E_{3\omega}^2 \right| dr. \tag{6.4}$$

If the medium is homogeneous, i.e., both the nonlinearity, $\chi_{1111}^{(3)}$, and refractive indices, $n_\omega, n_{3\omega}$, remain constant through the whole focal volume of the incident beam, the integral in Equation (6.3) vanishes, and no TH is generated. However, any inhomogeneity, such as an interface or a microscopic particle, will break this symmetry, leading to a strong TH signal (Shcheslavskiy et al. 2003, 2004; Barad et al. 1997). This effect can be used to quantify the solution flow in microfluidic devices. Indeed, many biologically relevant solutions and liquid chemicals have refractive indices very similar to the one of water, making it difficult to use conventional light microscopy to pick up the contrast between two liquids and/or to quantify the slow variation of optical refractive index occurring on the boundary between two liquids due to the interdiffusional processes. The third-order nonlinearity can vary in much larger range, leading to a more dramatic difference in the third harmonic signal. To make this methodology more quantitative and independent of the possible absorption and scattering in solution, we modified the experimental geometry, as shown in Figure 6.1. By focusing the laser beam first at the well-characterized interface (air-glass) and then at the glass-solution interface, one can avoid problems with light absorption/scattering, since the generated TH wave goes through the same distance of absorbing/scattering material for both focusing geometries. The ratio of those two signals can be used to extract the value of $\chi_{1111}^{(3)}$ for the solution under study

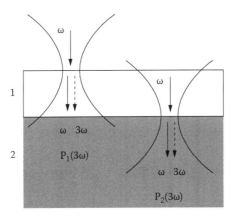

FIGURE 6.1 Schematic diagram of the experimental setting to measure the nonlinear optical constant of unknown solution (2). The power of the TH generated on the air-glass interface, $P_1(3\omega)$, is measured together with the power of the TH generated on the glass-liquid interface, $P_2(3\omega)$.

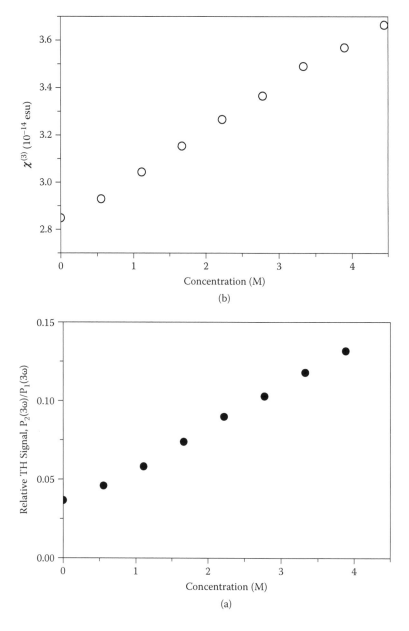

FIGURE 6.2 (a) The ratio of the third harmonic powers generated on two interfaces as a function of molar concentration of sodium chloride in distilled water. (b) The measured value of $\chi^{(3)}$ of solution as a function of molar concentration of sodium chloride in distilled water.

(Shcheslavskiy et al. 2003), if the values of refractive indices of this solution are known both for the incident and the TH wavelengths. An illustrative example of biologically important ionic solution is represented in Figure 6.2. In this case, sodium chloride water solution was characterized using TH microscopy using the approach depicted in Figure 6.1. The fundamental output of the Cr:forsterite laser ($\lambda = 1.25$ μm, pulse duration $\tau_p = 50$ fs) was used as an incident radiation. The TH signal measured at the glass-solution interface was normalized to the signal generated at the air-glass interface and is plotted in Figure 6.2a as a function of NaCl concentration. There is a negligible variation of the refractive indices across the whole range of NaCl concentrations; however, there is dramatic increase in the TH signal, which makes it possible to detect variations of NaCl concentration as small as 1 mM, if the noise of the input laser is kept below 1%. We also note that the change in nonlinear optical susceptibility does not have to be that large to see the appreciable effect on the TH signal. In this particular example, the largest difference in calculated values of $\chi^{(3)}_{1111}$ is only 35% (Figure 6.2b). While the presented methodology is not specific, i.e., the variations of the TH signal due to the change of $\chi^{(3)}_{1111}$ value can be attributed to many factors, such as variations of local concentrations, structural transformation of molecules, and chemical reactions, it provides a simple way of quantifying the dynamics of liquid flow occurring in a microfluidic device in space and time.

The same approach can be used to characterize solutions containing colloidal nanoparticles. To do this, we use the effective nonlinear media approach (Stroud and Wood 1989), i.e., express the effective nonlinear susceptibility of solution as

$$\chi^{(3)}_{\text{eff}} = \chi^{(3)}_{\text{medium}} + p\chi^{(3)}_{\text{particle}}, \tag{6.5}$$

where the first term is the nonlinear optical coefficient for the surrounding medium, which is, typically, water, and the second term takes into account the concentration (volume fraction, p) of nanoparticles, whose material is characterized with nonlinear susceptibility, $\chi^{(3)}_{\text{particle}}$. Equation (6.5) is only valid when the size of the nanoparticles is significantly smaller than the wavelength of the light. The refractive index of solution can be calculated in the same way (Stroud and Wood 1989). We tested the applicability of this approach using the solution of TiO_2 nanoparticles. Nanoparticles were about 10 nm in diameter and of approximately round shape. Only low concentrated solutions were used to avoid any possible aggregation. While all the prepared solutions showed rather significant scattering at the TH wavelength, it did not produce any problems with retrieving the effective values for third-order nonlinearity $\chi^{(3)}_{\text{eff}}$. When plotted against the concentration (Figure 6.3), the calculated value of the effective nonlinearity, $\chi^{(3)}_{\text{eff}}$, shows an excellent linear dependence, as predicted by Equation (6.5).

B. THIRD HARMONIC MICROSCOPY OF SUBMICRON PARTICLES

When the size of a particle in solution increases, the exact contribution to the TH signal from an individual particle has to be taken into account. Indeed, when such a particle is introduced into the focal volume of the laser beam, it can be treated as an inhomogeneity, which breaks the symmetry in Equation (6.3) and leads to a non-vanishing TH signal. To experimentally check this hypothesis, we moved the position of the focal point inside the solution filled with submicron-sized particles (Figure 6.4). If the solution were homogeneous, there would not be any TH signal; however, each particle passing through the focal volume creates conditions for

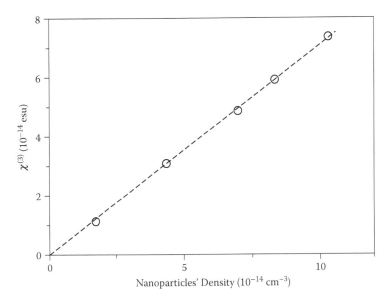

FIGURE 6.3 Nonlinear susceptibility of the solution as a function of TiO_2 nanoparticles concentration. (From Shcheslavskiy, V., Petrov, G., Yakovlev, V. V. *Appl. Phys. Lett.*, 82(22): 3982–3984, 2003. Used with permission.)

non-vanishing TH signal. Assuming that particles do not interact with each other, one should expect a linear concentration of the TH signal upon the concentration of particles in solution, reflecting the probability of a particle passing through the focal volume during the observation time. This is clearly shown in Figure 6.5, which shows the linear dependence of the power of the TH signal generated from the bulk

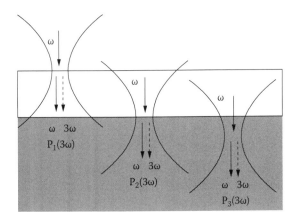

FIGURE 6.4 A slightly modified (as compared to Figure 6.1) geometry of the experimental setting to study the THG in solution containing micrscopic particles. The power of the TH signal, $P_3(3\omega)$, is measured by focusing the incident light radiation in the bulk of solution containing particles.

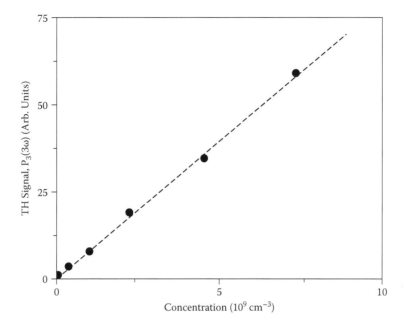

FIGURE 6.5 The measured power of the third harmonic power generated in the bulk of solution containing 0.5-μm-diameter fused silica microspheres as a function of microspheres' concentration. Circles are experimentally measured data points; the dashed line is a linear fit.

of solution of 0.5-μm-diameter polystyrene particles on the particles' concentration. The slope of this line is directly proportional to the TH signal from a single particle, and, when plotted against the particles' size, shows the fourth power dependence on the particles' size (Figure 6.6). This is a somewhat surprising result, since the TH signal from a very small particle should follow the sixth power dependence on the particles' size (Hill et al. 1993; Cheng and Xie 2002). We attribute this experimental observation to the existence of an intermediate region of particle sizes, where the particle's diameter becomes somewhat comparable with the wavelength of the TH radiation (in our experiments, it is $\lambda_\omega \cong 400$ nm). In fact, using the earlier developed approach (Hill et al. 1993), we calculated the particle size dependence of the TH signal as a function of the particle's diameter (Figure 6.7). It shows the sixth power dependence for smaller particles, which then turns to become close to the observed fourth power dependence for the particle's diameter of about 200 nm. For the larger size particles ($d > 1$ μm), the other effects, such as phase-matching and local field enhancement effects, become increasingly important (Shcheslavskiy et al. 2005a).

The ability of larger size particles to generate much stronger TH signals presents a great opportunity for detecting such submicroscopic particles in solution. We evaluated the possibility of detecting an individual particle in solution by trapping such a particle in a focused laser beam, and then used the same beam to generate the third harmonic in this trapped particle. We used Yb-doped continuous-wave fiber laser ($\lambda = 1.03$ μm, IPG Photonics, Inc.) and a low-concentrated solution of 200-nm-diameter polystyrene

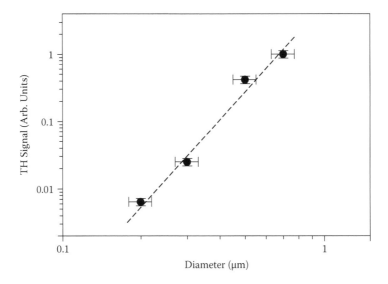

FIGURE 6.6 Solid circles: experimentally measured TH signal per particle as a function of the particles' diameter. Dashed line: linear fit exhibiting approximated fourth power dependence ($P(3\omega) \propto d^{4.1}$) of the TH signal on the particles' diameter.

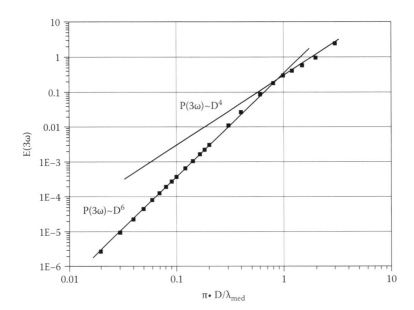

FIGURE 6.7 Squares: the calculated amplitude of the electric field of the TH wave generated by a particle of diameter, D. λ_{med} is the wavelength of the incident radiation in the particle's material. Solid line with ■: the sixth power dependence of the TH signal on the particles' diameter. Solid line: the fourth power dependence of the TH signal on the particles' diameter.

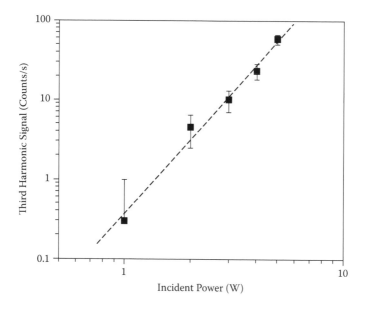

FIGURE 6.8 TH power generated from a 200-nm-diameter polystyrene particle trapped by a CW laser beam as a function of the incident power of the trapping beam. Dashed line shows a third-power dependence on the incident power.

spheres. An individual sphere was trapped in the focal plane of a high-numerical aperture (N.A. = 1.30) microscope objective, and the forward generated TH signal was collected using a fused silica lens (N.A. = 0.55) and directed to a photon-counting multiplier. As expected, when no particle is trapped, there is no measurable signal above the dark counts of the photomultiplier. In the presence of a particle in the focal volume, the signal is clearly detectable and is scaled as the third power of the incident power (Figure 6.8). At the maximum incident intensity (about 2×10^9 W $\cdot$ cm^{-2}), as much as 50 counts/s is detected on the photomultiplier. The continuous-wave laser used in those experiments was the major limitation in further increase of the output power of the TH. When this laser was replaced with a picosecond Nd:YVO$_4$ laser generating MHz-rate high-average power pulses, the efficiency of the nonlinear optical conversion was increased by more than seven orders of magnitude, reaching the level of 10^9 counts/s recorded by the detector, which is favorably compared with the second harmonic (SH) signal generated from submicroscopic particles under the same excitation conditions (Malmqvist and Hertz 1995). Clearly, such a strong signal level allows efficient real-time particle counting in solution. The advantage over conventional methods based on light scattering and/or fluorescence are obvious: there is no need for external labeling; the detected signal is spectrally distinguished from the incident radiation; and a strong localization of the TH signal allows exceptional spatial discrimination of the signal. Figure 6.9b shows a typical temporal trace of the TH signal, when a flow cell is placed in the focal plane of the laser beam (Figure 6.9a). Each individual spike of the TH signal corresponds to an individual particle passing through the focal volume. In principle, by knowing the time duration of such a peak and the

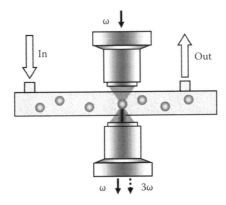

FIGURE 6.9a An experimental arrangement for observing the real-time flow of microscopic particle through the focal volume of the incident laser beam.

waist of the laser beam, one can also deduce the velocity of individual particles, which is always important to know for any microfluidic device.

The ability to perform simultaneous measurements of the TH on the interface and in the bulk of solution containing submicroscopic particles provides a very unique way of measuring the size of those particles and their nonlinear optical susceptibility (Shcheslavskiy et al. 2006a). Indeed, when such measurements are performed on the

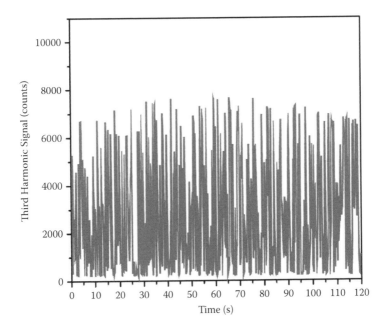

FIGURE 6.9b The power of the TH as a function of time (each individual peak of the TH signal corresponds to a particle, passing through the focal volume).

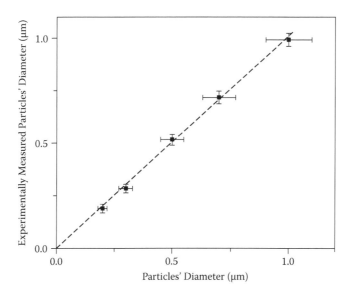

FIGURE 6.10 Experimentally measured diameter of submicron fused silica particles as a function of particles' diameter determined from scanning electron microscopy measurements. Dashed line corresponds to 1:1 size relationship. (From Shcheslavskiy, V., Petrov, G. I., Faustov, A., Yakovlev, V. V., and Saltiel, S. *Optics Lett.*, 31(10):1486–1488, 2006. Used with permission.)

interface, the effective nonlinear susceptibility [see Equation (6.5)] is calculated, which in turn depends on the volume fraction of those particles in solution (i.e., proportional to the product of their volume and concentration) and the nonlinear optical susceptibility of the particles' material. On the other hand, the TH signal generated from the bulk of solution is dependent upon the TH power generated from a single particle and its concentration in solution. Those two independent measurements provide the basis for simultaneously assessing the size and nonlinearity of the particles in solution for a given concentration of those particles in solution. However, the particles' concentration in solution can be independently measured using well-established photon correlation spectroscopy (Hess et al. 2002) using the same TH generation signal. Thus, the solution of submicroscopic particles can be fully characterized using a set of nonlinear optical measurements. To demonstrate the applicability and accuracy of this approach to characterization of particles in solution, we used an index-matching solution of fused silica submicroscopic particles. These particles are completely invisible in solution using conventional light scattering spectroscopy, but the TH signal is sufficiently strong to perform the procedure described above. The results are shown in Figures 6.10 and 6.11, where the same course of action was applied for different particles' sizes. As expected, the retrieved nonlinear susceptibility of fused silica agrees extremely well with the known literature data (Sheik-Bahae et al. 1989), while the extracted size particles (Figure 6.11) match the ones provided by the manufacturer with the accuracy of measurements.

The proposed procedure allows a rather accurate assessment of the size and optical properties of large molecular formations. In a recent example (Shcheslavskiy et al. 2006b), we attempted to interrogate the size of molecular assemblies of collagen

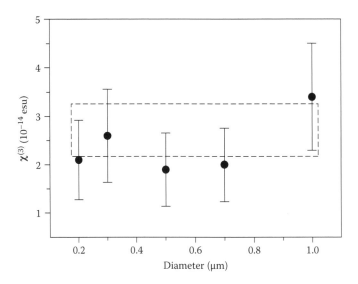

FIGURE 6.11 Third-order nonlinear susceptibility as a function of a diameter of fused silica spheres. The dashed rectangle represents the independently measured $\chi^{(3)}$ of fused silica with the appropriate error bars (Sheik-Bahae et al. 1989).

molecules in solution. While the exact shape of those assemblies is unknown, the earlier developed approach (Shcheslavskiy et al. 2005a) can be used to quantitatively describe the process of the third harmonic generation and accurately predict the formation of a new structural phase of collagen in solution (Shcheslavskiy et al. 2004; Shcheslavskiy et al. 2005b).

C. THIRD HARMONIC MICROSCOPY: FUTURE DIRECTIONS

In the original experimental setting, the TH signal was generated from the fundamental of the homemade Cr:forsterite laser (Shcheslavskiy et al. 2001). In our most recent work, a broadband continuum radiation generated from a fiber was used to excite nonlinear optical processes (Golovan et al. 2006b). The advantage of this approach is obvious—it allows the acquisition of spectral information about nonlinear response, which holds the keys to the chemically specific information (Clay et al. 2006). At the same time, the TH signal generated from interfaces is sufficiently strong and can be recorded with an adequate SNR using a relatively long (picosecond) pulse in a matter of seconds and milliseconds. However, there is a potential problem involved that has to be treated with caution. All the previous work on the TH imaging was done predominantly on the interface between two media, which have different values of $\chi^{(3)}$ and/or refractive index. In a microfluidic device there is still an interface between flowing liquid and glass, but this interface provides only a limited amount of useful information about the liquid flow. Much more important is the interface between two different liquids flowing in the same direction (Figure 6.12a). This interface is parallel to the optical axis of the optical system and shouldn't produce any TH signal.

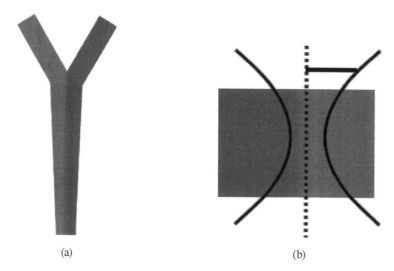

(a)   (b)

FIGURE 6.12 (a) The interface between two liquids being mixed in a microfluidic channel. (b) One of the possible geometrical arrangements allowing visualizing the mixing interface between different media by blocking half of the incident beam. Alternative geometry utilizes the spatial beam shaping (Krishnamachari and Potma, 2007).

However, if we consider the geometry of this boundary (Figure 6.12b), we see that if one half of the beam is blocked, the incident beam will come at an angle to this boundary, so that the incident light goes a different relative volume of two liquids before and after the focus, thus breaking the symmetry and generating the non-vanishing TH signal. The relative difference between the two signals created by blocking the opposite sides of the incident laser beam, provide enough information about the extension of the interface, which then can be analyzed and compared to the results of the previous studies.

The procedure to retrieve information about the properties of the liquid flow and inhomogeneities in solution relies on the set of consecutive measurements of the third harmonic signal along the direction of light propagation (i.e., along the z-axis). It would be more convenient to perform simultaneous measurements at several z-positions simultaneously. This can be done using a broadband ultrafast light source and a highly chromatic lens to focus the incident radiation onto the sample. Chromatic aberrations will disperse the focus along the direction of beam propagation, "color-encoding" the z-position (Garzon et al. 2004). The spectrum of the TH light will contain this distance information, which can be decoded back given the prior knowledge of those chromatic aberrations. The latter can be independently measured using the same approach and used for the exact z-position measurements.

As an alternate way to measure the gradients of structural and chemical composition, we intend to use the method of SH microscopy. In an isotropic environment, which has a center of symmetry, the second harmonic generation is not allowed in a dipole approximation, and is predominantly governed by weak quadrupole and surface dipole terms, which are orders of magnitude weaker. However, as we have already pointed out, the gradients of structural and chemical composition are present in a microfluidic device, and they break the inversion symmetry the same way as was first demonstrated in the late 1980s (Govorkov et al. 1989, 1990). The exact symmetry of $\chi^{(2)}$ tensor, which can be extracted from simple polarization measurements (see Golovan et al. 2006a,b; Petrov et al. 2006), contains information about the direction of those gradients, while the relative amplitude of the SH signal is directly related to the absolute value of this gradient (Govorkov et al. 1989).

III. NONLINEAR RAMAN MICROSPECTROSCOPY

A. Vibrational Microspectroscopy

Noninvasive microscopic imaging of biological systems remains a key problem in understanding the relationship between structure and function on the cellular and molecular levels. Optical spectroscopy, which uses information about the amplitude and frequencies of molecular vibrations, is typically considered one of the most chemically specific methods for in situ analytical detection of molecular species in solution. Vibrational spectroscopy is capable of providing some structural and functional information about molecular structures and their interactions. Raman spectroscopy and microscopy are particularly important, since they can provide submicron spatial resolution. Since the first introduction of the Raman microscope in 1973 (Hirschfeld 1973), optical and laser technology has made a tremendous step forward. The availability of inexpensive, energy efficient, stable, and reliable laser sources together with improved technology for spectral filtering and multichannel detection greatly increased our ability to study inorganic and organic materials in picoliter volumes. Raman confocal systems, which are now widely commercially available, permit rejections out of focus signal, making possible high-contrast high-resolution noninvasive imaging. However, Raman microscopy is still considered an up-and-coming technique for biological imaging. Despite the promise of providing chemically specific information about biological molecules in vivo, Raman spectroscopy of spontaneous Raman scattering suffers from a series of limitations, such as a fluorescent background and a low signal level. Thus, fluorescent spectroscopy is often used when real-time measurements are required (Lakowicz 1983). On the other hand, nonlinear Raman spectroscopy, and, in particular, spectroscopy of coherent anti-Stokes Raman scattering (CARS), can resolve most of the problems associated with conventional Raman spectroscopy (Hudson 1977). First, being a nonlinear optical method of spectroscopy, CARS spectroscopy relies on the interaction of high intensity laser pulses. Since the intensity of these pulses is highest in the focal point of a microscope, CARS spectroscopy potentially offers an excellent discrimination against the out-of-focus signal. A nonlinear optical Raman microscope would be an ideal confocal microscope without using an additional aperture (Wilson 1990; Duncan et al. 1982), thus greatly increasing the signal collection efficiency. Second, nonlinear Raman spectroscopy provides a way to increase a signal's level by increasing the intensity of incoming pump pulses. For ultrashort (picosecond, i.e., 10^{-12} s) pulses, a signal's level can be as much as several orders of magnitude higher than that for conventional Raman spectroscopy for a given average power of the incident laser beam (Hudson 1977). Another advantage of

CARS spectroscopy is that the detected signal is blue-shifted with respect to the excitation wavelengths; i.e., there is no problem of separating the signal from a fluorescent background. It should also be noted that short-pulse lasers, used for CARS spectroscopy, are naturally designed to study ultrafast processes on a molecular time scale, thus providing a unique opportunity to study molecular dynamics (such as protein folding, DNA transformations, etc.) in real-time (i.e., on the time scale of molecular processes). Last but not least, the light waves in the near-IR are much better suited for penetrating biological tissues, because fewer biological molecules absorb them, and because light scattering is less at longer wavelengths. All these features of CARS, a powerful spectroscopic tool, have been used extensively for flame, gas phase, plasma, and combustion diagnostics.

Since its first introduction (Duncan et al. 1982), nonlinear Raman microscopy based on CARS spectroscopy, driven by the progress of ultrafast laser technology, has made a tremendous step forward (Zumbusch et al. 1999; Potma et al. 2001; Hashimoto and Araki 2000; Wurpel et al. 2002; Dudovich et al. 2002; Lim et al. 2005; Petrov et al. 2007; Konorov et al. 2007; Kano and Hamaguchi 2007; Lozovoy and Dantus 2005; von Vocano and Motzkus 2006), making video-rate vibrational imaging possible (Evans et al. 2005; Pestov et al. 2008). However, despite the recent widespread of CARS microscopy and all the remarkable progress made over the past few years, the technique is still considered to be an emerging tool in microscopic imaging due to a number of technical difficulties, such as the complexity of the experimental setup for CARS microscopy, the high cost associated with lasers, and the so-called nonresonant background resulting from a nonspecific four-wave mixing process.

B. CARS MICROSPECTROSCOPY

CARS microspectroscopy is one of the examples of hyperspectral microscopy, which aims at obtaining a full vibrational spectrum from a given microscopic volume for the purpose of chemical analysis. In general, the complex chemical composition of a solution can complicate the vibrational spectrum. Most of the CARS imaging studies have been accomplished in a high-frequency vibrational range where the CARS signal from lipids is exceptionally strong. However, what is most interesting from a biochemical perspective is the so-called "fingerprint" region of vibrational frequencies from 500–1750 cm^{-1}. The strength of Raman lines is significantly reduced for most of the vibrations in this region, and a typical concentration of molecular species of interest (mostly proteins) is not as high as in lipid droplets. This type of imaging presents an apparent challenge for spectroscopists in terms of the detection and analysis of CARS signals.

CARS signal originates from a coherent excitation of vibrational level using a pair of optical pulses, ω_1 ("pump") and ω_2 ("Stokes"), separated by a frequency of this vibrational level, Ω; i.e.,

$$\omega_1 - \omega_2 = \Omega. \tag{6.6}$$

The third pulse at frequency ω_3 ("probe") is scattered off the coherently excited vibration to generate the signal at the CARS frequency, ω_{CARS} (Figure 6.13):

$$\omega_{CARS} = \omega_3 + \Omega = \omega_3 + (\omega_1 - \omega_2). \tag{6.7}$$

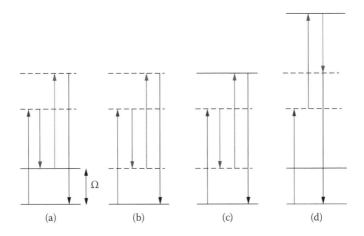

FIGURE 6.13 A schematic diagram of various four-wave mixing processes resulting in generation of $w_{CARS} = \omega_3 + (\omega_1 - \omega_2)$: (a) resonant CARS process, involving coherent excitation of a vibrational state (shown as a solid line); (b) non-resonant CARS process, involving transitions only through virtual states (shown as dashed lines); (c–d) different non-resonant CARS processes, which do not involve excitation of vibrational states, but involve electronic state transitions.

CARS signal is a coherent signal; i.e., the phase-matching conditions have to be fulfilled in order to achieve the efficient signal generation. Microscopic focusing conditions using a high-numerical-aperture lens significantly relax this restriction, allowing for simultaneous detection of the whole CARS spectrum without any special arrangements (Akhmanov and Koroteev 1981). This allows us to make an accurate analysis of the generated CARS signal and to directly compare it with the spontaneous Raman signal.

For simplicity of analysis, we consider just a single vibrational transition, which is characterized by the Raman cross section,

$$\frac{d\sigma}{d\Omega},$$

and the line width, Γ. The interaction volume is defined by the focal spot area,

$$A = \frac{\pi\omega_0^2}{2},$$

and the Rayleigh length,

$$l_R \cong \frac{2\pi\omega_0^2}{\lambda},$$

where ω_0 is the beam waist radius and λ is the incident wavelength, which is assumed to be approximately equal for all the light waves involved in this nonlinear optical interaction. This leads to the total power of CARS signal to be (Akhmanov and Koroteev 1981):

$$P_{CARS} \approx \left(\frac{8\pi c \omega_1 \omega_3}{\hbar \omega_2^4} \right)^4 \left(\frac{N}{\Gamma} \frac{d\sigma}{d\Omega} \right)^2 P_1 P_2 P_3, \tag{6.8}$$

which has to be compared with the power of the spontaneous Raman signal,

$$P_{Raman} \approx \Delta \Omega l_R N \frac{d\sigma}{d\Omega} P_3, \tag{6.9}$$

where N is the concentration of molecules, p_i is the incident power for the i-th frequency of the incident wave, and $\Delta\Omega$ is the collection angle for Raman microscopy. Most of the arguments for CARS microscopy are based on the direct comparison of Equations (6.8) and (6.9), which leads to the conclusion that for concentrated molecular species and high-power laser sources ($p_i > 10^3$ W), CARS signal is many orders of magnitude stronger than Raman signal. However, the reality is more complicated when this analysis, which was originally developed for diagnostics of gaseous systems, is applied to biochemical imaging.

The first important point we would like to make is related to the acceptable level of the incident power. Biological systems are not used to the high-intensity laser irradiation and impose a limit on the maximum incident intensity to be used without the induced cell damage. There is an ongoing debate about what the limiting intensity is for different incident excitation wavelengths, pulse durations, etc. (Squirrel et al. 1999; Konig et al. 1999; Yakovlev 2003; Schönle and Hell 1998); however, it is more or less accepted that the heating effects due to a linear absorption in the near-IR are rather insignificant (Schönle and Hell 1998), and the major contribution to cell damage comes from multiphoton absorption (Squirrel et al. 1999; Konig et al. 1999; Yakovlev 2003), which can be considerably reduced by implementing excitation sources with laser wavelength from 900–1300 nm (Squirrel et al. 1999; Konig et al. 1999; Yakovlev 2003). This allows the use of more than an order of magnitude higher incident laser powers, compared to 800-nm excitation.

The second important aspect of a fair comparison of nonlinear Raman microscopy and microscopy of spontaneous Raman scattering is related to evaluation of their signal-to-noise ratios, as was first demonstrated by Turner et al. (Tolles and Turner 1977; Eesley et al. 1978). Here, we simplify the analysis by assuming that highly stable laser sources are used and all the detectors are quantum limited (i.e., the only source of noise is the shot-noise). Both the laser and detection technologies made a significant step forward over the past 30 years, making the above assumption quite reasonable. Using Equations (6.8) and (6.9) and the procedure described by Eesley et al. (1978), we can express the improvement of the signal-to-noise ratio (SNR) for CARS spectroscopy with respect to spontaneous Raman spectroscopy as:

$$\frac{SNR_{CARS}}{SNR_{Raman}} \cong 50 \times \left(\frac{p_1}{kW} \right) \times \left(\frac{c}{1M} \right)^{1/2} \times \left(\frac{10\,cm^{-1}}{\Gamma} \right) \times \left(\frac{d\sigma}{d\Omega} \cdot \frac{1}{10^{-30}\,cm^2} \right)^{1/2} \times \left(\frac{\lambda_1}{1\,\mu m} \right)^{3/2}.$$

$$\tag{6.10}$$

In deriving Equation (6.10) we assumed that $p_1 = p_2$ and $\omega_1 \cong \omega_2 \cong \omega_3$, and expressed the concentration of molecules of interest in terms of the molar concentration, c, which is more convenient for biochemical analysis. Equation (6.10) provides a simple way of evaluating a possible use of CARS microspectroscopy. A typical value for the Raman cross section in the fingerprint region is

$$\frac{d\sigma}{d\Omega} \cong 10^{-30}\,\mathrm{cm}^2,$$

typical line width is $\Gamma \cong 10\,\mathrm{cm}^{-1}$. If one can afford using the incident power of $p_1 \cong 10$ kW, which should be possible for excitation wavelength longer than 1 μm, the resulted gain in the SNR will be a factor of 50 for the molecular concentration of $c = 10$ mM, which is considered to be rather high for most biological molecules. Since the SNR is scaled as a square root of the acquisition time, the total gain in the speed of recording CARS spectra is 2500, making CARS microspectroscopy extremely attractive. On the other hand, if the excitation wavelength is around 800 nm, biological structures demand the use of lower power excitation sources, and for the same experimental setting CARS microspectroscopy does not have any advantage with respect to spontaneous Raman spectroscopy.

The above formula does not take into account the fluorescence background, which is a common problem for Raman spectroscopy. This background reduces the SNR for Raman measurements (Yakovlev 2007) by a factor of

$$\frac{SNR_{\text{with fluorescence}}}{SNR_{\text{w/o fluorescence}}} = \sqrt{\frac{p_{\text{fluoresence}}}{p_{\text{Raman}}}}, \tag{6.11}$$

where $p_{\text{fluorescence}}$ is the fluorescence signal from the same volume. This is not the only factor, which is missing in Equation (6.10). CARS signal also suffers from a background that originates from a nonchemically specific four-wave mixing, which is considered to be one of the most prominent problems of CARS spectroscopy. The presence of this background reduces the SNR for CARS measurements (Akhmanov and Koroteev 1981; Eesley et al. 1978) by a factor of

$$\frac{SNR_{NR+R}}{SNR_R} = \sqrt{\frac{p_{NR}}{p_R}}, \tag{6.12}$$

where p_{NR}, p_R are the generated nonresonant and resonant powers of the CARS signal. While water molecules have a relatively low nonresonant Raman cross section, their concentration in solution is so high that the resonant contribution to the CARS signal is hardly noticeable if the concentration of molecular species of interest is smaller than 100 mM and no measures to reduce this background are taken. We discuss this issue separately, since it is the primary challenge in obtaining high-quality CARS spectral data, but for now we just assume that this background is not a problem, and

Equation (6.10) is the only fundamental limit imposed on CARS microspectroscopy. Most molecular species, e.g., proteins, rarely occur in a cell in concentrations higher than 1 mM and have a moderate Raman cross section in the fingerprint region. Earlier, we demonstrated that the use of the incident beams with a power in excess of 10 kW is essential to achieve significant SNR improvement of CARS spectra measurements. Since the incident light intensity is the major limiting factor, one has to evaluate the experimental conditions, which will allow the use of such high peak powers for cellular imaging. Recently, we did some systematic studies on cell and tissue damage using short-pulsed laser irradiation in the wavelength region around 1000 nm (Yakovlev et al. 2008). While the damage threshold somewhat varies for different cells and tissues, it can be approximately interpolated with a rather simple expression:

$$I_{max} \cong 1.5 \cdot 10^{11} \, \text{W/cm}^2 \sqrt{\frac{1 \, ps}{\tau_p}}, \tag{6.13}$$

where τ_p is the incident pulse duration of the light wave, which coherently excites molecular vibration. Using Equation (6.13) as a guideline, one can see that if the goal is to achieve microscopic spatial resolution, i.e., $\omega_0 \cong 0.5 \, \mu m$, the maximum allowed incident power is somewhere around $p_1 \cong 200 \, W$ for the incident pulse duration of $\tau_p = 4 \, ps$. Clearly, for microscopic imaging of low concentrated ($c \cong 1 \, mM$) molecular species, Raman microscopy does a much better job. CARS microscopy still provides better SNR, even at the highest possible spatial resolution, when used with higher concentrated molecular species and strong Raman transitions (vibrational imaging of lipid distribution using CH_2 stretch vibration provides one of the best examples of such an application [Hellerer et al. 2007]). For the most general case, one has to increase the incident beam diameter in order to accommodate higher incident power. Indeed, if the beam diameter is increased to about $\omega_0 \cong 3 \, \mu m$, which roughly corresponds to a typical cell, the incident power can be increased to about $p_1 \cong 10 \, kW$, making CARS spectral measurements highly favorable.

C. NONRESONANT BACKGROUND IN CARS MICROSPECTROSCOPY

The nonresonant background in CARS spectroscopy was always considered one of the biggest challenges to deployment of this spectroscopic technique for analytical measurements (Volkmer 2005). In brief, nonresonant contribution to the nonlinear optical polarization from surrounding molecules interferes with resonant contribution, producing an overall signal that has complex frequency dependence. The resulting line shape is given by the following equation:

$$R(\omega) \propto \left| \chi_{NR}^{(3)} + \sum_r \frac{A_r}{\omega_r - \omega - i\Gamma_r} \right|^2, \tag{6.14}$$

where A_r, ω_r, and Γ_r are the amplitude, the transition frequency, and the line width, respectively, of the rth Raman mode. The first term in Equation (6.14) is the nonresonant susceptibility, which is considered to be frequency independent. The second term in Equation (6.14) corresponds to the resonant contribution. If it is set to zero, the total signal will be just proportional to

$$\left| \chi_R^{(3)} \right|^2 = \left| \sum_r \frac{A_r}{\omega_r - \omega - i\Gamma_r} \right|^2, \qquad (6.15)$$

which is different from the spectral shape of a typical Raman spectrum taken for the same excitation wavelength and defined by

$$\left| \mathrm{Im}\, \chi_R^{(3)} \right|^2 = \left| \mathrm{Im} \sum_r \frac{A_r}{\omega_r - \omega - i\Gamma_r} \right|^2. \qquad (6.16)$$

Ideally, we want to retrieve the Raman spectrum, which would allow us to directly compare the experimentally measured spectral data with the Raman spectra. There are several ways of dealing with this problem.

1. Polarization Suppression

The idea of using polarization properties of $\chi^{(3)}$ tensor to suppress the nonresonant contribution was first suggested and experimentally demonstrated by Akhmanov et al. (1977). Later, Oudar et al. (1979) modified the earlier proposed ellipsometric technique to achieve either direct suppression of nonresonant background, or to use this nonresonant background for heterodyne mixing with the resonant contribution to extract either the real or imaginary part of the resonant susceptibility tensor $\chi_R^{(3)}$. Both approaches provide a substantial nonresonant background suppression, which is limited by the degree of depolarization introduced by high-numerical-aperture optics and the quality of broadband polarization components. Using polarization-preserving microscope objectives and specially selected polarizers and Fresnel rhombs for polarization rotation, we routinely achieved two to three orders of magnitude suppression of the nonresonant background (Figure 6.14). The problem comes, however, when a scattering medium is introduced in the focal plane, which can be, for example, a cell, an organelle, or a piece of tissue. The efficiency of the background suppression is substantially degraded. It gets even worse for typical tissue imaging, when the whole signal collected in the back-reflected geometry is produced through light scattering. An additional disadvantage of this polarization technique is that the useful signal in all configurations is also reduced.

2. Heterodyne Detection

Heterodyne detection is widely used to improve the SNR of optical measurements. The additional advantage of heterodyne detection for CARS measurements is that it allows direct extraction of either real or imaginary parts of the susceptibility tensor (Vinergoni et al. 2004; Evans et al. 2004). Being introduced for CARS spectros-copy about 30 years ago (Eesley et al. 1978), heterodyne detection was for a long

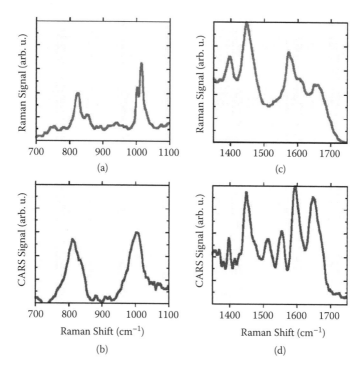

FIGURE 6.14 (a, c) Raman spectra of dried bacterial spores (*Bacillus subtilis*), and (b, d) the corresponding polarized CARS spectra of the same bacterial spores in aqueous solution. (From Petrov, G. I., Yakovlev, V. V., Sokolov, A. V., and Scully, M. O., *Optics Express*, 13(23): 9537–9542, 2005. Used with permission.)

time considered the best way of achieving the best SNR in CARS measurements (Akhmanov and Koroteev. 1981; Eesley et al. 1978). There are many ways of introducing weak signal mixing with a stronger reference beam. One of the numerous possibilities is to use the same polarization schemes, as described in a previous section, but to slightly misalign the analyzing polarizer in the detection channel (Oudar et al. 1979). This way the nonresonant signal will be mixed with the signal, and, if the signal is detected simultaneously in two orthogonal polarizations defined by the orientation of the analyzing polarizer, it leads to the direct measurement of the real part of $\chi_{res}^{(3)}$:

$$\frac{P_{parallel}^{CARS}}{P_{perpendicular}^{CARS}} = A + \mathrm{Re}\left(\chi_{res}^{(3)}\right). \tag{6.17}$$

While this signal is greatly magnified with respect to the polarized CARS signal, which is proportional to $\left|\chi_{res}^{(3)}\right|^2$, it is still difficult for interpretation. To resolve this problem, an additional quarter-wave plate is introduced into the detection channel.

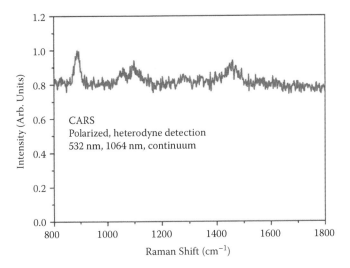

FIGURE 6.15 Experimentally measure heterodyne CARS spectrum of 5% (volume) ethanol solution in water.

The proper phase shift introduced to the signal wave allows direct measurements of the imaginary part of the resonant nonlinear optical susceptibility, i.e.,

$$\left. \frac{P^{CARS}_{parallel}}{P^{CARS}_{perpendicular}} \right|_{\text{with } 1/4-\text{waveplate}} = B + \text{Im}\left(\chi^{(3)}_{res} \right). \tag{6.18}$$

Figure 6.15 shows an illustrative example of the CARS spectrum, described by Equation (6.18), for a mixed solution of ethanol and water (1:20 in volume). The most important advantage of this type of heterodyne imaging is that the final CARS spectrum is directly related to the Raman spectrum, which is also proportional to the imaginary part of $\chi^{(3)}_{res}$. Polarization measurements are not the only way of making broadband heterodyne CARS measurements (Kee et al. 2006); however, all the proposed schemes suffer from light scattering in a sample, which affects the phase properties of the signal. It is also hardly applicable to epi-detection geometry, which is the most desirable arrangement for many biomedical applications.

3. Time-Delayed Methods

The nonresonant background in CARS spectroscopy originates from instantaneous four-mixing processes, while the resonant contribution involves real vibrational states. This provides a basis for possible discrimination against the nonresonant background. To do so, one has to come up with a pair of pulses that excite the vibrational state, and the third, time-delayed pulse will only contribute to the resonant part of the CARS signal. However, to make this scheme work efficiently, one has to overcome certain obstacles. To achieve high spectral resolution, the bandwidth of the third pulse should

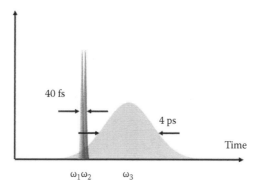

FIGURE 6.16 Temporal arrangement of three incident pulses, which requires two ultrashort pulses ("pump" and "Stokes") to be temporally overlapped and preceding the peak of the third (long) pulse.

be of the order or less than the line width of the Raman line, Γ, which requires the pulse duration of the third pulse to be longer than Γ^{-1}. On the other hand, Γ^{-1} roughly corresponds to the lifetime of the excited state; i.e., if the time delay is longer than Γ^{-1}, it will substantially degrade the amplitude of the resonance CARS signal. One of the possible solutions to this problem is to use a pair of time-overlapping short pulses to excite the transition and to use a longer probe pulse to scatter off the coherent excitation. This situation is illustrated in Figure 6.16. The time-overlap of excitation and probe pulses (i.e., the amplitude of the nonresonant contribution) is approximately proportional to the pulse duration of the excitation pulse(s), while the amplitude of the resonant contribution is roughly proportional to the pulse duration of the probe pulse. One can further suppress the overlap through some pulse shaping techniques, as was suggested by Pestov et al. (2007). In this approach, the probe pulse is passed through a spectral filter to produce a square-shaped spectrum, which in a time domain corresponds to the pulse, whose intensity profile looks like

$$I(t) \propto \left(\frac{\sin t}{t} \right)^2.$$

By making the excitation pulses overlap with the minimum of the probe pulse preceding its main maximum, the nonresonant background is further suppressed (Pestov et al. 2007). The same idea can be exploited with a single pulse excitation (Dudovich et al. 2003), when both pump pulses at frequencies ω_1, ω_2 are derived from a single ultra-broadband pulse.

   The great advantage of using time-delayed methods to reduce the nonresonant background is that they can work in highly scattering media and can be used for biomedical diagnostics of cells and tissues. The obscure disadvantage of this approach is the reduced amplitude of the CARS signal. Indeed, the amplitude of the resonant CARS signal is proportional to the energies of each of the excitation pulses (Pestov et al. 2008; Dudovich et al. 2003); the employment of shorter incident pulses in all the proposed schemes requires downscaling of the input energy as a square root of the pulse duration and, thus, diminution of the overall signal.

4. Phase Retrieval Algorithms

The spectral line shape in CARS spectroscopy is described by Equation (6.14). In order to investigate an unknown sample, one needs to extract the imaginary part of $\chi_R^{(3)}$ to be able to compare it with the known spontaneous Raman spectrum. To do so, one has to determine the phase of the resonant contribution with respect to the nonresonant one. This is a well-known problem of phase retrieval, which has been discussed in detail elsewhere (Lucarini et al. 2005). The basic idea is to use the whole CARS spectrum and the fact that the nonresonant background is approximately constant. The latter assumption is justified if there are no two-photon resonances in the molecular system (Akhmanov and Koroteev 1981). There are several approaches to retrieve the unknown phase (Lucarini et al. 2005), but the majority of those techniques are based on an iterative procedure, which often converges only for simple spectra and negligible noise. When dealing with real experimental data, such iterative procedures often fail to reproduce the spectroscopic data obtained by some other means.

The alternative approach is based on a non-iterative procedure using the maximum entropy model (MEM) to extract the complex dielectric susceptibility $\chi_R^{(3)}$ from the intensity measurements. This technique was first proposed 15 years ago (Vartiainen 1992), and recently was used for multiplexed CARS measurements (Petrov et al. 2007, Vartiainen et al. 2006).

The concept behind the maximum entropy model is to choose the spectrum in the form of a non-negative function of frequency, which corresponds to a time series with maximum entropy whose autocorrelation function is consistent with the set of known values.

The experimentally measured CARS signal is given by

$$I_{CARS}\left(\tilde{\omega}_1 - \tilde{\omega}_2\right) \propto \left|\chi^{(3)}\left(\tilde{\omega}_1 - \tilde{\omega}_2\right)\right|^2, \tag{6.17}$$

and is measured with in-frequency interval of $\tilde{\omega}_1 \le \tilde{\omega} \le \tilde{\omega}_2$. Let us assume that the real and imaginary parts of $\chi^{(3)}(\tilde{\omega}_1 - \tilde{\omega}_2)$ do not change sign in that frequency range and the nonresonant $\chi^{(3)}$ is real and does not depend on frequency. We define a normalized frequency as

$$v = \frac{\tilde{\omega} - \tilde{\omega}_1}{\tilde{\omega}_2 - \tilde{\omega}_1}. \tag{6.18}$$

The maximum entropy model estimate for $\left|\chi^{(3)}(v)\right|^2$ can be written as

$$\left|\chi^{(3)}(v)\right|^2 = \frac{|\beta|^2}{\left|1 + \sum_{n=1}^{M} a_n \exp(-i2\pi nv)\right|^2}, \tag{6.19}$$

where the unknown maximum entropy model coefficient a_n and $|\beta|^2$ are functions of estimated autocorrelation C_m. Those MEM coefficients can be found by solving a set of linear equations,

$$
\begin{pmatrix}
C_0 & C_{-1} & \cdots & C_{-M} \\
C_1 & C_0 & \cdots & C_{1-M} \\
\cdots & \cdots & \cdots & \cdots \\
C_M & C_{M-1} & \cdots & C_0
\end{pmatrix}
\begin{pmatrix}
1 \\
a_1 \\
\cdots \\
a_M
\end{pmatrix}
=
\begin{pmatrix}
|\beta|^2 \\
0 \\
\cdots \\
0
\end{pmatrix},
\tag{6.20}
$$

where C_m are the Fourier coefficients of signal $\left|\chi^{(3)}(v)\right|^2$ and are defined as

$$
C_m = \int_0^1 \left|\chi^{(3)}(v)\right|^2 \exp(i2\pi mv)\,dv, \quad m \le M.
$$

For discrete set of normalized frequencies:

$$
C_m = \frac{1}{K}\sum_{j=0}^{K-1}\left|\chi^{(3)}\right|^2 \exp\left(\frac{i2\pi mj}{K}\right), \quad m \le M, M \le \frac{K}{2}.
$$

Solution to the above Mth order equation always exists if the $M + 1$ by $M + 1$ Toeplitz matrix has a non-negative definite. In a typical experimental situation, $\left|\operatorname{Im}\chi_R^{(3)}\right| \ll \left|\chi_{NR}^{(3)} + \operatorname{Re}\chi_R^{(3)}\right|$, i.e., the nonresonant contribution, dominates the CARS signal, and the Raman line shape can be calculated using a simple expression:

$$
\operatorname{Im}\chi_R^{(3)}(v) = \sqrt{\chi^{(3)}(v)}\,\sin\theta(v), \tag{6.21}
$$

where $\theta(v)$ is the MEM's phase, defined as

$$
\theta(v) = \arg\left[1 + \sum_{n=1}^{M} a_n \exp(-i2\pi nv)\right].
$$

The above procedure of the Raman spectrum retrieval exhibits an amazing robustness with respect to the noise of the data; however, several empirically deduced conditions have to be satisfied for an accurate retrieval. First, the retrieved spectrum is always distorted on the edges, and, to get good precision, the CARS spectrum has to be taken in a very broad range (see, for example, Figure 6.17). Second, the presence of strong Raman lines, for which the above assumption of $\left|\operatorname{Im}\chi_R^{(3)}\right| \ll \left|\chi_{NR}^{(3)} + \operatorname{Re}\chi_R^{(3)}\right|$ is not fulfilled, might affect the retrieved amplitude and shape of this particular line; i.e., the developed algorithm works the best for relatively weak Raman lines. This is really important for CARS measurements. It appears that it is not necessary to attenuate the nonresonant background to complete extinction: some of the residual

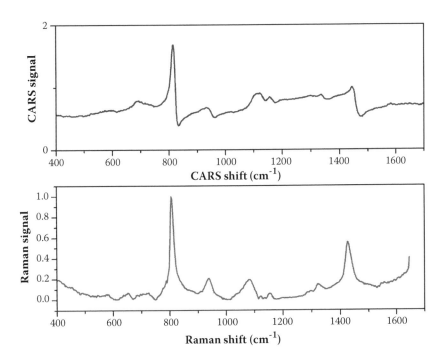

FIGURE 6.17 Top panel: experimentally measured CARS spectrum from an isopropyl solution. Bottom panel: Raman spectrum retrieved from the CARS spectrum.

background helps increase the signal amplitude. The additional surprising advantage of this approach is that it allows, unlike spontaneous Raman measurements, direct concentration measurements in solution. Indeed, the nonresonant background is predominantly due to the abundance of water molecules, whose concentration for most of the practical applications remains fixed. Thus, the ratio of the resonant signal to the nonresonant background provides a reliable measure of the relative concentration of molecules under study. Finally, we find that good agreement with independently measured Raman spectra is only possible if proper normalization of CARS signals is performed. To do so, we typically place a distilled water solution in the place of the sample and collect the reference CARS spectrum. This spectrum contains convoluted information about the spectral transmission of optical filters and spectrometer, the spectra response of a charge-coupled device (CCD) detector, and the spatial, wavelength-dependent overlap of all incident beams in the focal plane.

D. EXPERIMENTAL SETUP FOR BROADBAND CARS MICROSPECTROSCOPY

The concept of a broadband CARS microspectroscopy was first introduced in 2000 (Yakovlev 2001), and was further developed over the past few years (Yakovlev 2003; Petrov and Yakovlev 2005; Petrov et al. 2005, 2007). The major idea is

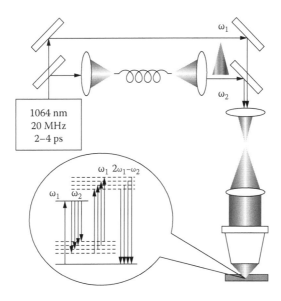

FIGURE 6.18 Schematic diagram illustrating the concept of ultrabroadband CARS microspectroscopy.

to develop a simple CARS microspectrometer that allows simultaneous recording of the whole vibrational spectrum. This concept is schematically presented in Figure 6.18. In the initial design (Figure 6.18), a fundamental high-energy, mode-locked Nd:YVO$_4$ oscillator (Petrov et al. 2003, 2004) was used to serve both as a pump and probe pulse, while the broadband continuum generated in a GeO$_2$ fiber was used as a broadband Stokes pulse. High spectral density generated through the process of stimulated Raman scattering in GeO$_2$ fiber (Petrov and Yakolvev 2005; Petrov et al. 2003) allows fast CARS spectral recording in the spectral range from 200–3000 cm^{-1}. The spectrum out of such fiber is shown in Figure 6.6, and it clearly shows that at least 100 nJ of the input energy is required to achieve a smooth and powerful continuum spectrum. Phase retrieval methodology was successfully adapted for CARS signal retrieval. Figures 6.19 and 6.20 illustrate the successful retrieval for several complex systems, which show congested Raman lines in the fingerprint region. Clearly, the described approach works for a variety of molecular systems and is capable of reproducing Raman spectra at a much faster acquisition rate.

The major challenge is to improve the sensitivity of CARS microspectroscopy, which requires a substantial increase of the signal and reduction of the nonresonant background. This can be achieved in a slightly modified setting, when the high-energy continuum generated in an optical fiber is compressed. While getting a transform-limited pulse is rather difficult, if possible, we can use a precompressed pulse, which can be as short as 50–70 fs. It is still two orders of magnitude shorter than the probe pulse and can be used for efficient vibrational excitation

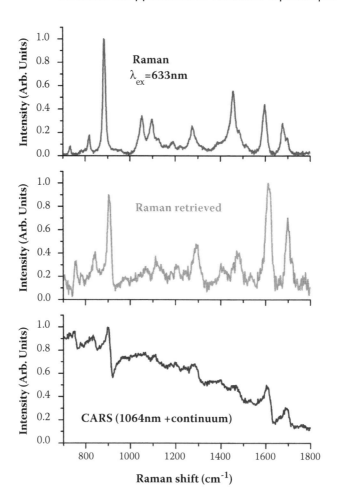

FIGURE 6.19 Vibrational spectra of vanillin in isopropanol solution. Bottom panel: experimentally measured CARS spectrum (1064 nm and continuum excitation, 1064 nm probe). Middle panel: retrieved Raman spectrum. Top panel: experimentally measured spontaneous Raman spectrum (excitation wavelength 532 nm). The acquisition time for CARS spectrum was 100 times shorter than for spontaneous Raman. The incident powers were set at approximately the same level.

(Dudovich et al. 2002; Lim et al. 2005) and nonresonant background suppression (Pestov et al. 2008). The latter is illustrated in the following example of glucose sensing using CARS microspectroscopy. In this particular example, a relatively low concentrated solution of glucose shows reproducible CARS signal (Figure 6.21a), which is retrieved using the algorithm described above (Figure 6.21b). An excellent agreement with the independently recorded Raman spectrum demonstrates the great promise of CARS microspectroscopy for noninvasive glucose concentration measurements.

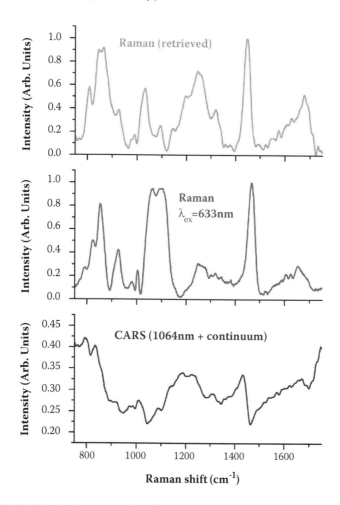

FIGURE 6.20 Vibrational spectra of collagen-rich tissue. Bottom panel: experimentally measured CARS spectrum (1064 nm and continuum excitation, 1064 nm probe). Middle panel: experimentally measured spontaneous Raman spectrum. Top panel: retrieved Raman spectrum. The acquisition time for CARS spectrum was 100 times shorter than for spontaneous Raman. The incident powers were set at approximately the same level. Autofluorescence background was digitally subtracted from experimentally measured Raman spectrum.

IV. APPLICATIONS TO MICROFLUIDIC DEVICES

A. PROTEIN CRYSTALLIZATION

We are currently pursuing protein crystallization imaging in microfluidic devices, which can use both THG and CARS microscopy. The use of the microfluidic platform is the newest and one of the most promising trends in the colossal project of obtaining the exact structure of more than 70,000 protein molecules. The goal is to use microfluidic devices as a multifunctional, multiparallel platform for trying out a

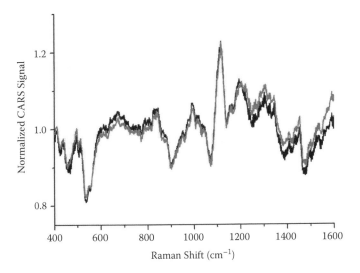

FIGURE 6.21a Typical CARS spectra from glucose solutions, showing spectra reproducibility for day-to-day operation. Black and red curves are CARS spectra from a buffer solution of glucose of slightly different concentrations (about 150 mg/dL). Spectra are shown on the same scale to emphasize the reproducibility.

very large number of protein crystallization conditions in order to identify which of them work best for a given protein (Anderson et al. 2006; Li et al. 2006). Microfluidic devices are ideally suited for screening thousands and millions of protein crystallization conditions, while minimizing the volume of a protein used (Anderson et al. 2006; Li et al. 2006). A single "reactor" is only 1 nL small; thus, the amount of proteins used is greatly minimized. Large (10–100 micron) sized crystals can be easily visualized in a microscope (Lounaci et al. 2006), while pre-crystallization conditions are typically correlated with light scattering measurements (Chaven and Saridakis 2008), which cannot distinguish small (submicron) sized crystals from protein and other molecular aggregates that often occur from membrane proteins. While at this moment most of the efforts are focused on developing microfluidic platforms capable of fulfilling this enormous task, sooner or later the need to identify pre-crystallization conditions in a submicron volume will be become increasingly important.

Our strategy involves the simultaneous use of THG and CARS microscopy to rapidly identify the presence of submicroscopic protein crystals in a nanoliter volume. Increase of the TH signal in the volume of solution is always associated with the formation of inhomogeneities, which can be either molecular aggregates or small crystals. However, vibrational spectroscopy can distinguish aggregates from microcrystals (Noda et al. 2007; Schwartz and Berglund 2000), which can be demonstrated using a model system of lysozyme crystallization. In brief, a highly concentrated protein solution (50 mg/mL) was prepared and micrometer-sized protein crystals were allowed to form (Figure 6.22a). CARS spectra were measured from both the solution and protein crystals. The retrieved Raman spectra from the bulk of the protein

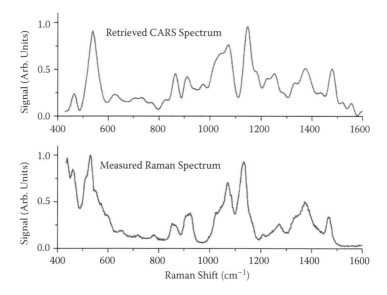

FIGURE 6.21b Top panel: Raman spectrum retrieved from the experimentally measured CARS spectrum (excitation wavelengths >1064 nm); bottom panel: experimentally measured Raman spectrum (excitation wavelength 532 nm). Glucose concentration is the same (150 mg/dL), and the incident average power is the same (50 mW). The CARS spectrum was collected within 1-s, while Raman spectrum took 100 s to collect. CARS spectrum was collected with a thermoelectrically cooled CCD and spectral resolution 1 cm^{-1}, Raman spectrum was collected with liquid nitrogen–cooled CCD and spectral resolution 5 cm^{-1}. Minor differences in spectral lines positions and their relative intensities are attributed to the very large difference in the incident wavelengths.

solution and an individual microcrystal are shown in Figure 6.22b and exhibit a dramatic difference in spectral shapes and amplitude, manifesting the great advantage of vibrational spectroscopy to identify the formation of microcrystals in solution.

B. Vibrational Flow Cytometry

Flow cytometry (Ormerod 2000) is yet another application of nonlinear optical spectroscopies. In a conventional flow cytometry, a stream of particles or cells is analyzed and assorted based on the electrical and/or optical signal generated from those particles or cells. The vast majority of applications involve fluorescence signal detection, which in turn requires external markers. While those fluorescent tags can be very specific, this procedure is invasive and often does not reflect the overall variation of biochemistry occurring in a cell due to a developing apoptosis or a response to a drug treatment (Nguyen et al. 2008). Vibrational spectroscopy has proven to be a valuable tool in diagnosing different stages of cancer (Andrus and Strickland 1998; Utzinger et al. 2001; Matthaus et al. 2006); however, most of the previous studies were performed on a large number of cells. Recently, we evaluated the approach based on a single-cell analysis using confocal Raman spectroscopy.

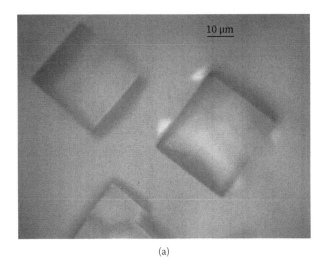

(a)

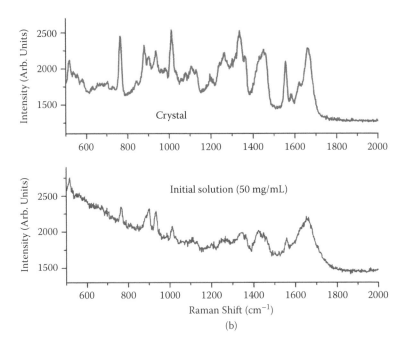

(b)

FIGURE 6.22 (a) The white-light image of lysozyme crystals. (b) The retrieved Raman spectrum of a crystalline lysozyme (top) and the retrieved Raman spectrum of lysozyme in solution (50 mg/mL; bottom).

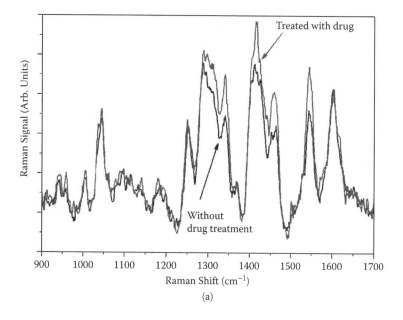

(a)

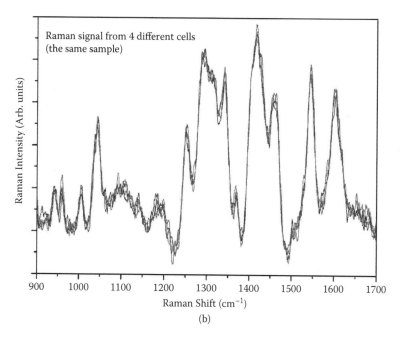

(b)

FIGURE 6.23 (a) Raman spectra of cancer cells with and without drug treatment. The observed difference in relative intensity of Raman lines is much larger than a standard variation with the same cell culture. (b) Raman spectra from four different cells from the same culture.

One of the promising directions in cancer drug development is the use of multiple drugs that bring the destructed biochemistry of a cell step-by-step to a normal state (Wallqvist et al. 2002). While microarrays, capable of analyzing those cells after each drug application, exist (Wallqvist et al. 2002), it would be highly desirable to analyze the variations of local biochemistry in those cells noninvasively. Raman spectroscopy is capable of accomplishing this task, which is illustrated by an example presented in Figure 6.23a, where cancer cells (PANC1 human prostate cells) were treated with a drug (hot water extract of *Nerium oleander*, 20 μg/mL). The difference in Raman spectra between different cells is quite remarkable, while the same cells show very little variation in their Raman spectra (Figure 6.23b). In this particular example, the laser beam at 632.8 nm (He-Ne laser) is focused into a spot size of about 2.5 μm, which was positioned at a cell nucleus, where the most changes in chemical/structural compositions are expected (Swain and Stevens 2007). While the presented results show definite promise, the major drawback was the extended acquisition time needed to collect high-quality data. If there are thousands of cells simultaneously developing and being treated with drugs on a microfluidic platform, one would prefer to shorten the acquisition time for such measurements, while keeping the superior SNR.

Our earlier estimations suggest that if the size of the incident laser beam is greater than 3 μm, CARS microspectroscopy is much more superior to conventional Raman spectroscopy, since it allows the use of high-power laser pulses benefiting the overall signal and SNR. Our previous results with bacterial spores (Petrov et al. 2007) indicate that even for tighter focusing, CARS microspectroscopy holds a significant advantage over Raman microspectroscopy. With the ongoing upgrades of our present experimental setting we expect that millisecond-rate acquisition times should be attainable for live-cell imaging. This will open a new window of opportunities to study cell cultures in vivo.

V. CONCLUSIONS

Nonlinear optical microspectroscopies based on harmonic generation and nonlinear Raman scattering are promising emerging techniques for noninvasive biochemical imaging of microfluidic devices. The developments in laser technology over the past few years made real-time measurements possible, matching the needs of high-throughput microfluidic devices. The advances in nonlinear laser spectroscopy have made a number of measurements possible, providing stiff competition for more traditional optical tools. Through the development of the phase retrieval algorithm, the direct comparison of CARS and Raman measurements becomes possible. The deployment of the CARS retrieval strategy with partially suppressed nonresonant background allows for the relative concentration measurements of molecular species of interest. At the same time, there are some limitations of CARS microspectroscopy, which are mostly related to imaging on a submicroscopic scale of low concentrated molecular species. However, there are many important applications where CARS microspectroscopy can be an indispensable tool for express chemical analysis.

ACKNOWLEDGMENTS

This work was partially supported by the NSF grant ECS-9984225, the NIH grants R21RR14257 and R03EB008535, and the University of Wisconsin Milwaukee RGI grant.

REFERENCES

Akhmanov, S. A., Bunkin, A. F., Ivanov, S. G., and Koroteev, N. I. 1977. Coherent ellipsometry of Raman scattering of light. *JETP Lett.* 25:444–49.

Akhmanov, S. A., and Koroteev, N. I. 1981. *Nonlinear optical methods in light scattering: Active spectroscopy of light scattering*. Moscow: Nauka.

Anderson, M. J., Hansen, C. L., and Quake, S. R. 2006. Phase knowledge enables rational screens for protein crystallization. *Proc. Natl. Acad. Sci. USA* 103:16746–51.

Andrus, P. G., and Strickland, R. D. 1998. Cancer grading by Fourier transform infrared spectroscopy. *Biospectroscopy* 4:37–46.

Barad, Y., Eisenberg, H., Horrowitz, M., and Silberberg, Y. 1997. Nonlinear laser scanning microscopy by third-harmonic generation. *Appl. Phys. Lett.* 70:922–24.

Barth, A., and Zscherp, C. 2002. What vibrations tell us about proteins. *Quart. Rev. Biophys.* 35:369–430.

Bjorklund, G. 1975. Effects of focusing on third-order nonlinear processes in isotropic media. *IEEE J. Quant. Electron.* 11:287–96.

Breslauer, D. N., Lee, P. J., and Lee, L. P. 2006. Microfluidics-based systems biology. *Mol. Biosys.* 2:97–112.

Chaven, N. E., and Saridakis, E. 2008. Protein crystallization: From purified protein to diffraction-quality crystal. *Nat. Meth.* 5:147–53.

Chen, D. L., and Ismagilov, R. F. 2006. Micrfluidic cartridges preloaded with nanoliter plugs of reagents: An alternative to 96-well plates for screening. *Curr. Opin. Chem. Biol.* 10:226–31.

Cheng, J. X., and Xie, X. S. 2002. Green's function formulation for third-harmonic generation microscopy. *J. Opt. Soc. Am. B* 19:1604–10.

Clay, G. O., Millard, A. C., Schaffer, C. B., Aus-der-Au, J., Tsai, P. S., Squier, J. A., and Kleinfeld, D. 2006. Spectroscopy of third-harmonic generation: Evidence for resonances in model compounds and ligated hemoglobin. *J. Opt. Soc. Am. B* 23:932–50.

Debarre, D., Pena, A. M., Supatto, W., Boulesteix, T., Strupler, M., Sauviat, M. P., Martin, J. L., Schanne-Klein, M. C., and Beaurepaire, E. 2007. Second- and third-harmonic generation microscopies for the structural imaging of intact tissues. *Med. Sci.* 22:845–50.

Debarre, D., Supatto, W., and Beaurepaire, E. 2005. Structure sensitivity in third-harmonic generation microscopy. *Opt. Lett.* 30:2134–36.

Debarre, D., Supatto, W., Farge, E., Moulia, B., Schanne-Klein, M. C., and Beaurepaire, E. 2004. Velocimetric third-harmonic generation microscopy: Micrometer-scale quantification of morphogenetic movements in unstained embryos. *Opt. Lett.* 29:2881–83.

Dittrich, P. S., and Manz, A. 2006. Lab-on-a-chip: Microfluidics in drug discovery. *Nature Rev. Drug Discov.* 5:210–18.

Dudovich, N., Oron, D., and Silberberg, Y. 2002. Single-pulse coherently controlled nonlinear Raman spectroscopy and microscopy. *Nature* 418:512–14.

Dudovich, N., Oron, D., and Silberberg, Y. 2003. Single-pulse coherent anti-Stokes Raman spectroscopy in the fingerprint spectral region. *J. Chem. Phys.* 118:9208–15.

Duncan, M. D., Reintjes, J., and Manuccia, T. J. 1982. Scanning coherent anti-Stokes Raman microscope. *Opt. Lett.* 7:350–52.

Eesley, G. L., Levenson, M. D., and Tolles, W. M. 1978. Optically heterodyned coherent Raman spectroscopy. *IEEE J. Quant. Electron.* 14:45–49.

Evans, C. L., Potma, E. O., Puorishaag, M., Cote, D., Lin, C. P., and Xie, X. S. 2005. Chemical imaging of tissue in vivo with vide rate coherent anti-Stokes Raman scattering microscopy. *Proc. Natl. Acad. Sci. USA* 102:16807–12.

Evans, C. L., Potma, E. O., and Xie, X. S. 2004. Coherent anti-Stokes Raman scattering spectral interferometry: Determination of the real and imaginary components of nonlinear susceptibility $\chi^{(3)}$ for vibrational microscopy. *Opt. Lett.* 29:2923–25.

Garzon, J., Meneses, J., Tribillon, G., Gharbi, T., and Plata, A. 2004. Chromatic confocal microscopy by means of continuum light generated through a standard single-mode fibre. *J. Opt. A* 6:544–48.

Golovan, L.A., Ivanov, D.A., Melnikov, V.A., Timoshenko, V.Yu., Zheltikov, A.M., Kashkarov, P.K., Petrov, G.I., and Yakovlev, V.V. 2006a. Form birefringence of oxidized porous silicon. *Appl. Phys. Lett.* 88:241113.

Golovan, L. A., Petrov, G. I., Fang, G. Y., Melnikov, V. A., Gavrilov, S. A., Zheltikov, A. M., Timoshenko, V. Y., Kashkarov, P. K., Yakovlev, V. V., and Li, C. F. 2006b. The role of phase-matching and nanocrystal-size effects in three-wave mixing and CARS processes in porous gallium phosphide. *Appl. Phys. B* 84:303–8.

Govorkov, S. V., Emelyanov, V. I., Koroteev, N. I., Petrov, G. I., Shumay, I. L., and Yakovlev, V. V. 1989. Inhomogeneous deformations of silicon layers probed by second harmonic generation in reflection. *J. Opt. Soc. Am. B* 6:111724.

Govorkov, S. V., Koroteev, N. I., Petrov, G. I., Shumay, I. L., and Yakovlev, V. V. 1990. Laser nonlinear-optical probing of Silicon-SiO$_2$ interfaces—Surface stress formation and relaxation. *Appl. Phys. A* 50:439–43.

Hansen, C., and Quake, S. R. 2003. Microfluidics in structural biology: Smaller, faster…better. *Curr. Opin. Struct. Biol.* 13:538–44.

Hashimoto, M., and Araki, T. 2000. Molecular vibration imaging in the fingerprint region by use of coherent anti-Stokes Raman scattering microscopy with a collinear configuration. *Opt. Lett.* 25:1768–70.

Hellerer, T., Axang, C., Brackmann, C., Hillertz, P., Pilon, M., and Enejder, A. 2007. Monitoring lipid storage in *Caenorhabditis elegans* using coherent anti-Stokes Raman scattering microscopy. *Proc. Natl. Acad. Sci. USA* 104:14658–63.

Hess, S. T., Huang, S. H., Heikal, A. A., and Webb, W. W. 2002. Biological and chemical applications of fluorescence correlation spectroscopy: A review. *Biochem.* 41:697–705.

Hill, S. C., Leach, D. H., and Chang, R. K. 1993. Third-order sum-frequency generation in droplets—Model with numerical results for third-harmonic generation. *J. Opt. Soc. Am. B* 10:16–33.

Hirschfeld, T. 1973. Raman microprobe: Vibrational spectroscopy in the femtogram range. *J. Opt. Soc. Am.* 63:476–83.

Hong, J. W., and Quake, S. R. 2003. Integrated nanoliter systems. *Nature Biotech.* 21:1179–83.

Huang, Y. S., Karashima, T., Yamamoto, M., and Hamaguchi, H. 2006. Molecular-level investigation of the structure, transformation, and bioactivity of single living fission yeast cells by time- and space-resolved Raman spectroscopy. *Biochemistry.* 44:10009–19.

Hudson, B. S. 1977. New laser techniques for biophysical studies. *Annual Review of Biophysics and Bioengineering.* 6:135–150.

Janasek, D., Franzke, J., and Manz, A. 2006. Scaling and the design of miniaturized chemical-analysis systems. *Nature* 442:374–80.

Kano, H., and Hamaguchi, H. O. 2007. Supercontinuum dynamically visualizes a dividing single cell. *Anal. Chem.* 79:8967–73.

Kee, T. W., Zhao, H. X., and Cicerone, M. T. 2006. One-laser interferometric broadband coherent anti-Stokes Raman scattering. *Opt. Express* 14:3631–40.

Kneipp, K., Kneipp, H., and Kneipp, J., 2006. Surface-enhanced Raman scattering in local optical fields of silver and gold nanoaggregates—From single-molecule Raman spectroscopy to ultrasensitive probing in live cells. *Acc. Chem. Res.* 39:443–50.

Konig, K., Becker, T. W., Fischer, P., Riemann, I., and Halbhuber, K. J. 1999. Pulse-length dependence of cellular response to intense near-infrared laser pulses in multiphoton microscopes. *Opt. Lett.* 24:113–15.

Konorov, S. O., Glover, C. H., Piret, J. M., Bryan, J., Schulze, H. G., Blades, M. W., and Turner, R. F. B. 2007. In situ analysis of living embryonic cells by coherent anti-Stokes Raman microscopy. *Anal. Chem.* 79:7221–25.

Lakowicz, J. R. 1983. *Principles of fluorescent spectroscopy.* New York: Plenum Press.

Li, L., Mustafi, D., Fu, Q., Tereshko, V., Chen, D. L., Tice, J. D., and Ismagilov, R. F. 2006. Nanoliter microfluidic hybrid method for simultaneous screening and optimization validated with crystallization of membrane proteins. *Proc. Natl. Acad. Sci. USA* 103:19243–48.

Lieber, C., and Mahadevan-Jansen, A. 2003. Automated method for subtraction of fluorescence from biological Raman spectra. *Appl. Spectros.* 57:1363–67.

Lim, S. H., Gaster, A. G., and Leone, S. R. 2005. Single-pulse coherently controlled nonlinear Raman scattering spectroscopy. *Phys. Rev. A* 72:0418031.

Lounaci, M., Rigolet, P., Casquillas, G. V., Huang, H. W., and Chen, Y. 2006. Toward a comparative study of protein crystallization in microfluidic chambers using vapor diffusion and batch techniques. *Microelect. Eng.* 83:1673–76.

Lowe, H., and Ehrfeld, W. 1999. State-of-the-art in microreaction technology: Concepts, manufacturing and applications. *Electrochim. Acta* 44:3679–89.

Lozovoy, V. V., and Dantus, M. 2005. Coherent control in femtochemistry. *Chem. Phys. Chem.* 6:1970–2000.

Lucarini, V., Saarinen, J. J., Peiponen, K. E., and Vartiainen, E. M. 2005. *Kramers-Kronig relations in optical material research.* Berlin: Springer.

Lucchertta, E. M., Lee, J. H., Fu, L. A., Patel, N. H., and Ismagilov, R. F. 2005. Dynamics of Drosophila embryonic patterning network perturbed in space and time using microfluidics. *Nature* 434:1134–38.

Malmqvist, L., and Hertz, H. M., 1995. Second-harmonic generation in optically trapped nonlinear particles with pulsed lasers. *Appl. Opt.* 34:3392–97.

Manz, A., Harrison, D. J., Verpoorte, E. M. J., Fettinger, J. C., Paulus, A., Ludi, H., and Widmer, H. M. 1992. Planar chips technology for miniaturization and integration of separation techniques into monitoring systems—Capability electrophoresis on a chip. *J. Chromatogr.* 593:253–58.

Matthaus, C., Boydston-White, S., Miljkovic, M., Romeo, M., and Diem, M. 2006. Raman and infrared microspectral imaging of mitotic cells. *Appl. Spectros.* 60:1–8.

Navratil, M., Mabbott, G. A., and Arriaga, E. A. 2006. Chemical microscopy applied to biological systems. *Anal. Chem.* 78:4005–19.

Nguyen, A., Marsaud, V., Bouclier, C., Top, S., Vessieres, A., Pigeon, P., Gref, E., Legrand, P., Jaouen, G., and Renoir, J. M. 2008. Nanoparticles loaded with ferrocenyl tamoxifen derivatives for breast cancer treatment. *Int. J. Pharmaceut.* 347:128–35.

Noda, K., Sato, H., Watanabe, S., Yokoyama, S., and Tashiro, H. 2007. Efficient characterization for protein crystals using confocal Raman spectroscopy. *Appl. Spectros.* 61:11–18.

Notingher, I., and Hench, L. L. 2006. Raman microspectroscopy: A noninvasive tool for studies of individual living cells in vitro. *Exp. Rev. Med. Dev.* 3:215–34.

Ormerod, M. G. 2000. *Flow cytometry—A practical approach.* Oxford, UK: Oxford University Press.

Oudar, J. L., Smith, R. W., and Shen, Y. R. 1979. Polarization-sensitive coherent anti-Stokes Raman spectroscopy. *Appl. Phys. Lett.* 34:758–60.

Park, T., Lee, S., Seong, G. H., Choo, J., Lee, E. K., Ji, W. H., Hwang, S. Y., Gweon, D. G., and Lee, S. 2005. Highly sensitive signal detection of duplex dye-labelled DNA oligonucleotides in a PDMS microfluidic chip: Confocal surface-enhanced Raman spectroscopic study. *Lab on a Chip* 5:437–42.

Pestov, D., Murawski, R. K., Aribunbold, G. O., Wang, X., Zhi, M.C., Sokolov, A. V., Sautenkov, V. A., et al. 2007. Optimizing the laser-pulse configuration for coherent Raman spectroscopy. *Science* 316:265–68.

Pestov, D., Wang, X., Ariunbold, G. O., Murawski, R. K., Sautenkov, V. A., Dogariu, A., Sokolov, A. V., and Scully, M. O. 2008. Single-shot detection of bacterial endospores via coherent Raman spectroscopy. *Proc. Natl. Acad. Sci. USA* 105:422–27.

Peticolas, W. L., Patapoff, T. W., Thomas, G. A., Postlewait, J., and Powell, J. W. 1996. Laser Raman microscopy of chromosomes in living eukaryotic cells: DNA polymorphism *in vivo*. *J. Raman Spectros.* 27:571–78.

Petrich, W. 2001. Mid-infrared and Raman spectroscopy for medical diagnostics. *Appl. Spectros. Rev.* 36:181–237.

Petrov, G. I., Arora, R., Yakovlev, V. V., Wang, X., Sokolov, A. V., and Scully, M. O. 2007. Comparison of coherent and spontaneous Raman microspectroscopies for noninvasive detection of single bacterial endospores. *Proc. Natl. Acad. Sci. USA* 104:7776–79.

Petrov, G. I., Shcheslavskiy, V. I., Yakovlev, V. V., Golovan, L. A., Krutkova, E. Y., Fedotov, A. B., Zheltikov, A. M., Timoshenko, V. Y., Kashkarov, P. K., and Stepovich, E. M. 2006. Effect of photonic crystal structure on the nonlinear optical anisotropy of birefringent porous silicon. *Opt. Lett.* 31:3152–54.

Petrov, G. I., and Yakovlev, V. V. 2005. Enhancing red-shifted white-light continuum generation in optical fibers for applications in nonlinear Raman microscopy. *Opt. Express* 13:1299–1306.

Petrov, G. I., Yakovlev, V. V., and Minkovski, N. I. 2003. Near-infrared continuum generation of femtosecond and picosecond pulses in doped optical fibers. *Appl. Phys. B* 77:219–25.

Petrov, G. I., Yakovlev, V. V., and Minkovski, N. I. 2004. Broadband nonlinear optical conversion of a high-energy diode-pumped picosecond laser. *Opt. Commun.* 229:441–45.

Petrov, G. I., Yakovlev, V. V., Sokolov, A., and Scully, M. O. 2005. Detection of *Bacillus subtilis* spores in water by means of broadband coherent anti-Stokes Raman spectroscopy. *Opt. Express* 13:9537–42.

Potma, E. O., de Boeij, W. P., van Haastert, P. J. M., and Wiersma, D. A. 2001. Real-time visualization of intracellular hydrodynamics in single living cells. *Proc. Natl. Acad. Sci.* 98:1577–82.

Schaller, R. D., Johnson, J. C., and Saykally, R. J. 2000. Nonlinear chemical imaging microscopy: Near-field third harmonic generation imaging of human red blood cells. *Anal. Chem.* 72:5361–64.

Schönle, A., and Hell, S. W. 1998. Heating by absorption in the focus of an objective lens. *Opt. Lett.* 23:325–27.

Schrotter, S. H., and Klockner, H. W. 1979. Raman cross-sections in liquids and gases. In *Raman spectroscopy of gases and liquids*, ed. A. Weber. Berlin: Springer-Verlag.

Schwartz, A. M., and Berglund, K. 2000. In situ monitoring and control of lysozyme concentration during crystallization in a hanging drop. *J. Cryst. Growth* 219:753–60.

Shcheslavskiy, V., Petrov, G. I., Faustov, A. R., Yakovlev, V. V., and Saltiel, S. 2005a. Third-harmonic Rayleigh scattering: Theory and experiment. *J. Opt. Soc. Am. B* 24:2402–08.

Shcheslavskiy, V., Petrov, G. I., Faustov, A. R., Yakovlev, V. V., and Saltiel, S. 2006a. How to measure $\chi^{(3)}$ of a nanoparticle. *Opt. Lett.* 31:1486–88.

Shcheslavskiy, V., Petrov, G. I., Saltiel, S., and Yakovlev, V. V. 2004. Quantitative imaging of aqueous solutions probed by the third-harmonic microscopy. *J. Struct. Biol.* 147:42–49.

Shcheslavskiy, V., Petrov, G. I., and Yakovlev, V. V. 2003. Nonlinear optical susceptibility measurements in solutions using third-harmonic generation at interfaces. *Appl. Phys. Lett.* 82:3982–84.

Shcheslavskiy, V., Petrov, G. I., and Yakovlev, V. V. 2005b. Nonlinear optical properties of collagen in solution. *Chem. Phys. Lett.* 402:170–74.

Shcheslavskiy, V., Saltiel, S., Ivanov, D. A., Ivanov, A. A., Petrussevich, V. Y., Petrov, G. I., and Yakovlev, V. V. 2006b. Nonlinear optics of molecular nanostructures in solution: Assessment of the size and nonlinear optical properties. *Chem. Phys. Lett.* 429:294–98.

Shcheslavskiy, V., Yakovlev, V. V., and Ivanov, A, 2001. High-energy self-starting femtosecond $Cr^{4+}:Mg_2SiO_4$ oscillator operating at a low repetition rate. *Opt. Lett.* 26:1999–2001.

Sheik-Bahae, M., Said, A. A., and Van Stryland, E. W. 1989. High-sensitivity, single-beam n_2 measurements. *Opt. Lett.* 14:955–57.

Shim, M. G., and Wilson, B. C. 1997. Development of an *in vivo* Raman spectroscopic system for diagnostic applications. *J. Raman Spectros.* 28:131–42.

Squirrell, J. M., Wokosin, D. L., White, J. G., and Bavister, B. D. 1999. Long-term two-photon fluorescence imaging of mammalian embryos without compromising viability. *Nat. Biotech.* 17:763–67.

Stroud, D., and Wood, V. 1989. Decoupling approximation for the nonlinear optical response of composite media. *J. Opt. Soc. Am. B* 6:778–86.

Swain, R. J., and Stevens, M. M. 2007. Raman microspectroscopy for non-invasive biochemical analysis of single cells. *Biochem. Soc. Tras.* 35:544–49.

Tolles, W. M., and Turner, R. D. 1977. A comparative analysis of the analytical capabilities of coherent antistokes Raman spectroscopy (CARS) relative to Raman scattering and absorption spectroscopy. *Appl. Spectros.* 31:96–103.

Utzinger, U., Heintzelman, D. L., Mahadevan-Jansen, A., Malpica, A., Follen, M., and Richards-Kortum, R. 2001. Near-infrared Raman spectroscopy for in vivo detection of cervical precancers. *Appl. Spectros.* 55:955–59.

Uzunbajakava, N., Lenferink, A., Kraan, Y., Volokhina, E., Vrensen, G., Greve, J., and Otto, C. 2003. Nonresonant confocal Raman imaging of DNA and protein distribution in apoptotic cells. *Biophys. J.* 84:3968–81.

Vartiainen, E. M. 1992. Phase retrieval approach for coherent anti-Stokes Raman scattering spectrum analysis. *J. Opt. Soc. Am. B* 9:1209–14.

Vartiainen, E. M., Rinia, H. A., Müller, M., and Bonn, M. 2006. Direct extraction of Raman line-shapes from congested CARS spectra. *Opt. Express* 14:3622–30.

Vinergoni, C., Bredfeldt, J. S., Marks, D. L., and Boppart, S. A. 2004. Nonlinear optical contrast enhancement for optical coherence tomography. *Opt. Express* 12:3310341.

Viskari, P. J., and Landers, J. P. 2006. Unconventional detection methods for microfluidic devices. *Electrophoresis* 27:1797–1810.

Volkmer, A. 2005. Vibrational imaging and microspectroscopies based on coherent anti-Stokes Raman scattering microscopy. *J. Phys. D* 38:R59–R81.

von Vocano, B., and Motzkus, M. 2006. Time-resolved two color single-beam CARS employing supercontinuum and femtosecond pulse shaping. *Opt. Commun.* 264:488–93.

Wallqvist, A., Rabow, A. A., Shoemaker, R. H., Sausville, E. A., and Covell, D. G. 2002. Establishing connections between microarray expression data and chemotherapeutic cancer pharmacology. *Mol. Cancer Therapeut.* 1:311–20.

Ward, J. W., and New, G. H. 1969. Optical third-harmonic generation in gases by a focused laser beam. *Phys. Rev.* 185:57–71.

Whitesides, G. M. 2006. The origins and the future of microfluidics. *Nature* 442:368–73.

Wilson, T. 1990. *Confocal microscopy*. London: Academic Press.

Wurpel, G. W. H., Schins, J. M., and Müller, M. 2002. Chemical specificity in three-dimensional imaging with multiplex coherent anti-Stokes Raman scattering microscopy. *Opt. Lett.* 27:1093–95.

Yakovlev, V. V. 2001. Advances in real-time nonlinear Raman microscopy. *Proceedings of SPIE: Biomedical Diagnostic, Guidance, and Surgical-Assist Systems III* 4254:7–105.

Yakovlev, V. V. 2003. Advanced instrumentation for non-linear Raman microscopy. *J. Raman Spectros.* 34:957–64.

Yakovlev, V. V. 2007. Time-gated confocal Raman microscopy. *Spectroscopy* 22:34–41.

Yakovlev, V. V., Noojin, G., Denton, M., and Thomas, R. J. 2008. Real-time monitoring of chemical and structural changes induced by ultrashort pulse interactions with tissues and cells. *SPIE Proc.* 6859:685914.

Zumbusch, A., Holton, G. R., and Xie, X. S. 1999. Three-dimensional vibrational imaging by coherent anti-Stokes Raman scattering. *Phys. Rev. Lett.* 82:4142–45.

7 Advanced Multiphoton and CARS* Microspectroscopy with Broadband-Shaped Femtosecond Laser Pulses

Bernhard von Vacano and Marcus Motzkus

CONTENTS

*CARS, coherent anti-Stokes Raman scattering

I. INTRODUCTION: NONLINEAR SPECTROSCOPIC SIGNALS FOR MICROSCOPIC IMAGING

In recent years, multiphoton microscopy has emerged as a new tool for bio-imaging, offering an unprecedented wealth of information (Zipfel et al. 2003; Xie et al. 2006). This includes fluorescence imaging with highly reduced photobleaching of the labeling dyes, bright images with very high three-dimensional resolution, deep-tissue imaging, and novel contrast mechanisms. In the most commonly employed two-photon fluorescence (TPF) microscopy, labeling fluorophores are excited simultaneously with two photons, which can roughly be understood as being equivalent to an excitation with a single photon of double energy, or half the wavelength (Figure 7.1a). The use of long excitation wavelengths in the near-infrared spectral region is the reason why multiphoton techniques usually achieve much higher penetration depths, as scattering is highly reduced. Also, the involvement of two photons in the excitation leads to a quadratic intensity dependence of the fluorescence. On the one hand, this requires pulsed laser sources for efficient excitation with high peak intensities but, on the other hand, it confines the signal generation to the focal volume, which is the reason for the highly beneficial 3D-imaging capabilities. Therefore, it is clear that at a given average laser power the shortest possible pulses achieve the highest multiphoton signal levels.

In practice, one uses femtosecond lasers, which necessarily feature a broader spectrum the shorter the pulses are. This provides another benefit for multiphoton imaging: ultrashort pulses (sub-20 femtoseconds [fs]) possess a laser bandwidth increased in such a way that several typical flurophores with different emission wavelengths can be excited at once. Usually, fluorescence signals in microscopy are recorded in a single

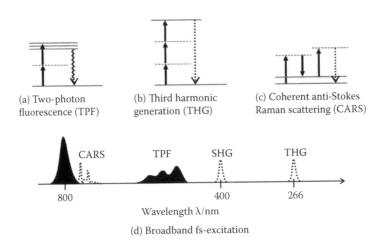

(a) Two-photon fluorescence (TPF)

(b) Third harmonic generation (THG)

(c) Coherent anti-Stokes Raman scattering (CARS)

CARS   TPF   SHG   THG

800   400   266

Wavelength λ/nm

(d) Broadband fs-excitation

FIGURE 7.1 Multiphoton spectroscopy schemes with corresponding energy level diagrams (a–c): Black solid arrows always indicate the energy equivalent of one excitation laser photon. (d): Sketch of possible excitation and signal spectra, demonstrating their relative spectral positions. Excitation with a fs-laser with a Gaussian spectrum around 800 nm has been assumed. Possible multiphoton signals include CARS, two-photon fluorescence (TPF), second harmonic generation (SHG), and third harmonic generation (THG).

channel by selecting a certain emission wavelength with an optical filter, which limits achievable image contrast between regions of a cell labeled with different fluorophors to the separation of the respective emission spectra of the dyes. If, however, the fluorescence signal that is detected is spectrally resolved, much more powerful data analysis can be performed, allowing even full separation of overlapping fluorescence spectra.

The considerations we have made here for two-photon fluorescence in principle hold for every multiphoton imaging technique. The signal is always proportional to a higher order of the intensity, requiring pulsed lasers and boasting 3D resolution. Other, still less commonly employed multiphoton techniques image native, untreated samples. One class is harmonic generation (Sun 2005), be it second harmonic generation (SHG) (Gauderon et al. 1998; Campagnola and Loew 2003) or third harmonic generation (THG, Figure 7.1b) (Barad et al. 1997). In SHG and THG imaging, two or three photons combine in the sample in a nonlinear optical process intrinsically occurring in some materials, yielding imaging contrast.

Despite the usefulness of fluorescence, SHG, and THG, another nonlinear optical spectroscopy method is even more powerful for functional imaging of unlabeled samples: CARS (Cheng and Xie 2004; Volkmer 2005). CARS refers to "coherent anti-Stokes Raman scattering" and is a three-photon process carrying information about the sample as a result of the involvement of a vibrational energy level (Figure 7.1c). In CARS, a pair of photons of different energy coherently excites a molecular vibration, which is then probed by the third photon. The signal created is a photon of higher energy, blue-shifted with respect to the probe photon by the energy of the vibrational level of the molecule (Figure 7.1d). Because the vibrational levels of a molecule are highly specific, CARS allows chemical identification and imaging based on the true chemical composition of the sample. At least, CARS requires photons at two different wavelengths, which can either be provided by the difficult combination and synchronization of two different lasers, or by a single beam of femtosecond laser pulses with sufficient bandwidth to contain both photons. This simplified approach (Dudovich et al. 2002, 2003; Lim et al. 2005; von Vacano et al. 2006a,b) again calls for excitation lasers with high optical bandwidth and, thus, ultrashort pulse duration.

Taken together, all the different methods introduced here are powerful nonlinear optical techniques with different capabilities for spectroscopy and imaging. In principle, a broadband laser pulse containing all the different photon energies (corresponding to the arrows in the level diagrams in Figures 7.1a–c) excites all the processes at once. From a fundamental point of view, they can be understood as light-matter interactions taking the system under study from an initial level via intermediate states to a certain target state. As light-matter interaction is a quantum mechanical phenomenon, all the different processes and thus "pathways" not only occur simultaneously, but give rise to quantum interferences (Figure 7.2).

The probability that a certain transition (for example, TPF, CARS, etc.) can take place is determined by the sum of the probability amplitudes of all indistinguishable pathways connecting the initial and target level (Meshulach and Silberberg 1998, 1999;). In Figure 7.2, this has been schematically depicted for a Raman-like transition and two-photon absorption. The indistinguishable pathways are given by all suitable combinations of photon energies in a broadband femtosecond pulse. If the phases

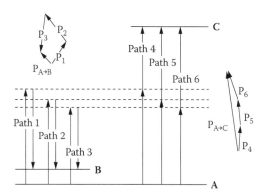

FIGURE 7.2 Coherent control of multiphoton processes. As an example, quantum interference between pathways leading from state A to state B in a Raman-type transition are shown (Paths 1–3), as well as pathways leading from the same initial state A to a target state C in a two-photon-absorption type transition (Paths 4–6). Depending on the quantum mechanical phases between the different pathways leading to the same target state, destructive or constructive interference can be induced, which is the foundation of coherent control. In this example, the probability amplitudes P_1, P_2, P_3 for the transition $A{\to}B$ add up almost completely destructively, while P_4, P_5, P_6 for the process $A{\to}C$ interfere constructively, leading to a large probability $P_{A{\to}C}$. Thus, coherent control allows selectively exciting a desired multiphoton process by applying correct phases for the different pathways.

of the pathways are arbitrarily controlled in order to select one target state over the other, one speaks of coherent control. Coherent control is a very powerful concept, aiming at selectively manipulating quantum states through the interaction with laser light (Bergmann et al. 1998; Rabitz et al. 2000; Rice and Zhao 2000; Shapiro and Brumer 2003). Methods of coherent control have found widespread applications in nonlinear optics (Broers et al. 1992; Omenetto et al. 2001; Zeidler et al. 2001b), atomic and molecular physics (Meshulach and Silberberg 1998; Hornung et al. 2000; Weiner et al. 1990), and even for steering chemical reactions (Brixner and Gerber 2003) and processes in photobiology (Herek et al. 2002). In addition, coherent control has proved to be a very valuable addition to spectroscopy (Wohlleben et al. 2005; Buist et al. 1999; Dela Cruz et al. 2005) and allows the extremely flexible implementation of nonlinear spectroscopy, where the desired multiphoton process can simply be selected by inducing the appropriate quantum interferences.

This latter application is the key to the broadband femtosecond approach to nonlinear microspectroscopy presented here. From the point of view of coherent control, it becomes clear that high spectral bandwidths are needed; providing a huge number of photon energies and, as such, interfering pathways is necessary to be able to achieve high interference contrast, and thus good controllability of optical processes.

In a most versatile and flexible scheme, all processes are therefore carried out in the same experimental setup: what is needed is a broadband laser source, the capability to deliver the femtosecond pulses in situ in the microscope with highest possible intensity, the ability to control the phase to select the desired process, and a multichannel detection unit. In a schematic way, this is shown in Figure 7.3.

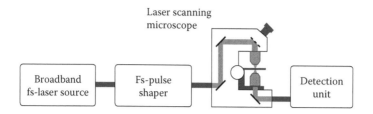

FIGURE 7.3 General scheme of a versatile next-generation broadband multiphoton laser-scanning microscope. A broadband femtosecond laser acts as the excitation source, which can be used for a multitude of nonlinear spectroscopic imaging techniques, such as TPF, SHG, THG, and CARS. The fs-pulse shaper ensures highest excitation efficiency in the microscope by compensating pulse distortions due to dispersion, and allows tailoring the excitation to improve contrast or enable spectroscopic acquisition despite the broadband excitation, as in the case of single-beam CARS. Detection can be performed in forward or backward (epi-) direction, depending on the application.

II. BROADBAND FEMTOSECOND PULSES IN MICROSCOPY

To introduce the application of ultrashort laser sources in microscopy, we want to review some properties of femtosecond pulses first; for a comprehensive introduction the reader may refer to one of the established textbooks on femtosecond lasers (Diels and Rudolph 2006). The most important notion is the Fourier transform relation between the temporal shape of a pulse and the spectrum necessary to create it. This leads to the well-known time-bandwidth product for the pulse temporal width (measured as full width at half maximum, FWHM) Δt and the pulse spectral width Δv.

$$\Delta t \cdot \Delta v = f \tag{7.1}$$

where f is a factor depending on the pulse profile (for the most commonly assumed Gaussian pulses, for example, f is 0.44): very short femtosecond pulses necessarily correspond to very broad excitation spectra (Figure 7.4a). However, a broad pulse spectrum alone does not guarantee short pulses: the time-bandwidth product [Equation (7.1)] only holds for "Fourier-transform limited" (FTL) pulses, meaning that all spectral components at frequencies v arrive simultaneously in time, creating the shortest pulse possible with this spectrum. The time at which a certain frequency component v from the laser beam passes at a given position is described by the group delay $GD(v)$, which can simply be calculated as the derivative of the spectral phase $\phi(v)$:

$$GD(v) = \frac{1}{2\pi} \frac{d}{dv} \phi(v) \tag{7.2}$$

Therefore, the spectral phase $\phi(v)$ and consequently the group delay $GD(v)$ for each color controls the temporal shape of the pulse. These important relations are summed up in Figure 7.4.

The "frequency sweep" resulting from a linearly varying $GD(v)$ of the pulse in Figure 7.4c is called a "chirp." This pulse is much longer that the respective 20 fs FTL pulse of the same bandwidth (Figure 7.4b). As any optical material shows dispersion

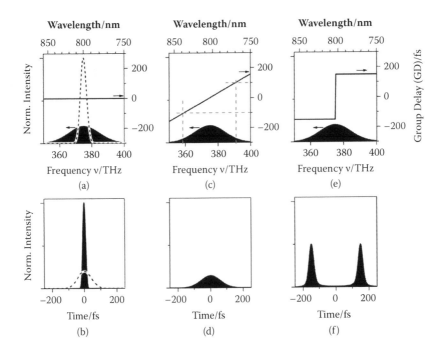

FIGURE 7.4 Properties of fs-pulses, showing the effect of bandwidth and group delay on the temporal shape. Shown are spectra (a) of two FTL pulses with a bandwidth corresponding to 100 fs (dashed curve) and 20 fs (hatched curve), as well as the resulting temporal intensity profiles (b). In panels (c) and (d), the effect of a linearly varying group delay on a pulse with an identical spectrum as the 20 fs pulse from panel (a) is shown, leading to an elongated "chirped" pulse in time (d). Deliberately modifying the group delay, e.g., in a step-function (e), allows pulses with a defined temporal shape, in this case (f) two-color double pulses.

of the refractive index, different colors of a laser pulse travel at different speed through the medium. Material dispersion thus leads to a varying group delay similar to the case shown in Figure 7.4c. Correspondingly, the temporal peak intensity drops dramatically. This is, of course, unwanted in nonlinear spectroscopic applications. Table 7.1 summarizes the dramatic effects of pulse duration, bandwidth, and dispersion on the different nonlinear spectroscopic signals. Constant pulse energy E_{pulse} (equivalent to the same number of photons in a pulse) is always assumed. The nonlinear signal can be calculated as follows:

$$I_{\text{Sig}} \propto \int dt \left[I(t) \right]^n \propto \left[\frac{E_{\text{pulse}}}{\Delta t} \right]^n \cdot \Delta t \qquad (7.3)$$

where I_{Sig} is the integrated signal intensity, n the order of the nonlinear process, and Δt the pulse duration.

It is evident from Table 7.1 that the shorter the effective pulse duration, the higher the nonlinear signal intensity. This is even more the case for third-order processes

TABLE 7.1

Influence of Laser Bandwidth and Effective Pulse Duration on Nonlinear Signal Intensity, Showing the Dramatic Effect of Dispersion on Ultrashort Pulses

Pulse	Effective Pulse Duration/fs	Relative Peak Intensity	Relative Second-Order Signal (TPF, SHG)	Relative Third-Order Signal (THG, CARS)	Optical Bandwidth (FWHM)/ cm^{-1}
100 fs FTL	100	1	1	1	147
100 fs after MO	149	0.67	0.67	0.45	
20 fs FTL	20	5	5	25	735
20 fs after MO	555	0.18	0.18	0.032	
10 fs FTL	10	10	10	100	1471
10 fs after MO	1109	0.09	0.09	0.008	

Note: For this table, a microscope objective with a group delay dispersion corresponding to 4 ps/PHz was assumed. Simulations have been performed with the Lab2 software package. (Schmidt, B., Hacker, M., Stobrawa, G., and Feurer, T. 2007. LAB2-A virtual femtosecond laser. lab.http://www.lab2.de.)

such as CARS and THG. Therefore, it seems ideal to use the shortest pulses possible for microscopy. However, as already cautioned before, this is not automatically true: the material dispersion of a typical microscope objective (MO) leads to a dramatic temporal broadening of ultrashort pulses. As such, the 10 fs laser pulse can turn out an effective picosecond pulse in the focus of the nonlinear microscope!

Thus, if such short pulses are to be used in microspectroscopy, a precompensation of the material dispersion introduced by the optical system is needed. Like that, all colors again arrive in focus at the same time. Experimental means to that end are discussed in the next section. If such additional control over the group delay is achieved, pulses with a defined temporal shape and phase can be crafted from a broadband ultrashort pulse, which will be a prerequisite for coherently controlled nonlinear spectroscopy schemes such as CARS. As an example, the creation of a two-color double pulse from the exemplary Gaussian 20 fs pulse is shown in Figures 7.4e and f. A group delay of −150 fs is introduced for the red spectral half, while a group delay of 150 fs is applied to the blue spectral half of the pulse. In time, this leads to a splitting of the pulse into a doublet (Figure 7.4f), with the first of the sub-pulses at time −150 fs being shifted in frequency to the red with respect to the second one at 150 fs.

The potential of broadband laser excitation and fs-pulse shaping for different microspectroscopy techniques ranges from pure dispersion compensation (in the case of SHG, THG), to highly functional pulse shaping (in the case of CARS), as summarized in Table 7.2. It is worth mentioning again that all techniques can be implemented in the same approach—broadband laser and pulse shaper. The detection technique of choice is just selected by the corresponding pulse shapes.

After having established the basics of femtosecond laser pulses in microspectroscopy, we now turn to more practical aspects of possible experimental implementations: laser sources and possibilities for pulse shaping.

III. EXPERIMENTAL IMPLEMENTATION: SOURCES AND PULSE SHAPERS

Pulsed ultrashort laser sources are commercially available nowadays in a wide variety. For nonlinear microscopy and microspectroscopy applications outside dedicated laser research facilities, robust systems that are easy to operate are necessary. This usually means mode-locked, unamplified lasers. Such systems typically emit pulses in the range of tens to hundreds of femtoseconds, with nanojoule pulse energy at megahertz repetition rates. The operating central wavelength and achievable pulse duration highly depends on the laser medium. The most widely used and up to now best-performing class of fs-oscillator laser systems is based on titanium sapphire (Ti:sapphire), emitting in a wavelength region around 800 nm and allowing pulse durations below 10 fs. These shortest pulse durations, well suited for the versatile nonlinear microspectroscopy schemes presented here, are only available with dedicated Ti:sapphire lasers, though. The majority of Ti:sapphire oscillators used for nonlinear microscopy operate in the 100 fs regime. The reasons for this are the maturity of the technology and the huge difficulties of using shorter pulses in microscopy without careful dispersion compensation, as demonstrated in Table 7.2. Disadvantages of Ti:sapphire lasers are their relatively high cost, their complex arrangement of free-space optics, and expenses for their continued operation, as they rely on relatively expensive pump sources at the Ti:sapphire absorption band around 532 nm. Therefore, a growing new class of fs-lasers, using doped optical fibers as laser medium, has received a lot of attention. Such lasers can be built entirely in fiber technology in a very robust way, and typically operate at telecommunication

TABLE 7.2
Benefits of Ultrashort Broadband Laser Pulses and Pulse Shaping for Nonlinear Microspectroscopy Comparing Different Techniques

Spectroscopic Technique	Benefits of Broad Bandwidth	Perspectives for Pulse Shaping
Harmonic generation (SHG, THG)	• Short pulses, corresponding high peak intensities for efficient signal generation	• Dispersion compensation to ensure highest peak intensities
TPF	• Short pulses, high efficiency • Simultaneous excitation of several fluorophors	• Dispersion compensation • Fluorophor selectivity for single channel detection
CARS	• Short pulses, high efficiency • Access to broad range of molecular vibrations	• Dispersion compensation • Enable spectroscopic resolution • Time-resolved measurements • Optical coherent "signal processing"

wavelengths, where a large number of reliable and competitively priced components are on hand. Fiber lasers thus promise to become ideal sources for nonlinear microspectroscopy applications. However, with pulses around 100 fs duration, they so far lack the ultrabroad bandwidths needed for the versatile spectroscopy schemes discussed here.

Another technological breakthrough in optical fiber technology, however, allows one to upgrade established 100 fs-class laser systems for broadband applications and even surpass the bandwidth of dedicated short-pulse Ti:sapphire lasers. Key to this is the use of novel microstructured optical fibers, which are designed to exhibit extremely high optical nonlinearities. If nanojoule femtosecond laser pulses are launched into such a fiber, the combination of different nonlinear optical processes leads to the creation of new frequency components. Therefore, the laser bandwidth can be increased dramatically by orders of magnitude.

Microstructured fibers can be designed with either very high flexibility as photonic crystal fibers (PCF), which consist of a solid micrometer sized core surrounded by a periodic arrangement of air-filled holes (Russel 2003) (Figure 7.5a), or as tapered fibers. For real-life applications, PCF are already commercially available with very different, highly reproducible properties. The dramatic bandwidth broadening effect (Figure 7.5b) is known as supercontinuum generation (Wadsworth et al. 2002; Dudley et al. 2006), and has already been demonstrated as a new excitation source for linear fluorescence laser scanning microscopy (McConnell 2004). The conversion is efficient; losses are mostly due to the fiber coupling, and transmission of around 50% is routinely achieved. Supercontinuum generated in a microstructured fiber does not, however, correspond to FTL pulses. The reason is, again, that the group delay of the different frequency components created in the supercontinuum varies, and is additionally modified by the intrinsic material dispersion of the fiber (as introduced in section II). Therefore, active group delay modification and pulse shaping is necessary for its efficient use in nonlinear microspectroscopy. This combination of a 100-fs class laser source with PCF and pulse shaping has successfully been demonstrated for Ti:sapphire laser systems (von Vacano et al. 2006a; Tada et al. 2007). The

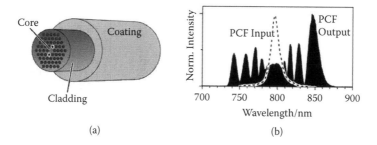

(a) (b)

FIGURE 7.5 Supercontinuum generation in a photonic crystal fiber (PCF). (a) Schematic drawing of a PCF, showing the solid micrometer sized core surrounded by a periodic array of air-filled holes. (b) Experimental demonstration of ultrabroad bandwidth generation from a standard 100 fs laser (PCF input, dashed curve) in only 12 mm of PCF. This spectrum (PCF output) has enough bandwidth to be compressed to sub-15 fs pulses (see text and Figure 7.7).

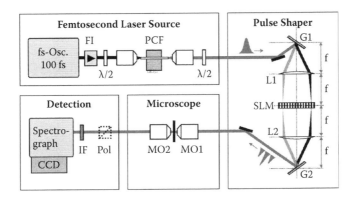

FIGURE 7.6 Schematic drawing of the optical setup including the 4*f* fs-pulse shaper, demonstrating the Fourier transform of the incoming pulses, modulated by the spatial light modulator (SLM) and finally leading to the desired shaped output pulses. FI: Faraday isolator, G1, G2: gratings, L1, L2: lenses (replaced by spherical focusing mirrors in reality), Pol: Polarizer for polarization shaping. In case of amplitude shaping, a polarizer is placed at the output port of the shaper. IF: Interference filter for excitation light rejection.

setup of such a broadband nonlinear spectroscopy source is sketched in Figure 7.6. This scheme should also provide a route to establishing fs-fiber lasers for broadband applications, which would constitute an all-fiber, very versatile excitation source for multiphoton spectroscopy.

In principle, femtosecond pulse shaping can be achieved in two different ways: The first possibility is an optical Fourier transform performed spatially (from time to frequency domain in a so-called 4*f* arrangement) using gratings and lenses (Weiner 2000); the second is the use of an acousto-optical programmable digital filter (AOPDF), which is based on a longitudinal acousto-optic interaction in an anisotropic birefringent crystal (Verluise et al. 2000). The AOPDF shaper is very compact, robust, and can be used with fs-oscillators, if some experimental measures are taken to account for the high MHz repetition rates (Ogilvie et al. 2006b). The alternative 4*f* setup (Weiner 2000; Stobrawa et al. 2001), being slightly more complex, offers direct access to the laser pulses in the frequency domain. A schematic view is shown in Figure 7.6. The colors of the incoming fs-laser beam are diffracted from the first grating in different angles. Placed at the focal distance *f* of the lens (which is usually a focusing mirror in the experiment to avoid dispersion), the dispersed colors are parallelized and focused each at a different position in the Fourier plane, at another distance *f* from the lens. On the other side of the Fourier plane, the setup is reproduced, leading to the characteristic 4*f* layout. Pulses are shaped by controlling each frequency component with a spatial light modulator (SLM). By different means, the phase (and therefore the group delay), amplitude, and polarization for each frequency component can be set as desired. The modification of the group delay only requires phase shaping, by applying the integrated $GD(v)$ as $\phi(v)$, introduced in section II. Like this, the complete bandwidth of the excitation source is maintained. Amplitude shaping, on the other hand, allows tailoring the spectrum, for example, to achieve more complex pulse shapes.

IV. *IN SITU* PULSE COMPRESSION AND CHARACTERIZATION

Being able to control the group delay is not yet sufficient for ensuring ultrashort pulses at the site of the nonlinear optical interaction in the sample: the necessary compensation for material dispersion or additional pulse distortions from supercontinuum generation has to be known precisely. One possibility in cases of moderate group delay deviations is to use an adaptive learning algorithm, which applies group delay compensations randomly that are generated in a given parameter space with the shaper, and measures a nonlinear signal. As any nonlinear signal is dependent on the pulse duration [Equation (7.3) and Table 7.1], it can be used as a feedback to guide the optimization algorithm toward ever better compensation, until, ideally, convergence is achieved for FTL pulses. This procedure has already been described in the earlier days of adaptive femtosecond pulse shaping (Baumert et al. 1997; Meshulach et al. 1997; Zeidler et al. 2000, 2001a), and recently also successfully applied to the compression of PCF supercontinuum for nonlinear microscopy (von Vacano et al. 2006a). The method is implemented very easily, but the optimization becomes more time consuming and difficult for complex pulses, and does not—with routinely justifiable effort—compensate all minor details of group delay distortions, which can be necessary in some cases to get the cleanest FTL pulses possible. Therefore, alternative schemes have been devised to directly measure the pulse shape *in situ*, and then use this information for compensation. One of them is based on scanning time-domain interferometry (Monmayrant et al. 2003), the other is based on the subsequent comparison of well-known reference compensation functions, introduced by the Dantus group as MIIPS (Lozovoy et al. 2004; Xu et al. 2006). The MIIPS method has successfully been demonstrated for microscopy, as have even faster alternative approaches based on the SPIDER (spectral shear interferometry for direct electric field reconstruction) technique (von Vacano et al. 2006a, 2007; Chen and Lim 2007). In SPIDER, two replicas of the pulse to be characterized have to be created, sum-frequency mixed with different laser colors, and brought to interfere. Huge simplification of SPIDER is possible if the task of creating exact copies of the pulse to be characterized is not performed by a complex interferometer, but by the pulse shaper itself which is later used for compression (Figure 7.7).

For microscopy with shaped pulses, the shaper-assisted collinear SPIDER (SAC-SPIDER) is ideally suited: it simply uses the pulse shaper present in the setup for the pulse measurement and, unlike MIIPS or FROG, is not intrinsically dependent on subsequent scanning in a measurement, but boasts real-time measurement capabilities (von Vacano et al. 2006b, 2007). As an example of the successful delivery of compressed ultrashort pulses, Figure 7.8 shows compressed 14.4 fs pulses, generated from PCF supercontinuum with an initial 100-fs laser source. In Figure 7.8a, the temporal profile of the uncompressed PCF supercontinuum (spectrum according to Figure 7.5b) is shown, as measured *in situ* by SAC-SPIDER within a fraction of a second. It can be seen that the pulse is stretched out in time due to its distortion from

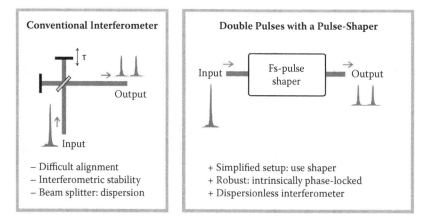

FIGURE 7.7 Usage of the pulse shaper to simplify SPIDER pulse compression, as employed in the SAC-SPIDER method (von Vacano et al. 2006b, 2007). See text for details.

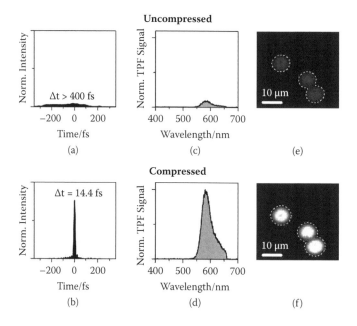

FIGURE 7.8 *In situ* pulse compression of a broadband laser source with fs-pulse shaper and the SAC-SPIDER technique. (a) Before compensation, pulses with more than 400 fs duration (FWHM) are impingent on the sample. (b) Compression with SAC-SPIDER yields 14.4 fs pulses with highly increased peak intensity. The effect of compression can be seen in the highly increased TPF signal of Rhodamine B, comparing panels (c) and (d). (e), (f): Equally, imaging of Rhodamine B-labeled polymer beads (size 8 μm), demonstrates the superior imaging contrast achieved with compressed ultrashort fs-pulses. The position of the beads is indicated by dashed white lines for orientation.

the nonlinear processes in the PCF and additional dispersion in the microscope. With this precise measurement, a group delay compensation and thus compression can immediately be achieved, resulting in ultrashort 14.4 fs pulses, as can be seen in Figure 7.8b.

As a demonstration of what this means for nonlinear spectroscopy, the TPF signal of a rhodamine B solution irradiated with uncompressed (Figure 7.8c) and compressed (Figure 7.8d) pulses is shown: the dramatic increase of signal intensity for the compressed pulses can immediately be discerned. (It has to be noted that the signal increase is a little bit lower than expected from the measured Δt values and Equation (7.3), but this is only due to the fact that the two-photon absorption cross section of the dye used is not constant across the pulse bandwidth of roughly 100 nm, as would be necessary for strict validity.) For an imaging application, this means highly increased image contrast, as can be seen from the comparison of Figures 7.8e and f. This is an experimental demonstration of the principles discussed in section II and laid out in Table 7.1, showing again that the use of ultrashort pulses in microscopy is only efficiently possible with very precise control over the pulse duration. As we have pointed out, the most powerful approach is the implementation of an fs-pulse shaper, which introduces so much flexibility that even the necessary measurements for pulse compression can be performed automatically in the same setup, using, e.g., SAC-SPIDER.

V. SINGLE-BEAM CARS MICROSPECTROSCOPY AND MICROSCOPY

With everything in place—broadband laser source, pulse shaper, and compressed ultrashort pulses in the microscope—we can now turn to CARS as a first important application of broadband nonlinear microspectroscopy. In contrast to TPF, SHG, or THG, CARS provides spectroscopic, chemically selective information without the need for any labeling dyes In combination with microscopy, this allows the study of whole new classes of systems, ranging from complex blended materials (von Vacano et al. 2006; Kee and Cicerone 2004) to drug-releasing polymers (Kang et al. 2006), and monitoring of the metabolism of living cells on a microscopic level (Xie et al. 2006). Therefore, CARS microscopy and microspectroscopy has received a lot of attention recently, and is a very fast growing, active field of research. A thorough introduction to theoretical aspects and applications can be found, among others, in reviews by Cheng and Xie (2004), Volkmer (2005), or Müller and Zumbusch (2007).

A. Introduction to Femtosecond and Single-Beam CARS

As already introduced in section I of this chapter, in a CARS process (Figures 7.9a–c; see also Figure 7.1c), a Raman transition between two vibrational energy levels of a molecule is coherently driven by two optical laser fields (frequencies ω_p' and ω_s) and subsequently probed by interaction with a third field at frequency ω_p. This generates the anti-Stokes signal at the blue-shifted frequency $\omega_{CARS} = \omega_p - \omega_s + \omega_p'$. The energy difference between pump and Stokes, $\omega_p - \omega_s$, defines which energy level is being probed. If it matches a transition Ω_R, the anti-Stokes signal reveals one data point of the vibrational spectrum.

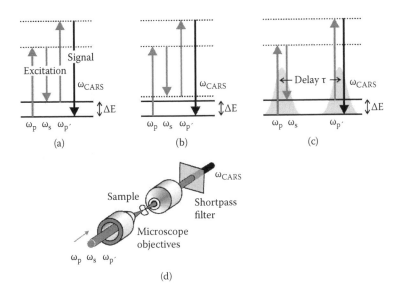

FIGURE 7.9 CARS energy level diagram showing the designation of the different frequency components taking part in the process: ω_p (pump), ω_s (Stokes), ω_p' (probe). (a) The blue-shifted signal ω_{CARS} is generated at the anti-Stokes frequency of ω_p'. This process is efficient, due to its resonance with a vibrational level at energy spacing ΔE. (b) Origin of nonresonant background: The CARS process can also occur via an intermediate virtual state that does not reflect the molecular energy levels. (c) Principle of time-resolved CARS experiments: Pump and Stokes fields from one or two ultrafast laser pulses excite a coherent molecular vibration, which is probed after a time delay τ by a probe pulse. (d) Schematic depiction of collinear, single-beam CARS: All frequency components are provided in the same beam, while the blue-shifted CARS signal can be extracted by spectral filtering.

Though generally considered this way in the framework of quantum states with discrete energies, molecular vibration can also be understood literally as atomic motion: a vibrational energy $\Delta E = 2\pi h \times \Omega_R$ of 1000 cm^{-1}, where h is the Planck constant, for example, corresponds to periodic oscillations of the atoms in the molecule within about 30 fs. With the development of ultrafast laser sources in the femtosecond regime, a direct observation of such motion has become possible. A prerequisite is the availability of pulses shorter than the vibrational period to be measured. Time-resolved four-wave mixing spectroscopies, such as femtosecond CARS, do not necessarily rely on a frequency dispersion of the generated signal, but on the measurement of the impulse response of the system (Zheltikov 2000). In the impulsive limit of extremely short laser pulses, this measured response reflects the complete dynamics of the molecule, and therefore shows oscillations at the frequencies of all molecular vibrations (Mukamel 1995). Experimentally, the condition for an impulsive measurement of a vibration with energy ΔE is that laser pulses employed have to be shorter than the vibrational period $T_{vib} = h/\Delta E$. In the frequency domain this translates into a pulse spectral width that is larger than $\Delta E/h$: within the ultrashort excitation pulse, photon pairs can be provided, allowing a coherent excitation of the vibration.

Time-resolved CARS offers a huge advantage over a pure frequency-dispersed approach: the nonresonant background can be suppressed completely. The reason for this becomes obvious from the scheme in Figure 7.9b: no more than virtual states are involved in the background generation, which only exist instantaneously during the interaction of the light fields with the molecule. Therefore, nonresonant signal can only be created when all interacting light fields overlap in time. Introducing the delay τ between the excitation with pump and Stokes and the probe pulse (Figure 7.9c) prevents this and does not allow generation of four-wave mixing signal without the resonant participation of an energy level. Such an approach, while not in the impulsive limit (Kamga and Sceats 1980; Volkmer et al. 2002), has been successfully implemented in CARS microscopy, where background suppression is vital for high-contrast chemical images (Cheng and Xie 2004; Volkmer 2005). In the impulsive limit of femtosecond CARS, however, the complete molecular information can be obtained from a spectrally integrated detection, which makes it a true Fourier-transform spectroscopic technique. Advantages are the simple single-channel detection, an almost arbitrarily high spectral resolution limited only by the length of the acquired delay series, and the possibility to even measure the relative phase between different molecular vibrations due to their acquisition in time domain (Zheltikov 2000). Microscopic imaging always requires collinear beam arrangement. It has become clear that phase matching for the CARS process can be fulfilled if the beams are very tightly focused with microscope objectives (Bjorklund 1975; Zumbusch et al. 1999).

Conventional CARS microscopy is typically performed in the frequency domain with pulses from distinct laser sources of different color, providing $\omega_p = \omega_{pr}$ and the red-shifted "Stokes" field ω_s, respectively. To obtain a complete spectrum, at least one laser and consequently $\omega_p - \omega_s$ have to be scanned. This approach makes CARS microspectroscopy experiments rather complicated and unpractical for non-specialists. Using ultrafast lasers, though, all frequencies can be contained in the broadband femtosecond pulse. Like this, the CARS setup can be extremely simplified: excitation is performed with a single laser beam of broadband femtosecond pulses, and the CARS signal is extracted spectrally, e.g., with a shortpass filter (Figure 7.9d). Because CARS is directly employed as a spectroscopy technique where the signal is probed by a part of the excitation laser itself, the spectral resolution is normally also determined by the laser spectral width. This is seemingly a contradiction: single-beam CARS excitation requires as broad spectra as possible to cover a wide range of frequency pairs (ω_p, ω_s), while the spectral resolution is lost with broadband probing. In 2002, however, the group of Silberberg presented an elegant scheme circumventing this problem by shaping the excitation pulse and thus coherently controlling the signal generation (Dudovich et al. 2002). With this, they demonstrated the feasibility of CARS incorporated within a single femtosecond laser pulse, with coherent control techniques restoring high spectral resolution and suppressing unwanted nonresonant background. The underlying key concept of "coherent control" was introduced at the beginning of this chapter. Several coherent control approaches have been proposed and realized for single-beam CARS: introducing a π-phase gate in the excitation pulses leads to interference between CARS signal and the nonresonant background and reveals vibrational levels (Oron et al. 2002; Dudovich et al. 2003). Resolution and sensitivity is further enhanced with additional polarization shaping of a narrow spectral feature (Oron et al. 2003; Lim et al. 2005). Such techniques are clearly rooted in the frequency domain, and their coherent control basis is thus understood in terms of interfering pathways. Another option is using periodic phase modulations to selectively excite Raman levels (Dudovich et al. 2002, 2003; von Vacano et al. 2006b), in analogy to a seminal control experiment on impulsive Raman scattering performed by Weiner et al. (1990). Again, this approach can be interpreted as interfering pathways control (Dudovich et al. 2003). However, in this case a time-domain picture is even more intuitive: periodic phase modulations lead to a splitting of the input pulse into a pulse train (Figure 7.10a). This allows a time-resolved measurement as depicted in Figure 7.9c for only two pulses. The problem due to the multipulse structure is that ambiguities arise, as pairs of sub-pulses with different temporal spacings, all being an integer multiple of the multipulse splitting, are present and act on the molecule under study (von Vacano and Motzkus 2006).

A direct microscopic time-resolved measurement would, however, simply be performed according to Figure 7.9c with two pulses. Using a pulse shaping apparatus, we have implemented a novel and flexible single-beam approach based on this, resolving the mentioned ambiguities of the multipulse method (Figure 7.10a) and unambiguously implementing impulsive time-resolved CARS (von Vacano and Motzkus 2006). The corresponding shaping strategy is the creation of two-color double pulses, by deliberately introducing different group delays for the blue and red part of the excitation spectrum (Figure 7.10b). This case has also been chosen as an example in Figures 7.4e and f. Without the flexibility of a pulse shaper, an implementation of impulsive time-resolved CARS in collinear geometry has successfully been shown by Ogilvie et al. using a Michelson-type interferometer (Ogilvie et al. 2006a; Cui et al. 2006).

To further improve the signal-to-noise ratio, we can use the shaper to set the polarization of the sub-pulses orthogonal to each other simply by polarization pulse shaping (Figure 7.10c), as we have shown recently (von Vacano and Motzkus 2007a). Which of the single-beam CARS schemes laid out in the previous section is eventually chosen for a microspectroscopy experiment depends on the application. In principle, they can all be implemented in software only, simply by programming the pulse shaper (Figures 7.4 and 7.6), respectively. Under favorable conditions, frequency-domain single-beam CARS schemes allow faster image acquisition, as the complete spectral information is detected simultaneously with a spectrograph. Time-resolved schemes typically require scanning of a pulse sequence and are hence inherently slower. However, they often prove more robust and intrinsically have advantages in the suppression of nonresonant background (von Vacano and Motzkus 2007b). The flexibility of single-beam CARS implemented with a pulse shaper, therefore, can be an important asset to adapt to special experimental needs. Shaping furthermore will allow the implementation of time-domain multiplexed acquisition, which combines the huge advantages of time-resolved CARS with rapid "one-shot" measurement.

In time domain, the CARS signal, $S_{CARS}(\tau)$, can be described mathematically as a function of the delay, τ, between pump/Stokes pulse, $E_1(t)$, and probe pulse, $E_2(t)$:

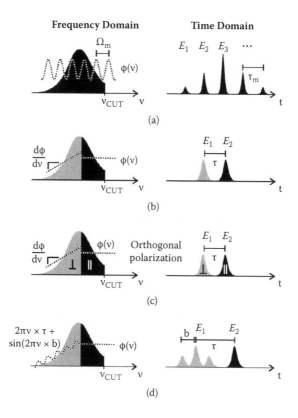

FIGURE 7.10 Single-beam CARS schemes in time domain. (a) Multipulse method, where a sine modulation of the spectral phase $\phi(\nu)$ with a period Ω_m leads to a splitting of the pulse train into a multipulse sequence with sub-pulse separations $\tau_m = 1/\Omega_m$. (b) For unambiguous time-resolved nonlinear spectroscopy, only two pulses have to be created. Additionally, to assign their role they can be made distinguishable by creating them from different spectral parts of the input pulse. This is done by applying a linear phase $\phi(\nu)$ for one part, leading to a group delay $\tau = GD(\nu) = 1/2\pi \times d\phi/d\nu$. See also Figure 7.4 for comparison. For conventional single-beam CARS measurements, the blue wing of the excitation spectrum is blocked for $\nu > \nu_{cut}$ to detect the much weaker, blue-shifted CARS signal in this spectral region. (c) In the two-color scheme, complete polarization control is possible. (d) Even more complex pulse sequences can also be created, e.g., with a multipulse sequence for selective excitation (sinusoidal phase) and a time-delayed probe.

$$S_{CARS}(\tau) \approx \int_{-\infty}^{\infty} dt \, \left| E_2(t) \right|^2 \left| \int_{-\infty}^{\infty} d\Omega \, \exp\left[-i\Omega(t+\tau) \right] A(\Omega) \cdot \chi^{(3)}(\Omega) \right|^2 \qquad (7.4)$$

where $A(\Omega)$ is the Raman population amplitude of the first pulse $E_1(t)$, and $\chi^{(3)}(\Omega)$ is the third-order susceptibility responsible for CARS. $A(\Omega)$ can be interpreted as the sum of all frequency pairs within the excitation pulse $E_1(t)$ having a frequency spacing of Ω, thus capable of populating Raman levels with this separation. Interpreted like this, it is easily calculated from the frequency-domain electric field $\tilde{E}_1(\omega)$ corresponding to $E_1(t)$ (Dudovich et al. 2003):

$$A(\Omega) = \int_0^\infty d\omega'\, \tilde{E}_1^*(\omega')\tilde{E}_1(\omega' + \Omega) \qquad (7.5)$$

Equation (7.5) is very important for single-beam CARS applications, because it quantifies the range of vibrational modes that can be accessed with a given laser pulse. Clearly, pulses with a broader spectrum and thus higher optical bandwidth allow covering a wider range of vibrational energies, as has also been pointed out in Table 7.1. It is, however, absolutely necessary that the pulses are perfectly compressed in the sample, so that all colors arrive at the same time.

B. SINGLE-BEAM TIME-RESOLVED CARS MICROSPECTROSCOPY

As an example of time-resolved single-beam CARS performed using orthogonally polarized two-color double pulses according to our scheme (Figures 7.4e and f and Figure 7.10c) with varied delay τ (von Vacano and Motzkus 2006), Figure 7.11 shows recorded time-dependent signal data, $S_{CARS}(\tau)$, and vibrational spectra recovered by

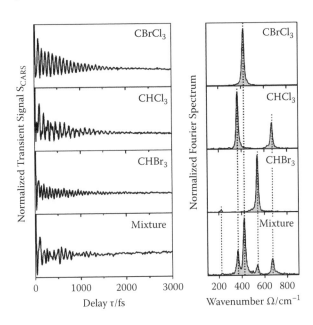

FIGURE 7.11 Time-resolved single-beam CARS transients (left column) and corresponding Fourier spectra (right column) for bromotrichloromethane (CBrCl$_3$), chloroform (CHCl$_3$), bromoform (CHBr$_3$) and a mixture of the three constituents. This result shows that the different components can clearly be distinguished by their characteristic vibrational resonances. (von Vacano and Motzkus 2007b). (From von Vacano and Motzkus, *Phys. Chem. Chem. Phys.*, 10:681–691, 2008. Used with permission.)

Fourier transform for different, chemically very similar halomethanes and a mixture thereof. The time-domain data in Figure 7.11 can be directly interpreted as an observation of molecular motion in real time, made possible by the compressed ultrashort pulses in the microscope. From the presence of different oscillatory patterns and beatings, it already becomes clear that the different molecules can be discriminated with high resolution. Correspondingly, the Fourier spectra in Figure 7.11 show markedly different vibrational resonances, which can also be discriminated in the ternary mixture of all components.

C. Time-Domain Control and Observation of Molecular Vibrations in CARS Microspectroscopy

Implementing the excitation pulse of a time-resolved single-beam CARS measurement as a deliberately shaped multipulse sequence (Figure 7.10d) allows the direct and intuitive time-domain observation and control of molecular vibrations. A mixture of molecules in the condensed phase is subjected to a single beam of shaped femtosecond laser pulses, which in a first step selectively excite defined molecular vibrations and in a second step interrogate the vibrational state of the molecules with a time-delayed probe pulse at a different color and polarization (von Vacano and Motzkus 2007a). The mixture here consisted of $CHCl_3$ and $CBrCl_3$.

As can be seen from the experimentally measured Fourier spectra of the time-resolved vibrations (Figure 7.12), shaping of the excitation pulse into a multipulse sequence with a temporal separation b (Figure 7.10d) allows the selective preparation

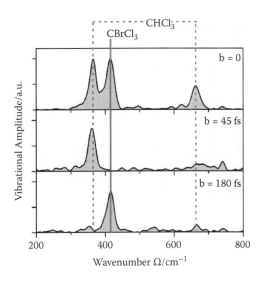

FIGURE 7.12 Single-beam Raman control of a mixture consisting of $CBrCl_3$ and $CHCl_3$. In the unshaped case ($b = 0$), all accessible modes from $CBrCl_3$ (solid vertical line) and $CHCl_3$ (dashed vertical lines) are visible. Selecting a suitable b parameter, only a distinct mode of either molecule can be excited.

of either the molecules in a desired vibrational state. This increased specificity in the excitation due to coherent control of the Raman excitation (Figure 7.2) is a valuable asset in further developing single-beam CARS for applications in high-contrast imaging or chemical microanalytics.

D. APPLICATION TO WHITE POWDER IDENTIFICATION AND THREAT ASSESSMENT

As an application example of broadband single-beam CARS microspectroscopy, we have studied a powder sample consisting of a mixture of potassium benzoate (KBenzoate) and the calcium salt of dipicolinic acid (CaDPA). While KBenzoate is a common food preservative, CaDPA is known as a marker molecule found in anthrax spores. Therefore, the rapid selective detection of CaDPA (Scully et al. 2002) and discrimination against other harmless compounds (such as KBenzoate, for example) can be an important step toward improving the assessment of "white powder" threats. With microspectroscopy, single microcrystals in the powder (or, in the case of a real biological agent, single bacterial spores) can be identified. Under a normal light microscope, KBenzoate and CaDPA look alike (Figure 7.13a).

Using our shaped broadband femtosecond laser beam a single crystal can be studied. The chemical constitution of KBenzoate and CaDPA is very similar (Figure 7.13b), both featuring an aromatic ring with carboxylate functions. The vibrational spectra are, however, markedly different (Figure 7.13c). While both compounds expectedly feature an aromatic ring breathing vibration around 1000 cm^{-1}, yet in the case of CaDPA at slightly lower wavenumbers, the other fingerprint vibrational bands at 750 cm^{-1} and 820 cm^{-1}, respectively, allow a clear distinction. As the time delay τ (or a corresponding temporal spacing in the multipulse approach) has to be scanned to obtain spectra such as in Figure 7.13c, presently the acquisition time for one spectrum is in the range of minutes. This can, however, be drastically sped up using faster pulse shaping apparatuses und somewhat higher laser pulse energies to increase the signal level. With the technology available today, spectral acquisition is likely to be cut down to tens of seconds. However, if a certain known molecule has to be detected, a different and faster approach can be pursued, such as recording the CARS signal for only a small number of specifically designed pulse shapes. This is especially attractive for chemical imaging, where each pixel of an image should be recorded in as little time as possible.

E. CHEMICAL IMAGING

To demonstrate this approach in principle, we have imaged the glass–liquid interface of our sample cell in a lateral and axial direction (Figure 7.14) by scanning the sample with a piezo stage. First, we recorded CARS transients for both regions, namely the $CHCl_3$ liquid phase and the glass cover slip sealing our sample cell (Figure 7.14a): for $CHCl_3$ (solid curve), the expected oscillations are seen, while the glass transient (dashed grey curve) remains constant for delays $\tau > 120$ fs. If the pulses $E_1(t)$ and $E_2(t)$ arrive simultaneously at the sample at $\tau = 0$, in both cases an elevated detected signal level S_{CARS} can be seen, which is mainly due to nonresonant CARS processes.

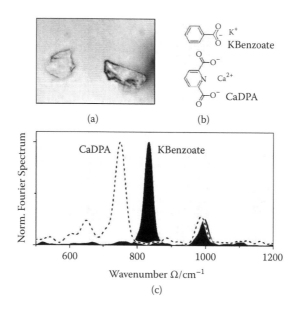

FIGURE 7.13 (a) Light microscopy transmission image (merged from two acquired camera frames) showing typical ~6 μm sized crystalites of a powder mixture prepared from potassium benzoate (KBenzoate) and the calcium salt of dipicolinic acid (CaDPA). (b) Chemical structures of the two compounds. (c) Single-beam CARS microspectroscopy performed on two microcrystals. These results have been obtained with an earlier version of our microscope setup, which did not yet allow simultaneous acquisition of light microscopy transmission images and CARS microspectroscopy. Therefore, different crystallites than shown in panel (a) have been measured. However, the clear differences in the spectral signature directly allow chemical identification of a microcrystal in the laser focus.

This information alone can be used for imaging, but due to its nonresonant origin does not reveal any chemical information.

A microscopic image of the glass-CHCl$_3$-interface based on the signal S_{CARS} at delay $\tau = 0$ can be seen in Figure 7.14b. There is only marginal contrast between the two layers, originating from the different level of nonresonant signal generated in the respective materials. Note that this contrast could vanish completely for different samples, or even invert if the nonresonant glass signal is stronger than that of the sample.

Therefore, such images are not useful for chemical mapping. If, on the other hand, the information in the oscillations of $S_{CARS}(\tau)$ is used for constructing the image, high-contrast chemical maps can be obtained, which selectively show only the presence of one chemical compound. As an example, we mapped a signal calculated from the oscillation amplitude of the CHCl$_3$-transient measured from four distinct points at τ_1, τ_2, τ_3, and τ_4 (indicated as dotted vertical lines in Figure 7.14a) as $|S_{CARS}(\tau_1) - S_{CARS}(\tau_2) + S_{CARS}(\tau_3) - S_{CARS}(\tau_4)|$. This simple procedure allows rapid acquisition of the image, as only four pulse sequences have to be applied and measured for each pixel. Acquiring each pulse sequence is performed in about 200 ms, keeping the

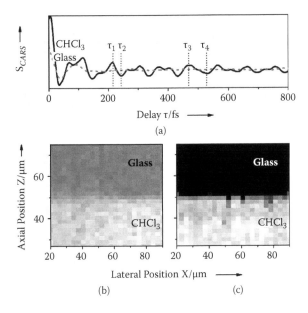

FIGURE 7.14 Microscopic imaging example, in which chemical contrast is obtained from the time-resolved CARS signal. (a) Measured CARS transients for chloroform ($CHCl_3$, solid curve) and a glass substrate (grey dashed curve). Selected delays $\tau_1 - \tau_4$ are indicated with dotted vertical lines, which are used for a chemically selective image of only $CHCl_3$. (b) CARS image of a glass/$CHCl_3$ interface. The dissatisfactory contrast is based only on the CARS signal $S_{CARS}(\tau)$ at zero delay $\tau = 0$. (c) High contrast chemically selective CARS image of the same sample, now based on the amplitude of the characteristic $CHCl_3$ oscillations measured at delays $\tau_1 - \tau_4$. The images were taken at identical scanning steps of $\Delta X = \Delta Z = 2$ μm. In both panels (b) and (c), the greyscale was normalized with white reflecting the maximum, and black zero signal.

measurement time of the complete image low. Alternatively, complete transients can be recorded and Fourier-transformed for each spatial position, but this, of course, leads to slower acquisition times per pixel. As mentioned earlier, using somewhat higher laser pulse energies and faster spatial light modulating schemes (Frumker et al. 2004; Frumker and Silberberg 2007) will generally drastically speed up time-resolved single-beam CARS measurements. Nonetheless, the present implementation already demonstrates very useful chemical imaging, as can be seen by the excellent contrast achieved in Figure 7.14c.

F. Improving Sensitivity with Heterodyne CARS Detection

In the previous sections, it was made clear that CARS indeed allows very high chemical specificity. Regarding the sensitivity, though, it faces a general problem: CARS is a coherent process, meaning that the signal waves created at each sample molecule add up with maximum constructive interference (given phase-matching is achieved). This leads to a quadratic dependence of the CARS signal on the concentration of the molecular species under study in the excitation volume (Levenson and Kano 1988). While this is beneficial for majority chemical species (for example, for lipid droplets in living cells [Nan et al. 2003]), it turns into a severe disadvantage at lower concentrations. However, the coherent nature of CARS can be used to mitigate this. Employing an auxiliary optical field as local oscillator (LO) with intensity I_{LO} brought to interference with the CARS signal I_{CARS}, linearization and amplification for sensitive detection can be achieved. The detected intensity is

$$I_{Det}(v) = I_{CARS}(v) + 2\sqrt{I_{CARS}(v)}\sqrt{I_{LO}(v)} \times \cos[\Delta\varphi(v)] + I_{LO}(v) \qquad (7.6)$$

with the characteristic interference term being proportional to the square roots of I_{CARS} and I_{LO} and dependent on the relative phase $\Delta\phi$. As $I_{CARS} \ll I_{LO}$, the interference term and hence the heterodyne CARS signal can be amplified by several orders of magnitude (Cui et al. 2006; von Vacano et al. 2006a). But most importantly, the sample concentration only enters linearly into the heterodyne signal, pushing CARS microspectroscopy toward analytical applications. Yet, it has to be stated that without further amplification by near-field effects at metal nanostructures (Koo et al. 2005), interferometric CARS also falls short in sensitivity compared to, e.g., fluorescence with its single-molecule sensitivity.

Heterodyne or interferometric detection has been implemented in CARS microscopy (Evans et al. 2004; Marks et al. 2004; Greve et al. 2005; Potma et al. 2006; Kee et al. 2006), also in order to suppress nonresonant background. Common to all interferometric approaches is a relatively complex experimental setup. In single-beam CARS it is much easier to introduce interferometric detection. In a frequency-domain scheme, the nonresonant background generated in the sample has deliberately been used as LO (Lim et al. 2005). One problem, however, is the dependence of the LO on the sample composition, causing difficult quantitative analysis in mixtures and a limited signal amplification. Recently, we demonstrated a novel, simplified approach to single-beam heterodyne CARS (von Vacano et al. 2006a). Instead of blocking the blue part of the broadband excitation spectrum for frequencies $v > v_{cut}$ in the pulse shaper (Figure 7.15a), we use it as strong and fully phase-controlled LO

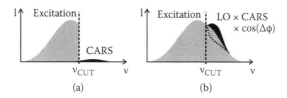

(a)   (b)

FIGURE 7.15 (a) "Conventional" single-beam CARS, where the blue wing of the excitation spectrum is suppressed in the pulse shaper in order to only detect the weaker CARS signal in this spectral position. (b) Single-beam heterodyne CARS approach. The blue wing of the excitation spectrum is not blocked, but used as a local oscillator (LO) field for interferometric detection of the CARS signal. The LO phase difference $\Delta\phi$ can be controlled precisely by the pulse shaper, in order to maximize the interference signal.

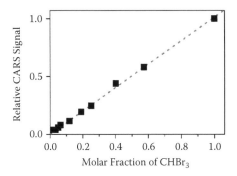

FIGURE 7.16 Linear concentration dependence achieved with single-beam heterodyne CARS. The sample used was $CHBr_3$ diluted in ethanol at different molar fractions ranging from 1 to 0.01. The indicated linear fit (slope 1.01 ± 0.01) through the origin confirms excellent agreement with the expected slope of 1.

(Figure 7.15b). Hence, sensitive heterodyne CARS detection is realized with intrinsic interferometric stability. Due to the interference with the strong external LO, signal amplification of a factor larger than 5000 was achieved (von Vacano et al. 2006a).

Figure 7.16 shows the linear concentration dependence resulting from interferometric detection and, consequently, the ability to measure CARS spectra of a diluted sample with concentrations in the ~100 mM range. For this data, single-beam heterodyne CARS microspectroscopy was performed on $CHBr_3$ diluted in ethanol for different concentrations in molar fractions ranging from 1–0.01 (von Vacano et al. 2006a). Even at the lowest concentration of $x(CHBr_3) = 0.01$ used in this study, corresponding to estimated 14 attomole (8×10^6 molecules) in the excitation volume of approximately 80 attoliters, the chemically selective CARS signal was detected. This increased sensitivity without added complexity in the experimental setup should prove useful in developing single-beam CARS for microanalytics and chemical imaging, e.g., of living biological cells.

The added benefit of intrinsic interferometric detection is only a further example of the great flexibility of using the pulse shaper in single-beam nonlinear microspectroscopy. The setup used here for CARS in its variants is, of course, also capable of immediately performing all the other nonlinear microspectroscopies simply by changing the shape of the excitation pulses with computer control. This is shown in the next section, where we discuss a broadband TPF application.

VI. BROADBAND MULTIPHOTON FLUORESCENCE MICROSPECTROSCOPY

To demonstrate the versatility of nonlinear microspectroscopy with shaped broadband laser pulses, TPF is chosen as a second example. Remember that TPF can be implemented only by programming the pulse shaper differently and sampling a different wavelength range of the signal spectrum in the same experimental setup. TPF is so far probably the most widely applied nonlinear optical spectroscopy method, considering its important application in microscopy. Due to the broad excitation bandwidth of our ultrashort fs-pulses, we are able to excite a variety of fluorescent labels at once, without having to tune the laser central wavelength, or even employing different lasers. The present scheme with integrated pulse shaper would also allow shaping the broadband pulses in such a way that the excitation is selective for a given dye. Therefore, the detection of the total fluorescence signal is sufficient for creating images showing the differently labeled areas in the specimen. This has successfully been shown for microscopic imaging (Pastirk et al. 2003; Ogilvie et al. 2006b). Here, we want to focus on a different approach that relies on simultaneous excitation and multichannel spectroscopic detection of the different fluorophors. Image contrast and selectivity is achieved by a decomposition of the microspectroscopy signal for each spatial position into the different contributions from all labelling dyes used (Dickinson et al. 2001; Lansford et al. 2001; Schultz et al. 2001).

The sample used for a demonstration of broadband TPF imaging with compressed PCF supercontinuum in Figure 7.17 was a commercially available test slide (Invitrogen FluoCells®, prepard slide #1, containing labeled bovine pulmonary artery endothelial cells). The conventional optical phase-contrast microscopy image

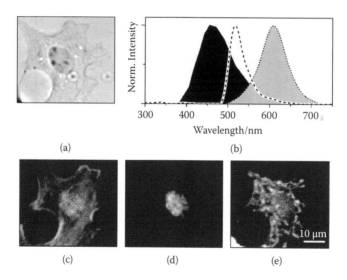

FIGURE 7.17 Broadband TPF microspectroscopy, demonstrated on BPAE cells (Invitrogen FluoCells® prepared slide #1) labeled with three different fluorophors. (a) A conventional phase-contrast image of a cell. (b) Normalized, smoothed emission spectra of the three dyes, as used for the linear decomposition of the TPF spectrum from each sample position: BODIPY® FL phallacidin (dashed curve), DAPI (black hatched curve), and MitoTracker® Red (dotted curve, hatched in grey). (c) Resulting partial image, showing the distribution of BODIPY® FL phallacidin and therefore the cytoskeleton. (d) Partial image of DAPI, revealing the nucleus. (e) Partial image of MitoTracker® Red signal, highlighting mitochondria. In panel (c), small residual crosstalk with the DAPI channel (d) can be seen, which is due to DAPI having a fluorescence signal in our sample an order of magnitude stronger.

(Figure 7.17a) shows the cell selected for spectroscopic TPF imaging. Different functional entities of the cell are labeled by different fluorescent dyes, which can all be excited in a two-photon transition within the broad excitation bandwidth provided. The normalized fluorescence spectra of the dyes are shown in Figure 7.17b, featuring different emission maxima but still strongly overlapping spectra. However, as the fluorescence signal from different dye molecules adds up incoherently in the detector, any TPF microspectroscopy signal will be a linear combination of these spectral profiles of the native dyes (note that this property of linear signal superposition is not the case in CARS spectroscopy, where signals and a nonresonant background add up coherently, leading to interferences).

Spectral decomposition can be performed in a variety of ways; here we used a powerful evolutionary algorithm (Zeidler et al. 2001a) to find the partial intensity of each dye spectrum for every spatial position in the sample. This immediately gives images showing only the distribution of one specific dye, which allows the location of the functional entities labeled (Figures 7.17c–e). The given example again shows how useful broadband excitation combined with spectrally resolved detection can be, if the pulse duration at the sample position can be controlled. The analysis of spectral profiles performed to discriminate the different dyes is much more powerful than simple single-channel detection at selected emission wavelengths.

VII. CONCLUSIONS AND OUTLOOK

This chapter has shown the great flexibility and potential of combining a broadband laser source, ultrafast pulse shaping, and microscopy for nonlinear spectroscopy. It allows tackling the quests of dispersion management and experimental simplification, and opens wholly new realms of application due to the combination with laser scanning microscopy. On the technical side, modern optical fiber technology and new simplified measurement techniques such as SAC-SPIDER have been shown to be key ingredients for success. The capability of performing all kinds of different spectroscopies implemented in the same setup only by coherent control and appropriate software-based pulse shaping cannot be underestimated. This will allow completely new insights into complex samples with multi-parameter spectroscopy at three-dimensional microscopic resolution.

ACKNOWLEDGMENTS

The authors would like to acknowledge the support of the "Fonds der chemischen Industrie" (by way of a scholarship to B. v. V.) and the Max-Planck Institute for Quantum Optics (Garching, Germany). Very helpful discussions concerning different aspects of the work presented with T. Buckup, J. Hauer, J. Möhring, and W. Wohlleben are highly appreciated.

REFERENCES

Barad, Y., Eisenberg, H., Horowitz, M., and Silberberg, Y. 1997. Nonlinear scanning laser microscopy by third harmonic generation. *Appl. Phys. Lett.* 70 (8):922–24.

Baumert, T., Brixner, T., Seyfried, V., Strehle, M., and Gerber, G. 1997. Femtosecond pulse shaping by an evolutionary algorithm with feedback. *Appl. Phys. B-Lasers Opt.* 65 (6):779–82.

Bergmann, K., Theuer, H., and Shore, B. W.. 1998. Coherent population transfer among quantum states of atoms and molecules. *Rev. Mod. Phys.* 7 (3):1003–25.

Bjorklund, G. C. 1975. Effects of focusing on third-order nonlinear processes in isotropic media. *IEEE J. Quantum Electron.* 11:287–96.

Brixner, T., and Gerber, G. 2003. Quantum control of gas-phase and liquid-phase femtochemistry. *Chem. Phys. Chem.* 4(5):418–38.

Broers, B., Vandenheuvell, H. B. V., and Noordam, L. D. 1992. Large interference effects of small chirp observed in two-photon absorption. *Opt. Comm.* 91(1–2):57–61.

Buist, A. H., Muller, M., Ghauharali, R. I., Brakenhoff, G. J., Squier, J. A., Bardeen, C. J., Yakovlev, V. V., and Wilson, K. R. 1999. Probing microscopic chemical environments with high-intensity chirped pulses. *Opt. Lett.* 24(4):244–46.

Campagnola, P. J., and Loew, L. M. 2003. Second-harmonic imaging microscopy for visualizing biomolecular arrays in cells, tissues and organisms. *Nat. Biotechnol.* 21(11):1356–60.

Chen, B.-C., and Lim, S.-H. 2007. Characterization of a broadband pulse for phase controlled multiphoton microscopy by single beam SPIDER. *Opt. Lett.* 32(16):2411–13.

Cheng, J. X., and Xie, X. S. 2004. Coherent anti-Stokes Raman scattering microscopy: Instrumentation, theory, and applications. *J. Phys. Chem. B* 108(3):827–40.

Cui, M., Joffre, M., Skodack, J., and Ogilvie, J. P. 2006. Interferometric Fourier transform coherent anti-Stokes Raman scattering. *Opt. Exp.* 14(18):8448–58.

Dela Cruz, J. M., Lozovoy, V. V., and Dantus, M. 2005. Quantitative mass spectrometric identification of isomers applying coherent laser control. *J. Phys. Chem. A* 109(38):8447–50.

Dickinson, M. E., Bearman, G., Tille, S., Lansford, R., and Fraser, S. E. 2001. Multi-spectral imaging and linear unmixing add a whole new dimension to laser scanning fluorescence microscopy. *Biotechniques* 31(6):1272–8.

Diels, J.-C., and Rudolph, W. 2006. *Ultrafast laser phenomena.* San Diego: Academic Press.

Dudley, J. M., Genty, G., and Coen, S. 2006. Supercontinuum generation in photonic crystal fiber. *Rev. Mod. Phys.* 78(4):1135–84.

Dudovich, N., Oron, D., and Silberberg, Y. 2002. Single-pulse coherently controlled nonlinear Raman spectroscopy and microscopy. *Nature* 418:512–14.

Dudovich, N., Oron, D., and Silberberg, Y. 2003. Single-pulse coherent anti-Stokes Raman spectroscopy in the fingerprint region. *J. Chem. Phys.* 118:9208–15.

Evans, C. L., Potma, E. O., and Xie, X. S. N. 2004. Coherent anti-Stokes Raman scattering spectral interferometry: Determination of the real and imaginary components of nonlinear susceptibility $\chi^{(3)}$ for vibrational microscopy. *Opt. Lett.* 29(24):2923–25.

Frumker, E., Oron, D., Mandelik, D., and Silberberg, Y. 2004. Femtosecond pulse-shape modulation at kilohertz rates. *Opt. Lett.* 29:890–92.

Frumker, E., and Silberberg, Y. 2007. Femtosecond pulse shaping using a two-dimensional liquid-crystal spatial light modulator. *Opt. Lett.* 32:1384–86.

Gauderon, R., Lukins, P. B., and Sheppard, C. J. R. 1998. Three-dimensional second-harmonic generation imaging with femtosecond laser pulses. *Opt. Lett.* 23(15):1209–11.

Greve, M., Bodermann, B., Telle, H. R., Baum, P., and Riedle, E. 2005. High-contrast chemical imaging with gated heterodyne coherent anti-Stokes Raman scattering microscopy. *Appl. Phys. B-Lasers Opt.* 81(7):875–79.

Herek, J. L., Wohlleben, W., Cogdell, R. J., Zeidler, D., and Motzkus, M. 2002. Quantum control of energy flow in light harvesting. *Nature* 417(6888):533–35.

Hornung, T., Meier, R., Zeidler, D., Kompa, K. L., Proch, D., and Motzkus, M. 2000. Optimal control of one- and two-photon transitions with shaped femtosecond pulses and feedback. *Appl. Phys. B-Lasers Opt.* 71(3):277–84.

Kamga, F. M., and Sceats, M. G. 1980. Pulse-sequenced coherent anti-Stokes Raman scattering spectroscopy: A method for suppression of the nonresonant background. *Opt. Lett.* 5(3):126–28.

Kang, E. N., Wang, H. F., Kwon, I. K., Robinson, J., Park, K., and Cheng, J. X. 2006. In situ visualization of paclitaxel distribution and release by coherent anti-stokes Raman scattering microscopy. *Anal. Chem.* 78(23):8036–43.

Kee, T. W., and Cicerone, M. T. 2004. Simple approach to one-laser, broadband coherent anti-Stokes Raman scattering microscopy. *Opt. Lett.* 29(23):2701–3.

Kee, T. W., Zhao, H., and Cicerone, M. T. 2006. One-laser interferometric broadband coherent anti-Stokes Raman scattering. *Opt. Exp.* 14(8):3631–40.

Koo, T. W., Chan, S., and Berlin, A. A. 2005. Single-molecule detection of biomolecules by surface-enhanced coherent anti-Stokes Raman scattering. *Opt. Lett.* 30(9):1024–26.

Lansford, R., Bearman, G., and Fraser, S. E. 2001. Resolution of multiple green fluorescent protein color variants and dyes using two-photon microscopy and imaging spectroscopy. *J. Biomed. Opt.* 6(3):311–18.

Levenson, M. D., and Kano, S. S. 1988. *Introduction to nonlinear lasers spectroscopy*, 2nd ed. San Diego: Academic Press.

Lim, S. H., Caster, A. G., and Leone, S. R. 2005. Single-pulse phase-control interferometric coherent anti-Stokes Raman scattering spectroscopy. *Phys. Rev. A* 72(4):041803.

Lozovoy, V. V., Pastirk, I., and Dantus, M. 2004. Multiphoton intrapulse interference. IV. Ultrashort laser pulse spectral phase characterization and compensation. *Opt. Lett.* 29(7):775–77.

Marks, D. L., Vinegoni, C., Bredfeldt, J. S., and Boppart, S. A. 2004. Interferometric differentiation between resonant coherent anti-Stokes Raman scattering and nonresonant four-wave-mixing processes. *Appl. Phys. Lett.* 85(23):5787–89.

McConnell, G. 2004. Confocal laser scanning fluorescence microscopy with a visible continuum source. *Opt. Exp.* 12(13):2844–50.

Meshulach, D., and Silberberg, Y. 1998. Coherent quantum control of two-photon transitions by a femtosecond laser pulse. *Nature* 396(6708):239–42.

Meshulach, D., and Silberberg, Y. 1999. Coherent quantum control of multiphoton transitions by shaped ultrashort optical pulses. *Phys. Rev. A* 60:1287–92.

Meshulach, D., Yelin, D., and Silberberg, Y. 1997. Adaptive ultrashort pulse compression and shaping. *Opt. Comm.* 138(4–6):345–48.

Monmayrant, A., Joffre, M., Oksenhendler, T., Herzog, R., Kaplan, D., and Tournois, P. 2003. Time-domain interferometry for direct electric-field reconstruction by use of an acousto-optic programmable filter and a two-photon detector. *Opt. Lett.* 28(4):278–80.

Mukamel, S. 1995. *Principles of nonlinear optical spectroscopy.* New York: Oxford University Press.

Müller, M., and Zumbusch, A. 2007. Coherent anti-Stokes Raman scattering microscopy. *ChemPhysChem* 8(15):2156–70.

Nan, X. L., Cheng, J. X., and Xie, X. S. 2003. Vibrational imaging of lipid droplets in live fibroblast cells with coherent anti-Stokes Raman scattering microscopy. *J. Lipid Res.* 44(11):2202–8.

Ogilvie, J. P., Beaurepaire, E., Alexandrou, A., and Joffre, M. 2006a. Fourier-transform coherent anti-Stokes Raman scattering microscopy. *Opt. Lett.* 31(4):480–82.

Ogilvie, J. P., Débarre, D., Solinas, X., Martin, J.-L., Beaurepaire, E., and Joffre, M. 2006b. Use of coherent control for selective two-photon fluorescence microscopy in live organisms. *Opt. Exp.* 14(2):759–66.

Omenetto, F. G., Taylor, A. J., Moores, M. D., and Reitze, D. H. 2001. Adaptive control of femtosecond pulse propagation in optical fibers. *Opt. Lett.* 26(12):938–40.

Oron, D., Dudovich, N., and Silberberg, Y. 2002. Single-pulse phase-contrast nonlinear Raman spectroscopy. *Phys. Rev. Lett.* 89(27):273001.

Oron, D., Dudovich, N., and Silberberg, Y. 2003. Femtosecond phase-and-polarization control for background-free coherent anti-Stokes Raman spectroscopy. *Phys. Rev. Lett.* 90(21):213902.

Pastirk, I., DelaCruz, J. M., Walowicz, K. A., Lozovoy, V. V., and Dantus, M. 2003. Selective two-photon microscopy with shaped femtosecond pulses. *Opt. Exp.* 11(14):1695–1701.

Potma, E. O., Evans, C. L., and Xie, X. S. 2006. Heterodyne coherent anti-Stokes Raman scattering (CARS) imaging. *Opt. Lett.* 31(2):241–43.

Rabitz, H., de Vivie-Riedle, R., Motzkus, M., and Kompa, K.-L. 2000. Whither the future of controlling quantum phenomena? *Science* 288(5. May):824–28.

Rice, S. A., and Zhao, M. 2000. *Optimal control of molecular dynamics.* New York: John Wiley & Sons.

Russel, P. 2003. Photonic crystal fibers. *Science* 299:358–362.

Schultz, R. A., Nielsen, T., Zavaleta, J. R., Ruch, R., Wyatt, R., and Garner, H. R. 2001. Hyperspectral imaging: A novel approach for microscopic analysis. *Cytometry* 43(4):239–47.

Scully, M. O., Kattawar, G. W., Lucht, R. P., Opatrny, T., Pilloff, H., Rebane, A., Sokolov, A. V., and Zubairy, M. S. 2002. FAST CARS: Engineering a laser spectroscopic technique for rapid identification of bacterial spores. *Proc. Natl. Acad. Sci. USA* 99(17):10994–11001.

Shapiro, M., and Brumer, P. 2003. *Principles of the Quantum Control of Molecular Processes.* Hoboken, NJ: John Wiley & Sons.

Stobrawa, G., Hacker, M., Feurer, T., Zeidler, D., Motzkus, M., and Reichel, F. 2001. A new high-resolution femtosecond pulse shaper. *Appl. Phys. B-Lasers Opt.* 72(5):627–30.

Sun, C. K. 2005. Higher harmonic generation microscopy. *Advances in Biochemical Engineering (Biotechnol., Microsc. Tech.)* 95:17–56.

Tada, J., Kono, T., Suda, A., Mizuno, H., Miyawaki, A., Midorikawa, K., and Kannari, F. 2007. Adaptively controlled supercontinuum pulse from a microstructure fiber for two-photon excited fluorescence microscopy. *Appl. Opt.* 46(15):3023–30.

Verluise, F., Laude, V., Cheng, Z., Spielmann, C., and Tournois, P. 2000. Amplitude and phase control of ultrashort pulses by use of an acousto-optic programmable dispersive filter: Pulse compression and shaping. *Opt. Lett.* 25(8):575–77.

Volkmer, A. 2005. Vibrational imaging and microspectroscopies based on coherent anti-Stokes Raman scattering microscopy. *J. Phys. D-Appl. Phys.* 38(5):R59–R81.

Volkmer, A., Book, L. D., and Xie, X. S.. 2002. Time-resolved coherent anti-Stokes Raman scattering microscopy: Imaging based on Raman free induction decay. *Appl. Phys. Lett.* 80(9):1505–7.

von Vacano, B., Buckup, T., and Motzkus, M. 2006a. Highly sensitive single-beam heterodyne coherent anti-Stokes Raman scattering. *Opt. Lett.* 31(16):2495–97.

von Vacano, B., Buckup, T., and Motzkus, M. 2006b. In-situ broadband pulse compression for multiphoton microscopy using a shaper-assisted collinear SPIDER. *Opt. Lett.* 31(8):1154–56.

von Vacano, B., Buckup, T., and Motzkus, M. 2007. Shaper-assisted collinear SPIDER: Fast and simple broadband pulse compression in nonlinear microscopy. *J. Opt. Soc. Am. B-Opt. Phys.* 24(5):1091–1100.

von Vacano, B., Meyer, L., and Motzkus, M.. 2006. Rapid polymer blend imaging with quantitative broadband multiplex-CARS microscopy. *J. Raman Spectrosc.* 38:916–26.

von Vacano, B., and Motzkus, M. 2006. Time-resolved two color single-beam CARS employing supercontinuum and femtosecond pulse shaping. *Opt. Comm.* 264:488–93.

von Vacano, B., and Motzkus, M. 2007a. Molecular discrimination of a mixture with single-beam Raman control. *J. Chem. Phys.* 127(14):144514.

von Vacano, B., and Motzkus, M. 2007b. Time-resolved single-beam CARS: Comparing approaches for simplified ultrafast vibrational spectroscopy in simulation and experiment. (In preparation.)

von Vacano, B., Wohlleben, W., and Motzkus, M. 2006a. Actively shaped supercontinuum from a photonic crystal fiber for nonlinear coherent microspectroscopy. *Opt. Lett.* 31(3):413–15.

von Vacano, B., Wohlleben, W., and Motzkus, M. 2006b. Single-beam CARS spectroscopy applied to low-wavenumber vibrational modes. *J. Raman Spectrosc.* 37(1–3):404–10.

Wadsworth, W. J., Ortigosa-Blanch, A., Knight, J. C., Birks, T. A., Man, T. P. M., and Russell, P. S. 2002. Supercontinuum generation in photonic crystal fibers and optical fiber tapers: A novel light source. *J. Opt. Soc. Am. B-Opt. Phys.* 19(9):2148–55.

Weiner, A. M. 2000. Femtosecond pulse shaping using spatial light modulators. *Rev. Sci. Instrum.* 71(5):1929–60.

Weiner, A. M., Leaird, D. E., Wiederrecht, G. P., and Nelson, K. A. 1990. Femtosecond pulse sequences used for optical manipulation of molecular motion. *Science* 247:1317–19.

Wohlleben, W., Buckup, T., Herek, J. L., and Motzkus, M. 2005. Coherent control for spectroscopy and manipulation of biological dynamics. *Chem. Phys. Chem.* 6(5):850–57.

Xie, X. S., Yu, J., and Yang, W. Y. 2006. Perspective—Living cells as test tubes. *Science* 312(5771):228–30.

Xu, B., Gunn, J. M., Dela Cruz, J. M., Lozovoy, V. V., and Dantus, M. 2006. Quantitative investigation of the multiphoton intrapulse interference phase scan method for simultaneous phase measurement and compensation of femtosecond laser pulses. *J. Opt. Soc. Am. B: Opt. Phys.* 23(4):750–59.

Zeidler, D., Frey, S., Kompa, K. L., and Motzkus, M. 2001a. Evolutionary algorithms and their application to optimal control studies. *Phys. Rev. A* 64(2):023420.

Zeidler, D., Hornung, T., Proch, D., and Motzkus, M. 2000. Adaptive compression of tunable pulses from a non-collinear type OPA to below 20fs by feedback-controlled pulse shaping. *Appl. Phys. B* 70:S125–S131.

Zeidler, D., Witte, T., Proch, D., and Motzkus, M. 2001b. Optical parametric amplification of a shaped white-light continuum. *Opt. Lett.* 26(23):1921–23.

Zheltikov, A. M. 2000. Coherent anti-Stokes Raman scattering: From proof-of-the-principle experiments to femtosecond CARS and higher order wave-mixing generalizations. *J. Raman Spectrosc.* 31(8-9):653–67.

Zipfel, W. R., Williams, R. M., and Webb, W. W. 2003. Nonlinear magic: Multiphoton microscopy in the biosciences. *Nat. Biotechnol.* 21:1369–77.

Zumbusch, A., Holtom, G. R., and Xie, X. S. 1999. Three-dimensional vibrational imaging by coherent anti-Stokes Raman scattering. *Phys. Rev. Lett.* 82(20):4142–45.

8 Nonlinear Optical Imaging with Sub-10 fs Pulses

Marcos Dantus, Dmitry Pestov, and Yair Andegeko

CONTENTS

I. INTRODUCTION

The fundamental goal of biomedical imaging is the retrieval of structural and dynamical information from the sample, with the greatest contrast, spatial, and temporal resolution possible. The workhorse of biomedical imaging is the optical microscope, which readily provides the resolution sufficient to observe subcellular organelles. The introduction of lasers and confocal optics brought the resolution scale down to diffraction-limited, i.e., on the order of the laser wavelength. Contrast enhancement came primarily from the use of fluorescent labeling compounds.

It is well known that at high enough electric field intensities, usually achieved by focusing the laser radiation, a number of nonlinear processes can come into play. However, it was not recognized until recently that nonlinear optical processes can be used to attain resolution well beyond the diffraction limit (Dyba and Hell 2002; Westphal and Hell 2005). The nonlinear effects have also permitted imaging through scattering biological tissue. In this chapter, we concentrate on two-photon microscopy (TPM) and explore the advantages and disadvantages of using ultrashort <10 fs laser pulses for this purpose.

TPM was introduced at a time when advanced optical microscopy involved the use of multiple laser wavelengths (Denk et al. 1990). For a given sample, one would

197

need to swap optics in order to excite different fluorescent labels and thus achieve the greatest contrast. With TPM, it became possible to excite a whole range of different laser dyes without having to change optics. Because of the quadratic dependence of the two-photon excitation fluorescence (TPEF) signal on the laser peak intensity, the use of two-photon-resonant processes reduced the interaction region to a small volume of the sample where the laser light was focused. It also allowed one to image as far as 1 mm through scattering tissue (Theer et al. 2003).

It is known that with the laser pulse energy kept constant, the TPEF signal magnitude is inversely proportional to the laser pulse duration. This is because the total amount of the collected signal not only depends on the peak field intensity, but also on the time duration that the laser field is acting upon the sample. The availability of femtosecond lasers and, in particular, Ti:sapphire lasers (Spence et al. 1991) fostered the development of TPM, but not for long. When TPM was introduced, the available laser sources produced 150-fs pulses. Now, in 2008, there are a number of companies offering lasers capable of generating sub-10 fs pulses. Despite the promise of 15-fold brighter signals, these new laser systems are not being used for TPM. Here we discuss reasons why and present a method that makes the utilization of these ultrashort pulse sources possible.

II. PULSE DISPERSION

When femtosecond pulses propagate through an optical medium they undergo chromatic dispersion. The cause for dispersion is that all optical media have a frequency-dependent index of refraction. Because of the uncertainty principle, femtosecond pulses have broad spectra, and this makes them orders of magnitude more sensitive to chromatic dispersion when compared to nano- or even picosecond pulses. If the phase retardation introduced by the medium is expanded in a Taylor series about the carrier frequency, the term that changes linearly with the frequency shift can be associated with a time delay, and therefore has no effect on the pulse shape and the TPM output. The subsequent terms, however, do affect the performance of TPM. The term that is quadratic in frequency is due to the second-order dispersion (SOD), also known as group delay dispersion (GDD). By definition, GDD is the derivative of the group delay with respect to the angular frequency, or the second derivative of the spectral phase. It is usually specified in fs^2 or ps^2. The GDD per unit length (in units of s^2/m) is referred to as group velocity dispersion (GVD).

If we consider a femtosecond pulse free of any phase distortion, such a pulse would satisfy the transform limit (TL) equation, which for Gaussian pulses is given by $\Delta v \times \Delta t = 2\ln(2)/\pi$ with Δv and Δt being the 1/e spectral bandwidth (in Hz) and the time duration of the pulse. When such a pulse acquires positive GDD, the high-frequency (blue-shifted) components are delayed with respect to the low-frequency (red-shifted) ones, and the pulse gets broadened by a factor of $(1 + (4\ln(2)\phi''/\tau_{in}^2)^2)^{1/2}$, where ϕ'' is the amount of GDD, and τ_{in} is the input pulse duration. The effect of negative GDD is the same, except for now the high-frequency components run ahead of low-frequency ones, see Figure 8.1. SOD lowers the peak intensity of the pulses and hence reduces the two-photon excitation efficiency.

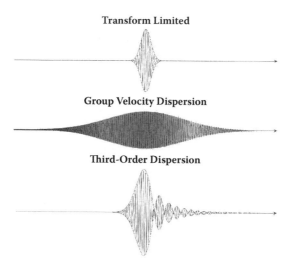

FIGURE 8.1 Time-domain effects of the second- and third-order dispersion. A TL pulse is one that is as short as possible given the available bandwidth. GDD causes time separation between different wavelengths of the pulse, and this broadens its duration. Third-order dispersion breaks the laser pulse into different sub-pulses in the time domain.

The term that depends on the third power of the frequency shift is known as third-order dispersion (TOD). When a TL pulse acquires a significant amount of TOD, the pulse envelope is distorted and a series of sub-pulses is produced, as shown in Figure 8.1. Unlike a pulse with SOD, a pulse with TOD leads to two-photon excitation with the same efficiency as a TL pulse but only for a particular two-photon frequency. At other frequencies, the amount of excitation is suppressed. The control over TOD would allow for preferential excitation in different spectral regions, while its correction would lead to efficient two-photon excitation over the whole accessed spectral range. Unfortunately, measuring and correcting TOD is not a simple task.

The reason two-photon excitation efficiency depends on the spectral phase of the laser pulse is beyond the scope of the peak intensity argument given above. It requires a more detailed consideration of how the pulse contributes to the nonlinear optical transitions. Let us consider, e.g., the case of two-photon excitation at wavelengths corresponding to $\lambda_{TPE} = 400$ nm. In addition to a pair of 800-nm photons, there are multiple combinations of spectral components within the bandwidth of an ultrashort pulse that can combine to cause excitation equivalent to 400 nm. Their wavelengths satisfy the relation $1/\lambda_{long} + 1/\lambda_{short} = 1/\lambda_{TPE}$. In particular, one can have photons at 790 nm combining with photons at 810 nm. Obviously, these channels will interfere constructively only if they all have the same phase. Mathematically, the requirement can be expressed as $\phi_{long} + \phi_{short} = $ const; i.e., the sum of the phases of the longer- and shorter-wavelength photons is independent of the choice of the wavelength pair. This condition is satisfied when all spectral components of the laser pulse have the same phase, i.e., for a TL pulse. Another interesting situation is when the sum is constant (e.g., zero) only for one specific value of λ_{TPE}. This can be accomplished by using a phase function that is antisymmetric with respect to $\lambda_{TPE}/2$. This arrangement is discussed later within the context of selective two-photon excitation. For pure TOD, an antisymmetric function, the condition above is satisfied for only one λ_{TPE} value for which TPE efficiency is maximum; at other wavelengths TPE efficiency is reduced.

III. MEASURING AND ELIMINATING PULSE DISPERSION

The main source of dispersion in microscopy is the microscope objective lens. Therefore, the first step is to obtain an accurate measurement of the spectral phase at the focal plane; the second step is to use adaptive optics to compensate the measured phase. A method that incorporates the indicated steps was invented by our research group in 2003 (Pastirk et al. 2003; Lozovoy et al. 2004; Xu et al. 2006; Dantus et al. 2007). This method, known as multiphoton intrapulse interference phase scan (MIIPS), uses a pulse shaper with a programmable spatial light modulator to introduce a reference phase function, $f(\lambda)$, that causes cancellation of the linear chirp at some wavelengths within the pulse spectrum. At those wavelengths, yet unknown, nonlinear processes such as second harmonic generation (SHG) are maximized. MIIPS uses this information to solve the equation $\phi''(\lambda) - f''(\lambda) = 0$ and find the second derivative of the original phase $\phi(\lambda)$ and its corresponding wavelength (Lozovoy et al. 2008). When the difference is zero, $f''(\lambda)$ is exactly the second derivative of the phase distortion to be measured. After double integration of the retrieved $\phi''(\lambda)$, the chromatic dispersion at the focal plane is known and the first step is complete. Note that the measured $\phi''(\lambda)$ is a function of wavelength, not a constant in the Taylor series expansion, which contains information about high-order phase distortions including SOD, TOD, and beyond. For the second step, the pulse shaper introduces a phase function that cancels the measured phase distortions at all wavelengths. This results in TL pulses at the focal plane. This process is fully automated and can be completed in less than one minute.

The ability to cancel all orders of phase distortion gives us an opportunity to evaluate the effect of partial dispersion correction on TPM. In particular, we focus on comparing SOD correction, which can be achieved with a prism pair arrangement, and correction of all orders of phase dispersion using MIIPS. For these measurements we used a pair of prisms in addition to our pulse shaper. With the aid of the pulse shaper, we found the condition for which SOD at the center wavelength was fully eliminated by the prism pair, and only higher-order dispersion was compensated by the pulse shaper.

As a first experiment, we wanted to return to the fundamental premise of TPM, which states that the efficiency of the two-photon excitation should be inversely proportional to the pulse duration. For TL pulses it is equivalent to having two-photon excitation efficiency proportional to the spectral bandwidth of the laser pulses.

The measured intensities of TPEF as a function of the spectral bandwidth, for GDD-compensated and TL pulses, are shown in Figure 8.2. The experiments were performed with the same average power at the sample (2.3 mW) for all bandwidths, using a red fluorescence standard slide (Chroma Technologies). We found that in the case of TL pulses, obtained via MIIPS compensation, the magnitude of TPEF

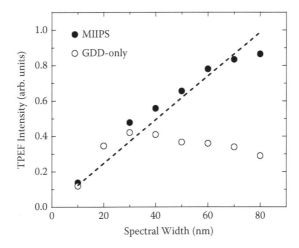

FIGURE 8.2 Dependence of the TPEF signal on the laser spectral bandwidth for TL (filled circles) and SOD-compensated (open circles) pulses of the laser pulse. The linear relation between TPEF and bandwidth is observed only for TL (i.e., MIIPS-compensated, pulses). (From Xi et al. *Opt. Comm.* 281(7):1841–1849, 2008. Used with permission.)

increased linearly almost over the whole studied range. It was enhanced by a factor of six as the laser bandwidth was tuned from 10 to 80 nm. In contrast, the laser pulses compensated only for SOD demonstrated a modest effect on the signal gain. Once the bandwidth reached 30 nm, no further increase in TPEF was observed. Therefore, we conclude that above 30-nm bandwidth, higher order dispersion, introduced by the prisms and the microscope objective, becomes dominant in the degradation of signal intensity in TPM.

IV. COMPENSATED PULSES RESULTS IN HIGHER SIGNAL IN TWO-PHOTON IMAGING OF BIOLOGICAL SAMPLES

The results shown above are in agreement with theory. However, it is important to demonstrate that they are applicable toward actual TPM of biological samples. We show here the effect of dispersion compensation on fixed biological samples and in living cells. TPM images on a fixed sample of mouse intestine (Molecular Probes, F-24631), given in Figure 8.3, were obtained with GDD-only compensation and after full MIIPS compensation for high-order dispersion. The average increase in the TPEF signal between these two conditions was ~5-fold. Figures obtained for GDD-only compensation are shown as obtained and also after a digital 6× increase in intensity. The increased two-photon efficiency achieved by using MIIPS results in an improvement of signal-to-noise ratio and contrast.

The next step was to test the difference between SOD compensation and compensation for high-order dispersion using MIIPS on living cells. Living U2OS cells stained with nuclear (blue) or mitochondrial (green) stain were imaged using GDD-only and MIIPS compensated pulses, as shown in Figure 8.4. The average increase in

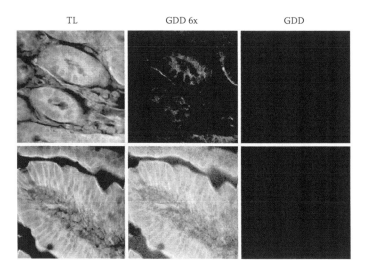

FIGURE 8.3 Greater signal with MIIPS compensated pulses (TL) compared to GDD compensated pulses on mouse intestine tissue. Cross-section of fixed intestine, stained with Alexa Fluor 350 wheat germ agglutinin (goblet cells), phalloidin Alexa Fluor 568 phalloidin (actin), and SYTOX Green nucleic acid stain (nuclei). The image is of a surface area expansion of villus showing epithelial cells that cover the villi (top) and microvilli (bottom) in the small intestine. The image obtained using TL pulses had at least five times greater signal intensity than that taken with GDD compensated pulses. The intensity with GDD compensated pulses was amplified 6× for viewing image detail. Image size is 75 μm (top) and 100 μm (bottom). The objective used is Zeiss 40×LD, 1.1 NA.

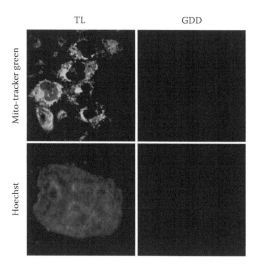

FIGURE 8.4 Greater signal with MIIPS compensated pulses compared to GDD compensated pulses is demonstrated in living U2OS cells. The cells were stained with Mito-Tracker 488 (mitochondria-green) and Hoechst 33258 (nuclei-blue). The image obtained using TL pulses had intensity of at least five times greater than that taken with GDD-only compensation. Image size is 100 μm (top) and 50 μm (bottom). The objective used is Zeiss 40×LD, 1.1 NA.

signal was ~5-fold, similar to the results shown in Figure 8.3. The observed increase of the excitation efficiency can be used to attenuate the incident laser power while maintaining satisfactory signal level. This attenuation would then reduce the exposure of the cells to the deleterious effects of the laser. We explore this effect later in the chapter.

V. TWO-PHOTON DEPTH IMAGING

One of the most important advantages of TPM over other microscopy modalities is the ability to image deep within biological tissue. Among the most impressive results we highlight the article by Denk and coworkers where they imaged neurons at a depth exceeding 1000 μm (Theer et al. 2003). Based on the increases in signal that we have registered through the use of dispersion compensated ultrashort pulses, we evaluated the effects of the increased signal on depth imaging. The motivation for this exploration is that deep tissue imaging will be essential to unravel biologically complex phenomena such as cancer. For example, the initial escape of metastatic cells from tumors has eluded studies in vivo, partly because of the relative inaccessibility of this process to direct observation. Recently, multiphoton microscopy has been applied to intravital-imaging studies (Peti-Peterdi et al. 2002; Sandoval et al. 2004; Molitoris and Sandoval 2005; Konig et al. 2007) and has been especially useful for characterizing primary-tumor properties, growth rates, and mechanisms of metastasis to target organs. Animal models of cancer that use the stable expression of green fluorescent protein (GFP) from tissue-specific promoters now make it possible to directly observe cell behavior in primary tumors of live animals, without worries that the labeling agent is not evenly taken up by the cells.

It is important to realize that dispersion compensation can eliminate the high-order phase distortions (in the spectral domain) introduced by the objective lens, as discussed above, but it cannot eliminate the scattering (in the spatial domain) that occurs in depth imaging. Here we explore the use of laser pulses that are dispersion compensated only before the medium. In principle, it is possible to compensate for dispersion at greater depths, but if the dispersion of tissues is similar to that of pure water, it should be insignificant. Finally, we could titrate the amount of laser power used, increasing the intensity as the focal plane moves deeper into the tissue.

To study the implications of spectral phase correction for two-photon depth-resolved imaging, we imaged a thick section of mouse kidney tissue, stained with DAPI (cell nuclei), Mitotracker-488 (mitochondria), and Phalloidin-568 (actin). The results, summarized in Figure 8.5, demonstrate increased penetration depth when higher-order dispersion is compensated via MIIPS, compared to GDD-only compensation. The images show the collagen wall components of a blood vessel in the mouse kidney at a depth of 40 μm. The collection duct region above it, at a depth of 50 μm, is seen only when MIIPS is applied.

The results from the depth-resolved imaging strongly support our hypothesis that dispersion compensation, especially when using pulses shorter than 20 fs, can

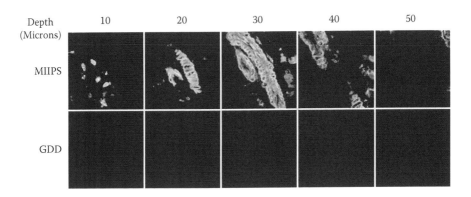

FIGURE 8.5 Effect of dispersion compensation in two-photon depth imaging. Image acquisition of thick mouse kidney sample collected in a series of z-sections at different depths, which demonstrates that MIIPS compensated pulses attain better signal, and deeper penetration than those with GDD-only compensation. The image size is about 100 µm. The objective used is Nikon 60 × 1.45 NA, with working distance of 130 µm. (From Xi et al. *Opt. Comm.* 281(7):1841–1849, 2008. Used with permission.)

make a very significant difference. Five-fold increase in the signal magnitude is typical, and some locations exhibit signal enhancement by as much as an order of magnitude.

VI. EFFECT OF DISPERSION COMPENSATION ON PHOTOBLEACHING

The results shown in Figures 8.3–8.5 indicate that the use of ultrashort, dispersion-compensated pulses can lead to essential improvements in TPM. However, one still needs to determine if those improvements are accompanied by a concomitant increase in photobleaching and other photo-induced damaging processes. Assuming that laser-induced effects are proportional to the number of molecules that are excited by the laser pulse, greater signal would necessarily imply greater damage. Greater damage could also be caused by three-photon processes, which are expected to be more probable when using ultrashort TL pulses. Conversely, greater efficiency of two-photon excitation implies that a lower photon flux can be used, and this may reduce the damage. Minimizing photobleaching and photodamage is of great importance when imaging fixed samples, because each time the sample is imaged, less signal is obtained. More importantly, when imaging living cells or imaging in vivo tissues, photodamage can alter the biology that is being observed. For these reasons we proceeded to evaluate photobleaching induced by SOD-compensated and by fully dispersion-compensated pulses using MIIPS.

For these experiments we picked a standard system that is easy to reproduce in other laboratories. We decided not to use a fluorescent dye solution because in

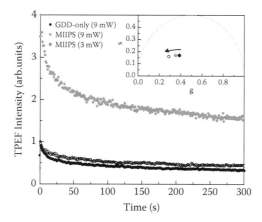

FIGURE 8.6 Effect of dispersion and pulse energy on photobleaching. Photobleaching curves for GDD-only (9 mW) and MIIPS compensated (9 mW and 3 mW) pulses. Inset: Phasor plot for the three considered cases. The modulation frequency, which is a free parameter in these measurements [for details, refer to (Digman et al. 2008)], is taken to be 0.03 rad/s. The counterclockwise shift of the phasor end point indicates the decrease of the decay rate.

microscopy, fluorescent tags are typically fixed. We also opted out a stained tissue as a sample since the experiment would be very difficult to reproduce by others. Based on these guidelines we decided to work with green fluorescent standard slides provided by Chroma Technologies. These slides are readily available to any microscopy laboratory and have the fluorophores evenly distributed within the sample. The only disadvantage is that the polymer matrix is different from a biological specimen; therefore, our experiments will need to be validated in fixed tissue and living cell samples.

For the first set of experiments we used a laser intensity of 9 mW for both GDD-compensated and TL pulses. The resulting fluorescence signal profiles for the two cases are shown in Figure 8.6 (black and grey solid circles, respectively). One can see that the use of TL pulses gives much higher magnitude of the TPEF signal, but the change in photobleaching rates is subtle. We obtain a somewhat more pronounced difference (Figure 8.6, empty circles) after attenuation of the TL pulse energy by a factor of three so that the TPEF signal magnitude is about the same as for GDD-compensated pulses. In our earlier publication (Xi et al. 2008) all three curves were fitted with a sum of four exponentials to retrieve the change in the relaxation rates. The inset in Figure 8.6 is our attempt to use the phasor analysis for data processing (Digman et al. 2008).

The phasor method associates the decay dynamics with a vector in a so-called phasor space. In particular, purely exponential decay corresponds to a phasor with its end point on a semicircle of radius 1/2 and centered at (1/2, 0). Tuning of the decay time from zero to infinity results in a counterclockwise displacement of the end point from (1,0) to (0,0) along the semicircle. Multi-exponential decay is equivalent to a point inside the semicircle, but its dependence on the weight-averaged decay rate remains the same. Namely, a counterclockwise shift around (1/2, 0) indicates an increase of the decay time (for details, see Digman et al. 2008). As one can see, the superposition of MIIPS compensation and lower pulse energies reduce the photobleaching rate. Let us note here though that the exposure time required for TPM imaging is much smaller than the measured photobleaching time. The laser-induced damage becomes an issue only for extensive real-time imaging experiments, when the sample is exposed to the laser radiation multiple times.

From the data shown in Figure 8.6 we also learn two important aspects of photobleaching. First, there is a component of the process that depends on a higher nonlinear process, which we assume here to be instantaneous three-photon excitation. For high intensities, when dispersion is fully compensated, third-order processes become comparable to two-photon excitation and greater photobleaching is observed. Once the laser intensity is attenuated, three-photon processes are no longer as probable. Under these conditions the photobleaching process is not dependent on the absorption due to the aforementioned instantaneous three-photon transitions, but rather on two-photon excitation followed by excitation with an additional photon arriving 10–100 fs later. This sequential excitation process is possible because the excited state of the chromophore is long lived compared with these delays. The greater efficiency of TL pulses allows the use of lower laser intensities and consequently reduces the damage the additional photons cause. For pulses compensated only for SOD, there are sub-pulses, as shown in Figure 8.1, that can cause photodamage.

VII. SELECTIVE TWO-PHOTON EXCITATION

As mentioned earlier, one of the great advantages of TPM was that a single laser could be used to excite a number of fluorescent molecules. However, this advantage is sometimes not desired. For example, there are several cases in which one would like to selectively excite one type and not all types of fluorescent molecules in the focal plane. One such situation would be the selective excitation of a particular fluorescent marker while preventing two-photon excitation of endogenous fluorescent molecules, which are abundant in living cells. Other situations could involve the determination of co-localization of different molecular species, and yet another application could be the selective activation of therapeutic photon reagents while preventing damage to healthy tissue. Some of these problems are already being addressed by using newly available tunable femtosecond laser sources. Here we discuss an alternative approach. Using ultrashort pulses implies that the laser pulse bandwidth is ultra-broad. By phase or amplitude manipulation of the pulse spectrum we are able to achieve the equivalent function of tuning the laser pulse without actually changing any of the laser source parameters, thus preventing a change in optical alignment.

The simplest approach to implement selective two-photon excitation is to permit excitation only from a portion of the spectrum that is necessary for a specific fluorophore while blocking the portions that might cause excitation of different ones. For example, one could block longer or shorter wavelengths. To do this experimentally, one uses a pulse shaper, which is similar to a spectrometer. At the Fourier plane, the position where all the frequencies are spread and focused, one can alter the phase or amplitude of some frequencies without affecting others. The pulses are then reflected out of the device, having acquired new spectral phase and amplitude properties. When the desired goal is to restrict excitation to shorter or longer wavelengths the undesired frequencies are simply blocked. However, if one desires highly selective two-photon excitation at one or more specific wavelengths, amplitude modulation leads to a significant loss of photons (Comstock et al. 2004). In other words, it is still effective but very inefficient.

Our first attempts at selective two-photon excitation made use of ~10 fs pulses modulated by a single sinusoidal phase function that stretched across the spectrum of the pulse (Pastirk et al. 2003; Dela Cruz et al. 2004b). These early attempts were far from optimum. Our group has since introduced an approach based on phase modula-tion to achieve highly selective nonlinear optical excitation (Comstock et al. 2004; Lozovoy et al. 2005, 2006). This method introduces binary phase functions (using only 0 and π values) to optimize two-photon excitation at the desired frequency while minimizing excitation elsewhere. When using binary phases, the resulting nonlinear fields, as all frequencies are coherently summed, acquire only two values, ± 1, making it relatively easy to maximize excitation at the desired frequency and minimize excitation elsewhere.

Following the successful demonstration of binary phase shaping for selective two-photon excitation in microscopy, we wanted to evaluate whether this type of approach could be used for deep tissue imaging and also for selective photodynamic therapy (PDT). We wondered if the binary phase modulation would withstand the scattering introduced by the biological tissue. For these experiments we used as the imaging target three capillary tubes with a fluorescent pH-sensitive dye, 8-hy-droxypyrene-1,3,6-trisulfonic acid (HPTS), submerged in a cell containing the same dye at the same concentration. The pH in the capillary solution was acidic, and the pH in the surrounding solution was basic. In front of the sample cell, we placed a 1-mm slice of biological tissue (chicken breast muscle). When using TL pulses, it was impossible to discriminate between the two solutions (Figures 8.7A [no tissue] and B [with tissue]). We then designed a binary phase to cause excitation of the acidic solu-tion, and a binary phase to cause excitation of the basic solution. The combination of these two excitation pulses was used to achieve selective functional imaging, shown in Figures 8.7C (no tissue) and D (with tissue). The reason for the remarkable success of this experiment, reported in (Dela Cruz et al. 2004a), is that two-photon imaging relies on ballistic photons. Ballistic photons are those that undergo minimal (or no) scattering as they reach their target. Therefore, ballistic photons preserve the binary phase imposed for selective excitation. This implies that binary phase shaping can be used for deep tissue functional imaging and selective activation of photodynamic therapy agents.

Selective two-photon excitation can provide additional contrast for single-cell imaging. It might be particularly useful for co-localization studies. In Figure 8.8, it is illustrated on HeLa cells. The cells were co-stained with Phalloidin-350 (actin filaments) and with MitoTracker 488 (mitochondria). An amplitude shaping was used

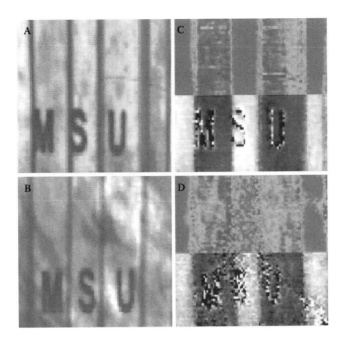

FIGURE 8.7 Selective two-photon imaging of HPTS. The sample consists of three 1-mm capillaries containing an acidic solution of HPTS submerged in a quartz cell filled with an alkaline solution of the same dye. To obtain the images, the sample was raster scanned (without and with scattering tissue) in the focal plane of the beam. (A and B) Two-photon excitation images obtained by raster scanning the sample without and with scattering tissue, respectively. The walls of the capillary tubes, which appear as vertical lines, are about 300 μm thick and are clearly visible in both images. Note that comparable image quality is obtained in the presence and absence of biological tissue. The images shown in (A) and (B) were obtained with TL pulses, which do not allow discriminating between the two different pH solutions. (C) A functional image highlighting the contrast obtained by using coherent control. This image was produced by taking the ratio of the data acquired with two binary phase masks, chosen for selective excitation of different parts of the SHG spectrum. (D) The functional image obtained after a slice of biological tissue was placed in front of the sample. Note that the presence of the tissue reduces the overall signal-to-noise ratio, but the discrimination between acidic and alkaline HPTS is conserved. The contrast in the functional images can be further enhanced by using false color (C Upper and D Upper). Higher values are shown in red, and lower values are shown in blue. (From Dela Cruz, J.M., Pastirk, I., Comstock M. et al. *Proc. Natt. Acad. Sci.* 10(49):16996–17001, 2004. Used with permission.)

to get selective excitation of the two dyes by transmitting only lower (red) or higher (blue) part of the pulse spectrum in frequency domain (Figure 8.9). The actin filaments (blue) were visualized using the blue pulses, while the mitochondria (green) was imaged using red pulses. The image of a single HeLa cell in Figure 8.8 clearly demonstrates typical subcellular localization of each structure.

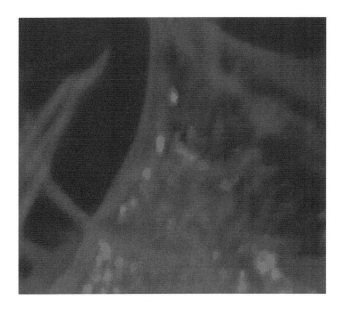

FIGURE 8.8 A section of a single HeLa (human cervical cancer) cell stained with Phalloidin 350 (actin filaments) and MitoTracker 488 (mitochondria) acquired using selective two-photon excitation, without fluorescence color filters. The image size is about 25 μm. (From Dantus, M., Lozovoy, V. V., and Pastirk, I. *Laser Focus World*, 43(5):101–104. 2007. Used with permission.)

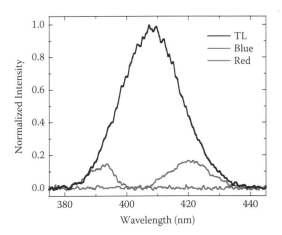

FIGURE 8.9 SHG spectra of the pulses used for selective excitation. (Adapted from [Xi et al. 2008]. Used with permission.)

VIII. CONCLUSION

We have identified high-order dispersion as the main reason why ultrashort, ~10 fs, pulses have rarely been used for nonlinear optical imaging. We discussed the MIIPS method for automated measurement and elimination of high-order dispersion. We provided quantitative analysis for the advantage of high-order dispersion as compared with correction limited to SOD. This enhancement was confirmed experimentally in fixed and living cells, as well as in depth imaging. Finally, we demonstrated that the broad bandwidth of ultrashort pulses can be used for selective two-photon excitation when appropriate phase or amplitude modulation is used.

In summary, nonlinear optical imaging methods, such as TPM, allow for an order of improvement in the signal simply by utilization of 10 times shorter pulses. The enhancement is even more dramatic for higher-order nonlinearities, such as third harmonic generation where two orders of magnitude increase is expected. These improvements, however, rely on successful measurement and elimination of phase distortions.

ACKNOWLEDGMENTS

We gratefully acknowledge funding for this research from the National Science Foundation, Major Research Instrumentation Grant CHE-0421047, and single investigator Grant CHE-0500661. We thank Dr. Vadim Lozovoy for stimulating discussions and help with the phasor analysis, and Dr. James H. Resau for technical assistance in the preparation of the mouse kidney sample. This work also benefited from the participation of Daniel Schlam through an undergraduate research opportunity. Some of the data was originally acquired by Lindsay R. Weisel, Dr. Peng Xi, and Dr. J. M. Dela Cruz. We are also grateful for the help from Nelson S. Winkler with the final preparation of the manuscript.

REFERENCES

Comstock, M., Lozovoy, V. V., Pastirk, I., and Dantus, M. 2004. Multiphoton intrapulse interference 6: Binary phase shaping. *Opt. Exp.* 12(6):1061–66.

Dantus, M., Lozovoy, V. V., and Pastirk, I. 2007. MIIPS characterizes and corrects femtosecond pulses. *Laser Focus World* 43(5):101–4.

Dela Cruz, J. M., Pastirk, I., Comstock, M., Lozovoy, V. V., and Dantus, M. 2004a. Use of coherent control methods through scattering biological tissue to achieve functional imaging. *Proc. Natl. Acad. Sci.* 10(49):16996–17001.

Dela Cruz, J. M., Pastirk, I., Lozovoy, V. V., Walowicz, K. A., and Dantus, M. 2004b. Multiphoton intrapulse interference 3: Probing microscopic chemical environments. *J. Phys. Chem. A* 108(1):53–58.

Denk, W., J. H. Strickler, and W. W. Webb. 1990. 2-Photon laser scanning fluorescence microscopy. *Science* 248(4951):73–76.

Digman, M. A., Caiolfa, V. R., Zamai, M., and Gratton, E. 2008. The phasor approach to fluorescence lifetime imaging analysis. *Biophys. J.* 94(2):L14–L16.

Dyba, M., and Hell, S. W. 2002. Focal spots of size lambda/23 open up far-field florescence microscopy at 33 nm axial resolution. *Phys. Rev. Lett.* 88(16):163901.

Konig, K., Ehlers, A., Riemann, I., Schenkl, S., Buckle, R., and Kaatz, M. 2007. Clinical two-photon microendoscopy. *Microsc. Res. Tech.* 70(5):398–402.

Lozovoy, V. V., Pastirk, I., and Dantus, M. 2004. Multiphoton intrapulse interference. IV. Ultrashort laser pulse spectral phase characterization and compensation. *Opt. Lett.* 29(7):775–77.

Lozovoy, V. V., Shane, J. C., Xu, B. W., and Dantus, M. 2005. Spectral phase optimization of femtosecond laser pulses for narrow-band, low-background nonlinear spectroscopy. *Opt. Exp.* 13(26):10882–87.

Lozovoy, V. V., Xu, B., Coello, Y., and Dantus, M. 2008. Direct measurement of spectral phase for ultrashort laser pulses. *Opt. Exp.* 16(2):592–97.

Lozovoy, V. V., Xu, B. W., Shane, J. C., and Dantus, M. 2006. Selective nonlinear optical excitation with pulses shaped by pseudorandom Galois fields. *Phys. Rev. A* 74(4):041805.

Molitoris, B. A., and Sandoval, R. M. 2005. Intravital multiphoton microscopy of dynamic renal processes. *Am. J. Physiol.-Renal Physiol.* 288(6):F1084–89.

Pastirk, I., Dela Cruz, J. M., Walowicz, K. A., Lozovoy, V. V., and Dantus, M. 2003. Selective two-photon microscopy with shaped femtosecond pulses. *Opt. Exp.* 11(14):1695–1701.

Peti-Peterdi, J., Morishima, S., Bell, P. D., and Okada, Y. 2002. Two-photon excitation fluorescence imaging of the living juxtaglomerular apparatus. *Am. J. Physiol. Renal Physiol.* 283(1):F197–F201.

Sandoval, R. M., Kennedy, M. D., Low, P. S., and Molitoris, B. A. 2004. Uptake and trafficking of fluorescent conjugates of folic acid in intact kidney determined using intravital two-photon microscopy. *Am. J. Physiol.-Cell Physiol.* 287(2):C517–26.

Spence, D. E., Kean, P. N., and Sibbett, W. 1991. 60-Fsec pulse generation from a self-mode-locked Ti-sapphire laser. *Opt. Lett.* 16(1):42–44.

Theer, P., Hasan, M. T, and Denk, W. 2003. Two-photon imaging to a depth of 1000 μm in living brains by use of a Ti:Al$_2$O$_3$ regenerative amplifier. *Opt. Lett.* 28(12):1022–24.

Westphal, V., and Hell, S. W. 2005. Nanoscale resolution in the focal plane of an optical microscope. *Phys. Rev. Lett.* 94(14):143903.

Xi, P., Andegeko, Y., Weisel, L. R., Lozovoy, V. V., and Dantus, M. 2008. Greater signal, increased depth, and less photobleaching in two-photon microscopy with 10 fs pulses. *Opt. Comm.* 281(7):1841–49.

Xu, B. W., Gunn, J. M., Dela Cruz, J. M., Lozovoy, V. V., and Dantus, M. 2006. Quantitative investigation of the multiphoton intrapulse interference phase scan method for simultaneous phase measurement and compensation of femtosecond laser pulses. *J. Opt. Soc. Am. B-Opt. Phys.* 23(4):750–59.

9 Imaging with Phase-Sensitive Narrowband Nonlinear Microscopy

Eric Olaf Potma and Vishnu Vardhan Krishnamachari

CONTENTS

I. THE IMPORTANCE OF PHASE

The signals generated in second harmonic generation (SHG) microscopy, third harmonic generation (THG) microscopy, and coherent anti-Stokes Raman scattering (CARS) microscopy are coherent. This implies that the oscillators in the focal volume, set in motion by the driving fields, radiate with a well-defined phase. In other words, there is a clear phase relationship between the waves that are emitted from the focal region. Because of this relationship, the resultant waves will interfere. The detected signal is thus highly dependent on the interference among waves emanating from different parts of the focal volume. Strong signals are expected if all the contributions are in phase, while much weaker signals are seen when the phases of the radiating contributions are not fully aligned.

Unlike in bulk nonlinear spectroscopy experiments, the signal in nonlinear microscopy is generated within a volume that is on the order of an optical wavelength. The axial extent of this volume is often referred to as the interaction length, which denotes the length within which the incident fields interact to produce a nonlinear polarization in the material. Such microscopic interaction lengths yield signal interference profiles that can differ markedly from those observed in macroscopic spectroscopy.

Phase is a key concept in nonlinear coherent microscopy. In a given detection direction, the occurrence of phase-shifts between waves leads to incomplete constructive interference, and thus to lower signals. Minimization of phase mismatch has, therefore, become synonymous with optimizing nonlinear coherent signals. In nonlinear microscopy, where a collinear excitation geometry is often used, a minimum phase mismatch is found along the propagation direction of the incoming beams. The strongest signals are indeed observed in the forward propagation direction and, vice versa, maximum phase mismatch is obtained in the backward propagation direction, where signals are typically weak in the absence of scattering.

In macroscopic nonlinear spectroscopy, dispersive effects are another important source of phase mismatch. Special beam geometries are required to maintain a minimum phase difference between the incident beams and the signal waves (Eckbreth 1978). In microscopy, however, the short interaction lengths render the dispersive phase-mismatch negligible (Cheng et al. 2002; Potma et al. 2000; Zumbusch et al. 1999). Even in THG, where the refractive index difference between the driving fields and the signal can be substantial, the dispersive phase mismatch accumulated in the focus of a high numerical objective is of little significance (Cheng and Xie 2002; Débarre et al. 2007).

While the dispersive phase mismatch is practically absent, the short interaction length produced by the high numerical aperture lens carries its own source of phase mismatch: the Gouy phase shift. The Gouy phase shift occurs in the focus of any lens and accounts for the π-phase inversion of the fields upon focusing (Feng and Winful 2001; Gouy 1890). In the tight focusing limit, however, the Gouy phase shift evolves over a distance comparable to an optical wavelength. Within the interaction length, the driving fields can now set up an effective Gouy phase shift that is much larger than π (Bjorklund 1975; Cheng et al. 2002; Cheng and Xie 2002; Kleinman et al. 1966; Moreaux et al. 2001). Under such conditions, the Gouy phase mismatch between the signal field and the incident waves can be significant, suppressing radiation in the phase-mismatched directions. The Gouy phase mismatch is responsible for the conical-shaped radiation profiles seen in THG microscopy (Cheng and Xie 2002; Débarre et al. 2007) and, for certain sample geometries, in SHG microscopy (Moreaux et al. 2000, 2001; Williams et al. 2005; Yew and Sheppard 2006).

The discussion above underlines the importance of phase effects in nonlinear coherent microscopy. Understanding the phase properties of the coherent signal generation process not only helps us appreciate the complexity of the imaging mechanism in nonlinear microscopes, but also offers opportunities to manipulate the phase of the beams in such a way that extra information about the sample can be obtained. Such form of interference control can, for instance, be realized by mixing an external beam coherently with the signal, opening the door to a true sampling of the electric field amplitude rather than the integrated intensity of the signal. In this form of interferometry, the sought-after information is engendered by the ability to control the temporal phase difference of the interfering waves. This approach has been applied before to bulk spectroscopic measurements based on the SHG, THG, and CARS techniques (Lüpke et al. 1989; Marowsky and Lüpke 1990; Yacoby et al. 1980). In microscopy, CARS interferometry has been applied to reject background contributions (Andresen et al. 2006; Potma et al. 2006), and SHG interferometry has been used for the purpose of tomographic imaging (Jiang et al. 2004; Vinegoni et al. 2004).

Another form of nonlinear interferometry is based on controlling the relative spatial phase of oscillating dipoles in the focal volume. By manipulating the phase properties of the incoming driving fields, oscillators at different points in the interaction volume can be engineered to oscillate with different phases. Upon emission, the waves from different points in focus may now interfere in a way substantially different from the interference seen in regular emission profiles. Spatial phase shaping, or focus engineering, introduces several ways to control the signal generation process, and offers new contrast mechanisms in nonlinear coherent microscopy. Focus-engineered CARS microscopy is an example of this approach (Krishnamachari and Potma, 2007a,b; Liu and Kim 2007).

In this chapter we explore several aspects of interferometric nonlinear microscopy. Our discussion is limited to methods that employ narrowband laser excitation; i.e., interferences in the spectral domain are beyond the scope of this chapter. Phase-controlled spectral interferometry has been used extensively in broadband CARS microspectroscopy (Cui et al. 2006; Dudovich et al. 2002; Kee et al. 2006; Lim et al. 2005; Marks and Boppart 2004; Oron et al. 2003; Vacano et al. 2006), in addition to several applications in SHG (Tang et al. 2006) and two-photon excited fluorescence microscopy (Ando et al. 2002; Chuntonov et al. 2008; Dudovich et al. 2001; Tang et al. 2006). Here, we focus on interferences in the temporal and spatial domains for the purpose of generating new contrast mechanisms in the nonlinear imaging microscope. Special emphasis is given to the CARS technique, because it is sensitive to the phase response of the sample caused by the presence of spectroscopic resonances.

We first review the essentials of the phase distribution of the electric fields at the focus of a high numerical aperture lens in Section II. After discussing the phase properties of the emitted signal, in Section III we zoom in on how the information carried by the emitted field can be detected with phase-sensitive detection methods. Interferometric CARS imaging is presented as a useful technique for background suppression and signal enhancement. In Section IV, the principles of spatial interferometry in coherent microscopy are laid out and applications are discussed. The influence of phase distortions in turbid samples on phase-sensitive nonlinear microscopy is considered in Section V. Finally, in Section VI, we conclude this chapter with a brief discussion on the utility of phase-sensitive approaches to coherent microscopy.

II. COHERENT SIGNALS FROM TIGHT FOCAL VOLUMES

To understand the physics behind phase-sensitive nonlinear microscopic methods, it is useful to first acquire a grasp of the mechanism of signal formation in nonlinear coherent imaging. The details of the tightly focused driving fields are essential ingredients for this understanding. We will focus on the CARS process and try to decipher how the properties of the focal fields bring about propagating coherent signal fields.

A. The Tightly Focused Field

To account for all possible phase-coherent effects involved in signal generation, in this chapter we use the full vectorial description of the focal fields. For a given spatial distribution of the light field incident at the back aperture of a high numerical aperture objective, the three-dimensional vectorial field distribution at its focus can be calculated based on the theory of angular spectrum representation by Richards and Wolf (1959). We assume that the incoming field at the back aperture plane of the lens has a Hermite-Gaussian (HG) mode. The simplest HG mode is HG00, which corresponds to the well-known Gaussian beam profile (Siegman 1986). Assuming no index mismatch and the incident light is x-polarized, the focal field for the HG00 mode can be written in the form shown by Novotny and Hecht (2006) as

$$E_{00} = E_0 e^{-ikf} \begin{pmatrix} I_{00} + I_{02} \cos 2\phi \\ I_{02} \sin 2\phi \\ -2iI_{01} \cos \phi \end{pmatrix}. \tag{9.1}$$

Here, E_0 is the field amplitude, k is the wavevector, f is the focal length of the lens, and ϕ corresponds to the azimuthal angle. The quantities I_{mn}, with $m = 0,1$ and $n = 0,...4$, are one-dimensional integrals with respect to the polar angle θ, and are written as

$$I_{mn}(\rho,z) = \int_0^{\theta'_{max}} f_w(\theta)\sqrt{\cos\theta}\, g_{mn}(\theta) J_l(k\rho\sin\theta) e^{ikz\cos\theta} \sin\theta\, d\theta, \tag{9.2}$$

where $l = n$ if $n \leq m$, and $l = n - m$ if $n > m$. The acceptance angle of the objective is indicated by θ'_{max}, and $\rho = (x^2 + y^2)^{1/2}$. The function $f_w(\theta)$ is the apodization function (Novotny and Hecht 2006), which describes the degree of the filling of the back aperture of the objective. Its value is one when the back aperture is completely filled and is less than one otherwise. The expressions for the functions $g_{mn}(\theta)$ are given in the appendix at the end of this chapter.

Figures 9.1a and 9.1b show the amplitude and the phase distributions in the focal plane of focused x-polarized HG00 mode of wavelength $\lambda = 800$ nm. The amplitude distribution corresponds to the well-known Airy disk of the focal plane. The successive rings are diffraction rings, which remind us that the tight focus is a direct result of wave interference. The interference nature is even more clear in the phase image of Figure 9.1b, which shows that each consecutive dark ring in the amplitude

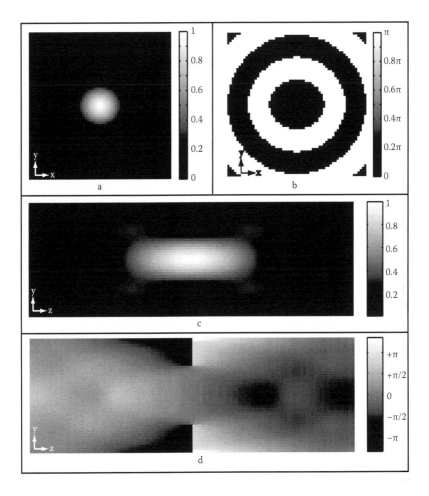

FIGURE 9.1 Amplitude and phase profiles of focal fields. Amplitude (a) and (b) phase in the focal xy-plane. Amplitude (c) and phase (d) in the yz-plane. The phase difference below and above the focal plane is a signature of the Gouy phase shift. The propagation phase was subtracted for clarity. Calculations are for a 1.1 NA water immersion lens and the wavelength is 800 nm. In the panels the lateral axis runs from −1.0 μm to 1.0 μm and the axial axis from −3 μm to 3 μm.

is accompanied by a π-phase jump of the field. In other words, successive rings are actually π out of phase, and the complete Airy pattern is certainly not uniform in terms of phase.

A cross-sectional view (yz) of the amplitude and phase of the focal field along the optical axis is depicted in Figure 9.1c and d, respectively. The phase distribution is plotted in its unwrapped representation and the linear phase due to propagation has been subtracted for clarity. It is evident that the phase before the focal plane and the phase after the focal plane ($z = 0$) are not the same. Along the optical axis, the phase undergoes a full π-phase swing. This phase swing is the Gouy phase shift (Feng and Winful 2001; Gouy 1890). The appearance of the Gouy phase shift is a natural property of any focal field. Without the π-phase swing, fields are unable to propagate through a focus. In the next subsection we will see that the Gouy phase shift difference between the interacting waves dramatically affects the phase-matching properties, i.e., interference, of the signal waves in SHG and THG microscopy techniques, but not in CARS.

B. NONLINEAR EXCITATION AT THE FOCUS

The focal fields set up a polarization in the material. In the case of CARS, we are interested in the polarization resulting from the combined action of the pump (of frequency ω_p) and Stokes (of frequency ω_S) beams, which induce motions in the electron clouds that oscillate at frequency $2\omega_p - \omega_S$, the anti-Stokes frequency. The ability of the material to oscillate at the anti-Stokes frequency when the pump and Stokes fields are present is given by the third-order nonlinear susceptibility $\chi^{(3)}$. The strength of the polarization is furthermore determined by the amplitude of the pump (E_p) and Stokes (E_S) driving fields. In the tensorial notation, where i, j, k, and l denote the polarization components of the nonlinear susceptibility, the third-order polarization in the polarization direction i is given as

$$P_i(r) = \sum_{j,k,l} \chi^{(3)}_{ijkl}(r)E_{pj}(r)E_{pk}(r)E^*_{Sl}(r). \tag{9.3}$$

Since both the incident beams are commonly polarized along the x-direction, the dominant polarization component is also along the same direction with negligible contributions from the off-axis tensor components. However, these neglected terms become important in techniques (such as polarization-sensitive CARS) where the incident polarization of the Stokes and the pump beams are different. Second, comparing the magnitudes of the orthogonal polarization components, we note that the magnitude of the y-component is negligible; the z-polarization component, however, becomes comparable to that of the x-component for higher-order incident beam modes. While this longitudinal component plays a major role in near-field detection methods (Liu et al. 2006), its contribution is of limited importance in case the CARS signal is detected in the far-field.

We will assume that all the incident fields are polarized in the x-direction. The pump and Stokes focal fields, which each look like the complex fields depicted in Figure 9.1, set up an effective CARS excitation field $E_{ex} = E^2_p(r)E^*_S(r)$. The anti-Stokes polarization in the excitation volume induced by the effective excitation field can be considered as a collection of radiating dipoles. The resulting CARS amplitude at any far-field point Q with coordinates $\boldsymbol{R} \equiv (R, \theta, \phi)$ is a sum of the amplitude contributions form all these dipoles and is given by

$$E(\boldsymbol{R}) = -\int_v \frac{e^{ik_{as}|\boldsymbol{R}-\boldsymbol{r}|}}{4\pi|\boldsymbol{R}-\boldsymbol{r}|^3}(\boldsymbol{R}-\boldsymbol{r})\times[(\boldsymbol{R}-\boldsymbol{r})\times \boldsymbol{P}(\boldsymbol{r})]d^3\boldsymbol{r}, \tag{9.4}$$

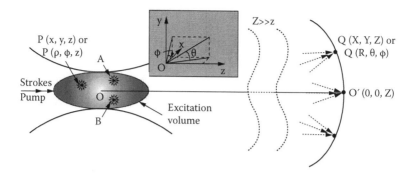

FIGURE 9.2 Sketch of far-field detection of CARS waves generated in the focal volume. The figure shows the yz cross-section of the excitation volume. O is the origin of the coordinate system and also the center of the focal volume and O′ is a far-field point on the optical axis. The coordinates of the points in the near-field are represented by (x,y,z) and those of the points in the far-field by (X,Y,Z) in cartesian coordinate system. The outgoing arrows around the near-field points represent the CARS waves generated in the focal volume. The incoming arrows at the far-field points represent the contributions from the individual points in the focal volume.

where k_{as} is the magnitude of the wave-vector of the CARS field and V is the interaction volume. In Equation (9.4), the integrand corresponds to the electric field at location $\boldsymbol{R}$ in the far field due to the joint action of radiating dipoles located at points r in the focal volume. Figure 9.2 illustrates the geometry of far-field signal collection. The total CARS intensity generated from the focal volume is obtained by integrating the intensities at the far-field plane within the detection angle of the collecting lens.

For a focal volume that is uniform and with HG00 pump and Stokes incident fields, the anti-Stokes radiation at the far-field is reminiscent of an HG00 propagating beam as well. To understand the phase effects in the interaction volume on the resulting far-field intensity, much insight can be obtained by comparing the phase properties of E_{ex} with those of an HG00 focal field at the anti-Stokes frequency. In Figures 9.3a and b, the CARS excitation field is compared to a propagating anti-Stokes field in the vicinity of the focal field. Both fields display the characteristic Gouy phase swing. The phase difference between these fields is plotted in Figure 9.3c. It is seen that the phase difference along the optical axis is very small. This implies that the excitation field is capable of driving dipoles at the anti-Stokes frequency in a way that facilitates radiation toward the far-field; i.e., the oscillators are in phase in the direction of propagation. Hence, in CARS the oscillating dipoles constructively interfere in the far-field because the effective phase swing of E_{ex} mimics the Gouy phase shift of a propagating anti-Stokes field.

In the next sections, we shall see that by manipulating the spatial as well as temporal phase of the third-order polarization at the anti-Stokes frequency the properties of the far-field CARS signal can be favorably influenced. We will first discuss phase manipulations in terms of temporal interference in Section III and then zoom into spatial phase manipulations in Section IV.

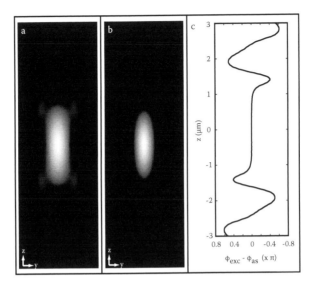

FIGURE 9.3 Comparison of a focal field at the anti-Stokes frequency (a) with the CARS excitation field (b). In (c) the phase difference between (a) and (b) along the optical axis is given. Note the limited phase difference within the interaction length. Calculations are for a 1.1 NA water immersion lens. The pump wavelength is 800 nm, the Stokes wavelength is 1064 nm. In the panels the lateral axis runs from −1.0 μm to 1.0 μm.

III. INTERFEROMETRIC NONLINEAR MICROSCOPY

A. Why Interferometry Is Useful

An interferometer is a useful device. In interferometry, two fields are mixed and allowed to coherently interfere. In its simplest form, an interferometer combines two phase-coherent fields by means of a beam splitter. If E_1 and E_2 are the two fields combined at the beam splitter with reflectivity r, and ϕ is the relative phase between the fields, the intensity in the reflected direction of E_1 after the beam splitter is

$$I = r\left|E_1\right|^2 + (1-r)\left|E_2\right|^2 + 2\sqrt{r(1-r)}\left|E_1 E_2\right|\cos\phi. \tag{9.5}$$

At the beam splitter, the phase difference between the fields dictates which way the energy will flow. In the case of $E_1 = E_2$ and $r = 0.5$, all the energy is found in the reflected channel and none in the transmission channel when $\phi = 0$, and vice versa, when $\phi = \pi$ the total energy is now directed in the transmission direction. Hence, at the beam splitter, a redistribution of the total energy takes place contingent on the phase difference between the two fields.

In nonlinear interferometry, a similar scheme can be used. What makes interferometry attractive to nonlinear coherent microscopy is the last term of Equation (9.5). First, this term is not sensitive to the intensity of each of the fields, but to the product of the actual field amplitudes. Moreover, direct information on the phase difference between

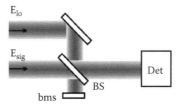

FIGURE 9.4 Schematic of interferometric combination of a local oscillator field with the signal field by means of a beam splitter (BS) with reflectivity r. A beamstop is indicated by bms.

the two fields can be obtained. Hence, when the amplitude and the phase of one of the fields, say E_1, are well known, then a measurement may reveal the amplitude and the phase of the unknown field E_2. The known field is commonly called the local oscillator field, or E_{lo}, and the unknown field is typically the signal field E_{sig} of interest. In a typical nonlinear interferometric measurement, the signal field is mixed with a pre-generated and well-characterized local oscillator field, as sketched in Figure 9.4 (Bredfeldt et al. 2005; Evans et al. 2004; Vinegoni et al. 2004). The advantage of interferometric detection of the signal is that phase information about the electric field, information that is lost when only signal intensity is detected, can be obtained. Detailed knowledge of the phase of the electric field is particularly relevant to CARS microscopy, where phase shifts may occur as a consequence of spectroscopic resonances.

A second advantage of interferometric detection in nonlinear microscopy is that information carried by weak signal fields can be amplified with a stronger local oscillator field. It is evident from the last term of Equation (9.5) that a stronger E_{lo} will increase the overall contribution of this mixing term on the detector. Field amplification of a signal field offers the opportunity to lift weak signals to detectable levels. Analysis shows that the interferometric detection scheme can increase detection sensitivity by three orders of magnitude by raising otherwise buried signals above the detection noise (Eesley et al. 1978; Jurna et al. 2007). An example of field amplification is shown for CARS interferometric imaging in Figure 9.5a, but this principle applies equally to SHG and THG microscopy.

A third advantage of nonlinear interferometry is that the mixing term scales linearly with the sample concentration through E_{sig}. In the case of CARS microscopy, the anti-Stokes field can be expressed as

$$E_{sig} \propto E_{ex} \chi^{(3)}. \tag{9.6}$$

The nonlinear susceptibility is linearly proportional to the number of dipole oscillators (N), and thus scales linearly with the concentration of Raman scatterers in the sample. A linear concentration dependence is attractive for imaging applications, as the image intensities are now a direct reflection of the number of target compounds in focus. In regular CARS imaging, the signal depends quadratically on the number of scatterers, which often complicates analysis. Similar arguments apply to SHG and THG signals. The retrieved linear concentration dependence of the CARS signal is demonstrated in Figure 9.5b.

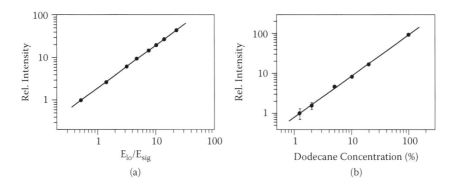

FIGURE 9.5 (a) Field amplification with a local oscillator field. A linear dependence of the CARS interferometric mixing contribution is found with increasing field strength of the local oscillator. CARS signals were measured from a dodecane solution at 2845 cm^{-1} using a 1.15 NA water immersion lens. (b) Concentration dependence of interferometric CARS contribution in a dodecane:deuterated-dodecane solution, measured at 2845 cm^{-1}. A linear concentration dependence is found.

B. NONLINEAR INTERFEROMETRY AND SPECTRAL RESONANCES

All the nonlinear coherent imaging techniques discussed thus far are typically considered to be electronically off-resonant techniques. This implies that the signals can be generated without exciting molecular electronic eigenstates. In other words, SHG, THG, and CARS contrast is not commonly used to probe electronic spectroscopic contrast in microscopy. On the other hand, CARS is the only method of the pack that is sensitive to vibrational resonances. This spectroscopic sensitivity has direct implications for the phase properties of the resulting anti-Stokes signal. In this section, we discuss what kind of spectral phase information is carried by the CARS signal and how nonlinear interferometry can be employed to detect this information.

The components of the third-order nonlinear susceptibility relevant to the CARS process are conveniently subdivided into two terms: vibrationally resonant ($\chi_r^{(3)}$) and vibrationally nonresonant ($\chi_{nr}^{(3)}$) components. The total response of the material depends on the sum of these two terms:

$$\chi_{CARS}^{(3)} = \chi_{nr}^{(3)} + \chi_r^{(3)}. \tag{9.7}$$

The nonresonant contributions pertain to electron cloud oscillations that oscillate at the anti-Stokes frequency but do not couple to the nuclear eigenfrequencies. These oscillatory motions follow the driving fields without retardation at all frequencies. The material response can, therefore, be described by a susceptibility that is purely real and does not depend on the frequencies of the driving fields. The resonant contributions, on the other hand, are induced by electron cloud oscillations that are enhanced by the presence of Raman active nuclear modes. The presence of nuclear oscillatory motion introduces retardation effects relative to the driving fields; i.e., there is phase shift between the driving fields and the material oscillatory response.

This phase shift is akin to the phase shift experienced by a damped harmonic oscillator driven in the vicinity of the oscillator's eigenfrequency ω_v. Hence, the presence of a vibrational resonance not only changes the amplitude of the signal field, but also its phase. To incorporate this effect, the resonant nonlinear susceptibility is no longer real as it contains imaginary contributions:

$$\chi_r^{(3)} = \text{Re}\{\chi_r^{(3)}\} + i\,\text{Im}\{\chi_r^{(3)}\}. \tag{9.8}$$

A formal expression for the resonant nonlinear susceptibility can be obtained by describing the light-matter interactions in a density matrix formalism (Boyd 2003; Mukamel 1995), which is beyond the scope of this chapter. A third-order perturbative expansion of the system's density matrix yields the following form for the nonlinear susceptibility:

$$\chi_r^{(3)} = \frac{N}{V}\sum_R \frac{A_R}{\omega_R - (\omega_p - \omega_s) - i\Gamma_R}, \tag{9.9}$$

where A_R is the amplitude associated with eigenfrequency ω_R, Γ_R is the corresponding vibrational decay rate of the mode, and V denotes the unit volume. Equation (9) predicts Lorentzian lineshapes for the Raman resonances.

The amplitude and phase of $\chi_r^{(3)}$ are plotted in Figure 9.6a, whereas in Figure 9.6b the same function is depicted in terms of real and imaginary parts. It is clear from Figure 9.6a that the phase of the material's resonant oscillatory response undergoes a π phase shift relative to the nonresonant response in the vicinity of the spectral resonance. This is a direct manifestation of the retardation observed when driving the oscillators near their Raman resonances. In nonlinear interferometry, the $\chi_r^{(3)}$ and $\chi_{nr}^{(3)}$ contributions can be discriminated from each other based on this spectral phase shift.

When the CARS field is mixed with an external local oscillator anti-Stokes field, the detected signal can be written from Equations (9.5)–(9.8) as

$$S \propto r\left|E_{lo}\right|^2 + (1-r)\left|E_{sig}\right|^2 + 2\sqrt{r(1-r)}\left|E_{lo}E_{ex}\right|\left(\left[\chi_{nr}^{(3)} + \text{Re}\{\chi_r^{(3)}\}\right]\cos\phi + \text{Im}\{\chi_r^{(3)}\}\sin\phi\right).$$

$$\tag{9.10}$$

It is seen that the mixing term now contains two components with different ϕ dependences. When $\phi = 0$ the first component, which consists of the real part of $\chi_r^{(3)}$, is maximized, while the second contribution is zero. When $\phi = \pi$ the second term, which depends solely on the imaginary part of $\chi_r^{(3)}$, is maximized. For this setting of the phase, the mixing term yields only resonant contributions to the signal. This is a very attractive feature, because with the aid of lock-in techniques the $|E_{lo}|^2$ and $|E_{sig}|^2$ contributions can be electronically filtered out such that the resulting signal scales with $\text{Im}\{\chi_r^{(3)}\}$. The CARS signal is now free of nonresonant background contributions and directly proportional to the Raman differential cross section (Hellwarth 1977; Mukamel 1995). In principle, with the aid of interferometric detection, the CARS

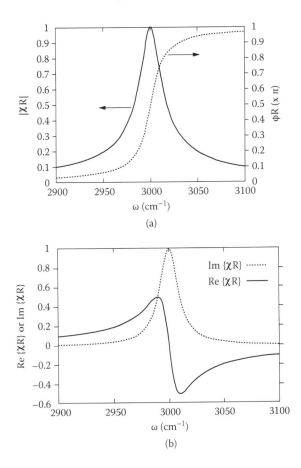

FIGURE 9.6 (a) Amplitude and phase of a vibrational transition modeled with $\omega_R = 3000$ cm^{-1} and spectral width (half-width at a half max) of 10 cm^{-1}. (b) Same transition represented in terms of real and imaginary part.

microscope can produce the same contrast as a Raman microscope, with the added advantage that the CARS signal levels are up to five orders of magnitude higher than the levels observed in spontaneous Raman scattering.

The advantages offered by interferometric detection are compromised by practical limitations related to phase noise and loss of phase quality due to scattering. We will briefly discuss these issues in Section V.

C. INTERFERENCE IN FOCUS

In the previous subsection the advantages of interferometric CARS detection were summed up. However, to benefit from these advantages, a rather complicated setup has to be constructed. In particular, mixing the CARS field generated in the microscope with the local oscillator field may pose some problems. In laser scanning microscopes, the angular direction of the CARS field changes upon raster-scanning

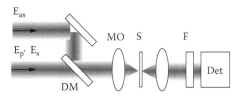

FIGURE 9.7 Schematic of an in-line interferometer. The anti-Stokes local oscillator field is collinearly overlapped with the pump and Stokes beams on a dichroic mirror (DM). All fields are focused by a microscope objective (MO) into the sample (S), and the total signal at the anti-Stokes frequency is detected through a spectral bandpass filter (F) at the photodetector.

the sample, which makes interferometric mixing with an external field a daunting task. In addition, the phase stability between the two fields will be compromised as the beams travel through different parts of the optical setup.

Such problems can be circumvented if the local oscillator co-travels with the pump and Stokes excitation beams into the microscope, as depicted in Figure 9.7. In this configuration, the local oscillator is generated either in-line (Andresen et al. 2006; Lee et al. 2007; Yacoby et al. 1980) or in a compact interferometer (Potma et al. 2006), which minimizes phase fluctuations. An active stabilization scheme is relatively easily incorporated into the interferometer (Krishnamachari et al. 2006). Because the local oscillator propagates through the same optics as the pump and Stokes beams, no temporal phase differences are accumulated beyond the interferometer.

Figure 9.8 shows the CARS signal of a bulk sample in the field of view of a laser-scanning microscope when a local oscillator field is co-propagated with the pump and Stokes beams. The apparent interference pattern is the result of the relative phase shift between the CARS polarization and the local oscillator field as a function of scan angle (Lee et al. 2007). Such a pattern is typical of a laser-scanning microscope and is an indication of the different dispersive phase shifts experienced by the beams when collinearly traveling through the imaging system at different angles. When the beams are kept fixed and the sample is scanned instead, the phase shift between the participating beams would remain constant and no interference pattern would be observed.

The interference process in this collinear approach is, however, different from the interference realized by mixing the local oscillator and the CARS field on a beam splitter. Interference takes place in the sample, which, in the presence of multiple frequencies, mediates the transfer of energy between the beams that participate in the nonlinear process. The local oscillator mixes with the anti-Stokes polarization in the focal volume, and is thus coherently coupled with the pump and Stokes beams in the sample through the third-order polarization of the material. In other words, the material's polarization, and its ability to radiate, is directly controlled in this collinear interferometric scheme. Under these conditions, energy from the local oscillator may flow to the pump and Stokes fields, and vice versa. For instance, when the local oscillator field is π out of phase with the pump/Stokes-induced anti-Stokes polarization in the focal interaction volume, complete depletion of the local oscillator may occur. The energy of the local oscillator field is not redistributed in terms

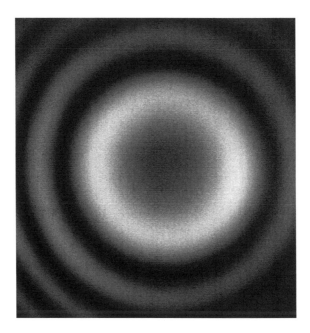

FIGURE 9.8 CARS fringe pattern observed in the field of view of a laser-scanning microscope using a co-propagating local oscillator. Fringes result from dispersion-related phase-shifts for different angles of propagation through the microscope optics. Sample is a homogeneous solution of DMSO. Raman shift is 2913 cm^{-1}. Lens is a 1.15 NA water immersion objective lens.

of direction, as would be the case on a beam splitter, but in terms of frequency; i.e., color conversion takes place through the coupled fields in the material. In the same spirit, with in-focus interference the anti-Stokes radiation can be completely "switched off" by controlling the phase and amplitude of the local oscillator. The ability to switch off CARS signals with a low power beam (E_{lo}) may have interesting applications in imaging experiments.

D. INTERFEROMETRIC IMAGING

The ability to generate contrast based on $\mathrm{Im}\,\chi_r^{(3)}$ in the CARS microscope is very attractive for many imaging applications. In particular, CARS interferometry can resolve weaker bands that are often hidden in the shadow of strong Raman peaks. Whenever the tails on the blue side of strong Raman bands overlap with weaker bands, the resonant contribution of the weak spectral feature is regularly out of phase with the overlapping blue tail of the strong Raman line. Consequently, destructive interference occurs between the spectral features, which often suppresses the weak spectral lines. In such cases, imaging contrast based on a weak band is notably absent. CARS interferometry circumvents this problem by interfering constructively with only the Raman band of interest.

In Figure 9.9a, a CARS image of a protein flake of casein suspended in water is taken at 2930 cm^{-1}, a Raman shift that overlaps with the CH_3-symmetric stretch

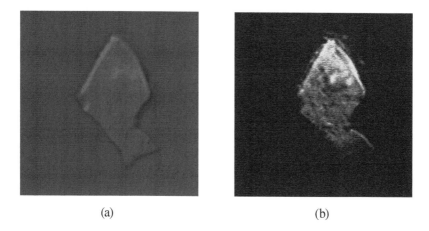

(a)   (b)

FIGURE 9.9 CARS interferometric imaging of a casein protein flake suspended in water. In (a) a regular CARS image is shown, displaying limited contrast. The signal strength in the interferometric image in (b) is proportional on $\chi^{(3)}$, showing much improved contrast. Image size is 50 µm × 50 µm. Raman shift is 2930 cm^{-1}.

spectral region. However, the CH$_3$-symmetric stretch is situated on the blue side of the strong CH$_2$-symmetric stretch. The contrast from the CH$_3$-band from proteins is, therefore, not very pronounced in regular CARS imaging. With CARS interferometry, the CH$_3$-resonant field can be filtered out of the blur comprised of nonresonant background and CH$_2$-resonant contributions. In Figure 9.9b, the same protein flake is shown when recorded with interferometric CARS. The image is devoid of nonresonant background, while the Im$\{\chi_r^{(3)}\}$ contrast is attributed predominantly to the CH$_3$ mode.

IV. SPATIAL PHASE SHAPING IN COHERENT MICROSCOPY

We have seen that the phase with which the radiating dipoles in the focal volume oscillate is a key factor that shapes the resulting anti-Stokes radiation. The phase of the Raman oscillator at a given point in the focal volume can be controlled by manipulating the phase of the driving fields. Hence, through focus engineering (i.e., phase shaping of the incident radiation) the CARS radiation can be controlled. Anti-Stokes generation can be controlled in such a way that better contrast is obtained in the microscope. In the following sections, we briefly discuss the principle of FE-CARS and we outline several mechanisms of focus engineering that improve image contrast.

A. Phase Shaped Excitation Fields

The driving fields in nonlinear coherent microscopy can be dressed with alternative phase profiles. One of the simplest phase masks is a one-dimensional π-phase step across the transverse Gaussian beam profile. The resulting phase pattern resembles that of a HG01 (step along y) or HG10 (step along x) beam modes. These fields are modeled as (Novotny and Hecht 2006)

$$E_{01} = E_0 e^{-ikf} \begin{pmatrix} iI_{11} \cos\phi + iI_{14} \cos 3\phi \\ -iI_{12} \sin\phi + iI_{14} \sin 3\phi \\ -2I_{10} - 2I_{13} \cos 2\phi \end{pmatrix}$$
$$E_{10} = E_0 e^{-ikf} \begin{pmatrix} i(I_{11} + 2I_{12}) \sin\phi + iI_{14} \sin 3\phi \\ -iI_{12} \cos\phi - iI_{14} \cos 3\phi \\ 2I_{13} \sin 2\phi \end{pmatrix}$$

$$\tag{9.11}$$

Figure 9.10 displays the amplitude and phase of the focused HG01 field. It is clear that the phase step not only leads to a split of the focal field amplitude into two lobes, but also to a sharp π-phase step across the focus in the lateral direction. When the Stokes field in the CARS process is shaped according to an HG01 phase step, the resulting CARS excitation field also exhibits a sharp lateral π-phase step in focus through the relation $\phi_{ex}(r) = 2\phi_p(r) - \phi_S(r)$, where $\phi_p(r)$ and $\phi_p(r)$ are the spatial phase of the pump and Stokes beams, respectively. The corresponding CARS excitation field is sketched in Figure 9.10c.

Under regular (HG00) excitation conditions the strongest CARS signal is observed along the optical axis (Figure 9.11a). The observed radiation profile confirms that the dipole oscillators radiate in-phase in the direction of propagation, while the phase matching diminishes for larger angles θ. When the dipoles in the focal volume are driven with a phase associated with an HG01 CARS excitation field, the radiation is no longer phase matched along the optical axis. In this direction, no CARS is observed. Instead, a lobed emission is produced, indicative of the altered phase-matching conditions (Figure 9.11b). The example above illustrates that emission from the bulk, detected along the propagation direction of the beams, can be selectively switched on and off by applying a simple π-phase step in the CARS excitation volume. Similar effects can be obtained with more advanced phase profiles as well. This recipe can be used to suppress the signal from the bulk in forward-detected CARS. The signal from small objects and interfaces with a $\chi^{(3)}$ distinctly different from the bulk, however, shows a different trend under such an excitation condition. In particular, when the object is addressed at one of its vibrational resonances, the object's CARS emission will have a different amplitude and phase relative to its surroundings. Consequently, the CARS field from the resonant object will not interfere in the same fashion as the field of the surrounding bulk. In fact, as shown in the next subsection, the signal from resonant interfaces is unaffected, whereas the bulk signal is suppressed through destructive interference. This is the main form of contrast enhancement in focus-engineered (FE)-CARS: phase control allows highlighting of select objects while the signal from the bulk is suppressed.

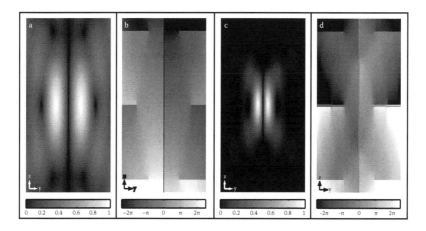

FIGURE 9.10 Calculated amplitude and phase profiles of HG01 focal fields. Amplitude (a) and phase (b) of Stokes focal field at 1064 nm. Amplitude (c) and phase (d) of the CARS excitation field using a pump field at 800 nm. Calculations are for a 1.1 NA water immersion lens. In all panels the axial axis runs from −3 μm to 3 μm and the lateral axis from −1.5 μm to 1.5 μm. Phase values beyond the interval −2π to 2π are discarded to better represent the contrast near the focal plane.

B. Spectral Dependence

The main source of contrast in FE-CARS is based on differences in the amplitude and phase of $\chi^{(3)}$. The spectral phase plays an important role in FE-CARS. While the phase of the nonresonant CARS signal is independent of ω, the resonant part of $\chi^{(3)}$ exhibits a characteristic π-jump in the vicinity of a vibrational resonance ω_R. In the presence of a spatial π-step in focus, the nonresonant background destructively

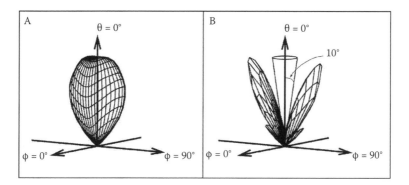

FIGURE 9.11 Calculated angle-resolved CARS radiation intensity in the case of a Stokes beam with a HG00 phase pattern (a) and in the case of a Stokes beam with a HG01 phase profile (b). Note that no CARS intensity along the optical axis is seen.

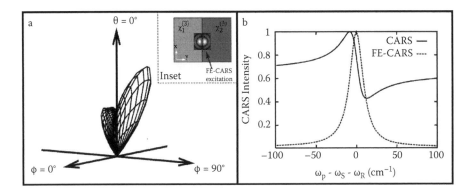

FIGURE 9.12 (a) Calculated FE-CARS radiation profile when a HG01 excitation field overlaps with a lateral interface between a resonant and a nonresonant material. Note that the intensity along the optical axis is no longer zero due to partial lifting of the phase step by the interface. The inset shows the excitation field relative to the orientation of the interface. (b) Comparison of the calculated spectral dependence of CARS in a bulk material with a weak resonance and FE-CARS measured at an interface similar to the one considered in (a). Note the Raman-like spectral dependence of the FE-CARS signal.

interferes. In case the focal volume contains a resonant object in a nonresonant medium, the spectral phase-shift of the resonant object can counterbalance the spatial phase-step of the driving fields. Consequently, destructive interference no longer occurs and a strong signal is detected along the propagation axis, as depicted in Figure 9.12a. Hence, strong signals are expected at the interfaces of resonant objects. Because FE-CARS is sensitive to spectral phase-shifts, a property of only the resonant signal, the magnitude of the contrast follows the spontaneous Raman spectrum. This principle is illustrated in Figure 9.12b, where the spectral dependence of a weak resonance is plotted in the case of regular CARS and FE-CARS. It is seen that the FE-CARS spectrum is not affected by interferences with the nonresonant background, and a clean Raman-like spectral dependence is retrieved.

C. INTERFACE HIGHLIGHTING

As shown in the previous subsection, FE-CARS can be used to accentuate resonant object interfaces. Such an imaging capability can be a very useful complement to a regular CARS image. For instance, when an HG10 Stokes beam is used, interfaces aligned in the y-direction will be highlighted in this imaging mode. An example is shown in Figure 9.13a, where a simple interface between d-DMSO and paraffin oil is imaged at 2977 cm^{-1}. At this Raman shift, the paraffin oil is resonantly driven on the blue side of the CH$_2$ stretching band, whereas d-DMSO is nonresonant. In regular CARS, the interface appears as a dip, which is the result of the destructive interference of the resonant paraffin signal with the nonresonant d-DMSO contribution. In other words, the field generated on the paraffin side of the interface is π out of phase with the field generated on the nonresonant d-DMSO side. In FE-CARS, using the HG10 shaped Stokes beam, the interface is highlighted instead. The strong signal

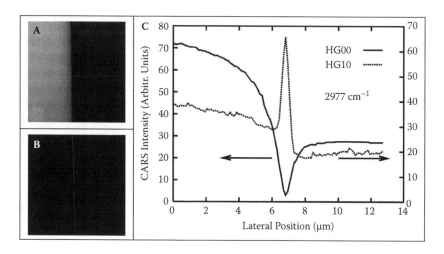

FIGURE 9.13 Experimental demonstration of differential imaging of a lateral edge using HG10 CARS excitation. The interface is between d-DMSO and paraffin oil diluted with 6% dodecane. Both sides of the interface have a refractive index of 1.478 at room temperature. (a) Regular HG00 CARS imaging. (b) FE-CARS imaging of the same interface. (c) Cross section taken perpendicular to the interface, showing the marked different contrast between the two excitation modes.

at the interface is the direct result of counterbalancing between the applied spatial phase step and the spectral phase step across the interface. The submicron width of the interface signal corresponds to the lateral resolution of the microscope. Away from the interface, the signal from the bulk is significantly reduced due to destructive interference between the out-of-phase segments of the focal volume.

Different interfaces can be visualized using alternative phase-shaped beams. For instance, a phase pattern that resembles the transverse phase of a Laguerre-Gaussian beam will highlight interfaces along both the x- and y-directions (Krishnamachari and Potma 2007b). Interfaces perpendicular to the axial direction can be accentuated by shaping the Stokes beam as an "optical bottle beam" (OBB) (Arlt and Padgett 2000). An OBB beam exhibits a sharp π-phase step along the longitudinal direction, in addition to the natural Gouy phase shift, introducing destructive interference between the CARS fields generated above and below the focal plane of a bulk sample. At the axial interface, similar to the situation at the lateral interface, the spatial phase step can offset the spectral phase jump in the case of OBB excitation, effectively highlighting the interface (Krishnamachari and Potma 2007c). Consequently, axial interfaces can be seen in forward-detected CARS with contrast that is reminiscent of epi-CARS detection. Indeed, the axial-edge sensitivity of epi-CARS also results from constructive interference across the interface, albeit in a different fashion. Nonetheless, the epi-CARS signal suffers from intrinsically low signals and substantial contributions of back-scattered forward CARS (Evans et al. 2005), problems that are less relevant to forward-detected FE-CARS.

Whereas special beam-shapes can be employed to highlight axial edges, the same interface can be accentuated with a lateral phase step. This somewhat fortuitous

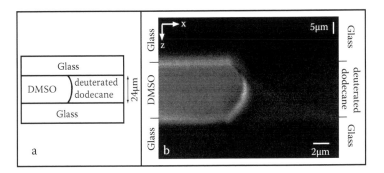

FIGURE 9.14 Illustration of multi-dimensional differential imaging using HG10 CARS excitation. (a) Schematic of the sample configuration; (b) *xz* cross-section of the glass/DMSO/deuterated-dodecane/glass interface.

effect is a result of the altered phase-matching conditions at the axial interface in the case of a CARS excitation field dressed with a lateral π-phase step (Krishnamachari and Potma 2008). In Figure 9.14, a cross-sectional view is shown of an interface between DMSO and d-dodecane sandwiched between two coverslips. At 2933 cm⁻¹, the DMSO layer is moderately vibrationally resonant, whereas the d-dodecane layer is far from vibrational resonance. As expected, in regular CARS the DMSO layer appears bright, while signal from the d-dodecane layer is significantly lower. In FE-CARS using an HG10 shaped Stokes beam, not only is the interface between the liquids highlighted, but differential contrast from interface between the DMSO and the glass is seen as well. Consequently, a simple π-step phase mask is sufficient for differential imaging of both lateral and axial interfaces.

In the examples above, we have discussed differential CARS imaging contrast based on phase shaping of the Stokes beam. However, an HG10 focus-engineered Stokes beam displays suboptimal overlap with a non-shaped (Gaussian) pump beam. Consequently, the effective CARS excitation amplitude is reduced, producing signals that are less than optimal. A solution is found by applying an HG10 phase mask to both the pump and Stokes (Liu and Kim 2007). In this case, maximum overlap between the driving fields is obtained, while the effective phase of the CARS excitation field is virtually identical to the case of Stokes-only shaping.

D. IMAGING WITH PHASE-SHAPED BEAMS

The differential contrast provided by FE-CARS can be useful for elucidating additional features not observed in regular CARS images. In case the phase pattern on the excitation beams is applied with a spatial light modulator, simply switching between CARS and FE-CARS improves the resolving power of a CARS microscope. The contrast mechanisms introduced through focus engineering reveal new information of biological samples. An example is shown in Figure 9.15, where the connective tissue in the vicinity of a mouse aorta is imaged with multimodal focus-engineered microscopy. The CARS image in (a) reveals fibrous structures

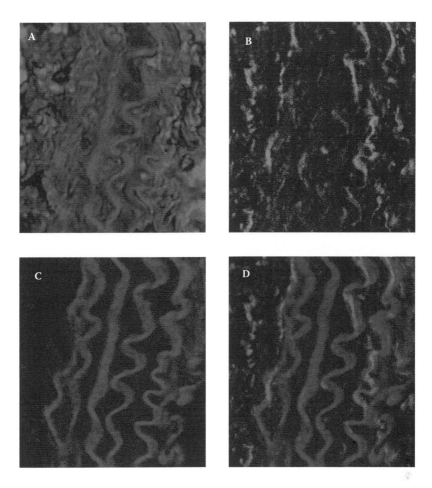

FIGURE 9.15 Multimodal imaging of a mouse aorta with FE-CARS. (a) CARS image at 2911 cm^{-1}; (b) FE-CARS image; (c) Two-photon excited fluorescence image showing elastin; (d) overlay of FE-CARS image (light gray) and fluorescence image (dark gray). Images are 40×45 μm.

that overlap with the TPEF signal from elastin (c). In (b) an HG10 phase mask is applied to acquire an FE-CARS image. The contrast is completely different, as can be seen from the overlay as shown in (d). Several vibrationally resonant interfaces are highlighted in the FE-CARS image, providing additional dimensions for image analysis.

E. COMBINING FE-CARS WITH INTERFEROMETRIC TECHNIQUES

The ability to control the phase of focal fields at different locations in the interaction volume has interesting applications when combined with interferometric mixing techniques. When a local oscillator field is focus engineered and overlapped with the CARS polarization in focus, certain parts of the local oscillator focal field can be chosen to be in-phase with the CARS polarization, while others are selectively tuned out-of-phase. In this scheme, the CARS radiation of the out-of-phase portions can be effectively switched off if the amplitude of the local oscillator field is chosen adequately. In other words, the CARS radiation from certain parts of the focal volume can be suppressed, while other parts are retained. Such an imaging mode is reminiscent of stimulated emission depletion (STED) microscopy, which achieves focal volume reduction through suppressing fluorescence from certain parts of the focus with the aid of stimulated emission (Hell and Wichmann 1994; Klar and Hell 1999; Klar et al. 2000). A possible CARS analog could thus consist of focal volume reduction through interference only. Note that such a scheme is very favorable in terms of the applied power of incident light, as the additional local oscillator is typically less than a nW, which is negligible compared to the pump and Stokes excitation powers. Hence, potential CARS focal volume reduction can be achieved without applying additional high power incident beams.

V.   PHASE DISTORTIONS

All the phase-sensitive microscopy techniques discussed in this chapter depend on the ability to synthesize and retain well-defined focal phase profiles. Naturally, phase distortions are the mortal enemies of this phase-sensitive approach. The primary source of phase distortions is the variation of the linear refractive index ($\chi^{(1)}$) in the sample. In CARS interferometry, unwanted phase shifts between the CARS excitation fields and the local oscillator may occur when the beams encounter $\chi^{(1)}$ objects with a chromatic dispersion that is significantly different from their surrounding. Because the local oscillator and the CARS excitation fields are of different color, any dispersive object will introduce a relative phase shift between the fields. While their influence is rather small in out-of-focus regions, dispersive effects are most notable when such objects are located in focus. Amplitude effects may also play a role, in case the local oscillator field is scattered off an object with a different efficiency than the CARS excitation fields. Biological materials often exhibit a refractive index that varies significantly on the micrometer scale, and consequently scatter light. Application of index matching fluids, such as the use of glycerol in tissues (Tuchin 2005), has the potential to reduce these effects considerably.

In focus-engineered nonlinear coherent microscopy, scattering of the excitation beams in turbid media also leads to compromised imaging contrast. Nonetheless, while scattered portions of the phase-shaped beam will display randomized phases, these contributions will play a minor role in the generation of the nonlinear coherent signal. Only the coherent portion of the incident beam will be able to effectively generate a coherent signal. Hence, phase scrambling will bring down the relative magnitude of the coherent signal by reducing the effective amplitude of the coherent driving field. This effect can be compensated for when adaptive optics techniques are used, which have shown to significantly improve the CARS signal intensity recorded in tissue samples (Wright et al. 2007). While scattering will diminish the signal intensity in the image, the resolution is not necessarily compromised because the shape of the coherent excitation volume is essentially unaffected. Note that in regular CARS imaging, the signal strength is affected by partial phase randomization of the excitation fields in a similar fashion as in FE-CARS.

It is useful to consider the extent of focal quality reduction in terms of phase due to tissue scattering. How much does a focus-engineered beam suffer from tissue scattering? An example is found in STED microscopy, a technique that also relies on the quality of alternative focal beam profiles. To reduce the axial resolution of the fluorescence microscope, STED employs an OBB depletion beam (Klar et al. 2000). The sharp π-step of the OBB beam along the optical axis introduces a dark spot at the location of the focal plane. The pitch of this dark spot, crucial to STED performance, depends on the ability to maintain a sharp axial π-step across the focal volume. The success of STED is testament to the sufficient preservation of sharp focal phase steps in biological materials, and to the possibility of applying focus-engineered nonlinear coherent imaging to turbid media.

VI. CONCLUSION

In this chapter, we discussed several approaches to interferometric nonlinear micros-copy. The fascinating aspect of interferometric microscopy is that it takes full advan-tage of the intrinsic property of coherence in CARS, SHG, and THG. Here, we have zoomed into possible coherent manipulations schemes in the narrowband excitation regime. The ability to selectively control the radiation from the focal interaction vol-ume through either interferometric mixing or focus engineering has opened up new ways to investigate the sample with nonlinear coherent imaging techniques. As we have seen in the case of CARS, radiation control not only allows background suppres-sion, $\mathrm{Im}\{\chi_r^{(3)}\}$ imaging, and differential contrast imaging, it also offers opportunities for imaging with super-resolution. As such, this approach may have far-reaching implications for biological imaging studies.

REFERENCES

Ando, T., Urakami, T., Itoh, H., and Tsuchiya, Y. 2002. Optimization of resonant two-photon absorption with adaptive quantum control. *Appl. Phys. Lett.* 80:4265–67.

Andresen, E. R., Keiding, S. R., and Potma, E. O. 2006. Picosecond anti-Stokes genera-tion in a photonic-crystal fiber for interferometric CARS microscopy. *Opt. Express* 14:7246–51.

Arlt, J., and Padgett, M. J. 2000. Generation of a beam with a dark focus surrounded by regions of higher intensity: The optical bottle beam. *Opt. Lett.* 25:191–93.

Bjorklund, G. C. 1975. Effects of focusing on the third-order nonlinear process in isotropic media. *IEEE J. Quant. Electron.* QE-11:287–96.

Boyd, R. W. 2003. *Nonlinear optics.* San Diego: Academic Press.

Bredfeldt, J. S., Vinegoni, C., Marks, D. L., and Boppart, S. A. 2005. Molecularly sensitive optical coherence tomography. *Opt. Lett.* 30:495–97.

Cheng, J.-X., Volkmer, A., and Xie, X. S. 2002. Theoretical and experimental characterization of coherent anti-Stokes Raman scattering microscopy. *J. Opt. Soc. Am. B* 19:1363–75.

Cheng, J.-X., and Xie, X. S. 2002. Green's function formulation for third-harmonic generation microscopy. *J. Opt. Soc. Am. B* 19:1604–10.

Chuntonov, L., Rybak, L., Gandman, A., and Amitay, Z. 2008. Enhancement of intermediate-field two-photon absorption by rationally shaped femtosecond pulses. *Phys. Rev. A* 77:021403.

Cui, M., Joffre, M., Skodack, J., and Ogilvie, J. P. 2006. Interferometric Fourier transform coherent anti-Stokes Raman scattering. *Opt. Express* 14:8448–58.

Débarre, D., Olivier, N. and Beaurepaire, E. 2007. Signal epidetection in third-harmonic generation microscopy of turbid media. *Opt. Express* 15:8913–24.

Dudovich, N., Dayan, B., Faeder, S. M. G., and Silberberg, Y. 2001. Transform-limited pulses are not optimal for resonant multiphoton transitions. *Phys. Rev. Lett.* 86:47–50.

Dudovich, N., Oron, D., and Silberberg, Y. 2002. Single-pulse coherently controlled nonlinear Raman spectroscopy and microscopy. *Nature* 418:512–14.

Eckbreth, A. C. 1978. BOXCARS: Crossed-beam phase-matched CARS generation in gases. *Appl. Phys. Lett.* 32:421–23.

Eesley, G. L., Levenson, M. D. and Tolles, W. M. 1978. Optically heterodyned coherent raman spectroscopy. *IEEE J. Quant. Electron.* QE-14:45–49.

Evans, C. L., Potma, E. O., Puoris'haag, M., Cote, D., Lin, C., and Xie, X. S. 2005. Chemical imaging of tissue *in vivo* with video-rate coherent anti-Stokes Raman scattering (CARS) microscopy. *Proc. Natl. Acad. Sci. USA* 102:16807–12.

Evans, C. L., Potma, E. O., and Xie, X. S. 2004. Coherent anti-Stokes Raman scattering interferometry: Determination of the real and imaginary components of the nonlinear susceptibility for vibrational microscopy. *Opt. Lett.* 29:2930–32.

Feng, S., and Winful, H. G. 2001. Physical origin of the Gouy phase shift. *Opt. Lett.* 26:485–87.

Gouy, C. R. 1890. Sur une propriete nouvelle des ondes lumineuses. *Compt. Rendue Acad. Sci. Paris* 110:1251–53.

Hell, S., and Wichmann, J. 1994. Breaking the diffraction limit by stimulated emission: Stimulated-emission depletion fluorescence microscopy. *Opt. Lett.* 19:780–82.

Hellwarth, R. W. 1977. Third-order optical susceptibilities of liquids and solids. In *Progress in quantum electronics*, eds. J. H. Sanders, S. Stenholm, 1–68. New York: Pergamon.

Jiang, Y., Tomov, I., Wang, Y., and Chen, Z. 2004. Second-harmonic optical coherence tomography. *Opt. Lett.* 29:1090–92.

Jurna, M., Korterik, J. P., Otto, C., and Offerhaus, H. L. 2007. Shot noise limited heterodyne detection of CARS signals. *Opt. Express* 15:15207–13.

Kee, T. W., Zhao, H., and Cicerone, M. T. 2006. One-laser interferometric broadband coherent anti-Stokes Raman scattering. *Opt. Express* 14:3631–36.

Klar, T. A., and Hell, S. 1999. Subdiffraction resolution in far-field fluorescence microscopy. *Opt. Lett.* 24:954–56.

Klar, T. A., Jakobs, S., Dyba, M., Egner, A., and Hell, S. W. 2000. Fluorescence microscopy with diffraction resolution barrier broken by stimulated emission. *Proc. Natl. Acad. Sci. USA* 97:8206–10.

Kleinman, D. A., Ashkin, A., and Boyd, G. D. 1966. Second-harmonic generation of light by focused laser beams. *Phys. Rev.* 145:338–79.

Krishnamachari, V. V., Andresen, E. R., Keiding, S. R., and Potma, E. O. 2006. An active interferometer-stabilization scheme with linear phase control. *Optics Express* 14:5210–15.

Krishnamachari, V. V., and Potma, E. O. 2007a. Detecting lateral interfaces with focus-engineered coherent anti-Stokes Raman scattering microscopy. *J. Raman Spectrosc.* 39:593–98.

Krishnamachari, V. V., and Potma, E. O. 2007b. Focus-engineered coherent anti-Stokes Raman scattering: A numerical investigation. *J. Opt. Soc. Am. A* 24:1138–47.

Krishnamachari, V. V., and Potma, E. O. 2007c. Imaging chemical interfaces perpendicular to the optical axis with phase-shaped coherent anti-Stokes Raman scattering microscopy. *Chem. Phys.* 341:81–88.

Krishnamachari, V. V., and Potma, E. O. 2008. Multi-dimensional differential imaging with FE-CARS microscopy. *Vib. Spectrosc.:* DOI 10.1016/j.vibspec.2008.07.009.

Lee, E. S., Lee, J. Y., and Yoo, Y. S. 2007. Nonlinear optical interference of two successive coherent anti-Stokes Raman scattering signals for biological imaging applications. *J. Biomed. Opt.* 12:024010.

Lim, S. H., Caster, A., and Leone, S. R. 2005. Single pulse phase-control interferometric coherent anti-Stokes Raman scattering (CARS) spectroscopy. *Phys. Rev. A* 72:041803.

Liu, C., Huang, Z., Lu, F.. Zheng, W., Hutmacher, D. W., and Sheppard, C. 2006. Near-field effects on coherent anti-Stokes Raman scattering microscopy imaging. *Opt. Express* 15:4118–31.

Liu, C., and Kim, D. Y. 2007. Differential imaging in coherent anti-Stokes Raman scattering microscopy with Laguerre-Gaussian excitation beams. *Opt. Express* 15:10123–34.

Lüpke, G., Marowsky, G., and Steinhoff, R. 1989. Phase-controlled nonlinear interferometry. *Appl. Phys. B* 49:283–39.

Marks, D. L., and Boppart, S. A. 2004. Nonlinear interferometric vibrational imaging. *Phys. Rev. Lett.* 92(12):123901–123905.

Marowsky, G., and Lüpke, G. 1990. CARS-background suppression by phase-controlled nonlinear interferometry. *Appl. Phys. B* 51:49–51.

Moreaux, L., Sandre, O., Charpak, S., Blanchard-Desce, M., and Mertz, J. 2001. Coherent scattering in multi-harmonic light microscopy. *Biophys. J.* 80:1568–74.

Moreaux, L., Sandre, O., and Mertz, J. 2000. Membrane imaging by second-harmonic generation microscopy. *J. Opt. Soc. Am. B* 17:1685–94.

Mukamel, S. 1995. *Principles of nonlinear optical spectroscopy.* New York: Oxford.

Novotny, L., and Hecht, B. 2006. *Principles of nano-optics.* New York: Cambridge University Press.

Oron, D., Dudovich, N., and Silberberg, Y. 2003. Femtosecond phase-and-polarization control for background-free coherent anti-Stokes Raman spectroscopy. *Phys. Rev. Lett.* 90: 213901–213904.

Potma, E. O., Boeij, W. P. d., and Wiersma, D. A. 2000. Nonlinear coherent four-wave mixing in optical microscopy. *J. Opt. Soc. Am. B* 17:1678–84.

Potma, E. O., Evans, C. L., and Xie, X. S. 2006. Heterodyne coherent anti-Stokes Raman scattering (CARS) imaging. *Opt. Lett.* 31:241–43.

Richards, B., and Wolf, E. 1959. Electromagnetic diffraction in optical systems II: Structure of the image field in an aplanatic system. *Proc. Roy. Soc. A* 253:358–79.

Siegman, A. E. 1986. *Lasers.* Sausolito, CA: University Science Books.

Tang, S., Krasieva, T. B., Chen, Z., Tempea, G., and Tromberg, B. J. 2006. Effect of pulse duration on two-photon excited flourescence and second harmonic generation in nonlinear optical microscopy. *J. Biomed. Opt.* 11:020501.

Tuchin, V. V. 2005. Optical clearing of tissues and blood using the immersion method. *J. Phys. D* 38:2497–2518.

Vacano, B. V., Buckup, T., and Motzkus, M. 2006. Highly sensitive single-beam heterodyne coherent anti-Stokes Raman scattering. *Opt. Lett.* 31:2495–97.

Vinegoni, C., Bredfeldt, J., Marks, D., and Boppart, S. 2004. Nonlinear optical contrast enhancement for optical coherent tomography. *Opt. Express* 12:331–41.

Williams, R. M., Zipfel, W. R., and Webb, W. W. 2005. Interpreting second-harmonic generation images of collagen I fibrils. *Biophys. J.* 88:1377–86.

Wright, A. J., Poland, S. P., Girkin, J. M., Freudiger, C. W., Evans, C. L., and Xie, X. S. 2007. Adaptive optics for enhanced signal in CARS microscopy. *Opt. Express* 15:18209–19.

Yacoby, Y., Fitzgibbon, R., and Lax, B. 1980. Coherent cancellation of backgroud in four-wave mixing spectroscopy. *J. Appl. Phys.* 51:3072–77.

Yew, E., and Sheppard, C. J. R. 2006. Effects of axial field components on second harmonic generation microscopy. *Opt. Express* 14:1167–74.

Zumbusch, A., Holtom, G., and Xie, X. S. 1999. Vibrational microscopy using coherent anti-Stokes Raman scattering. *Phys. Rev. Lett.* 82:4142–45.

APPENDIX

The factors $g_{mn}(\theta)$, used in the calculation of the focal fields, are given as

$$g_{00} = 1 + \cos\theta$$

$$g_{01} = \sin\theta$$

$$g_{02} = 1 - \cos\theta$$

$$g_{10} = g_{13} = \sin^2\theta$$

$$g_{11} = \sin\theta(1 + 3\cos\theta)$$

$$g_{12} = g_{14} = \sin\theta(1 - \cos\theta)$$

10 Biomolecular Imaging by Near-Field Nonlinear Spectroscopy and Microscopy

Norihiko Hayazawa, Taro Ichimura,
Katsuyoshi Ikeda, and Satoshi Kawata

CONTENTS

I. INTRODUCTION

For biochemical characterization, fluorescence spectroscopy has been widely used due to the advanced labeling technologies of dye molecules to specific biochemical sites of the target molecules (Tsien 1998). Specifically, the finding of green fluorescent protein from the luminescent jellyfish (Shimomura et al. 1962; Prasher et al. 1992) allowed for genetic encoding of strong fluorescence and led to a wide variety of fluorescent proteins for the application of molecular and cellular biology (Kogure et al. 2006; Ando et al. 2002). Three-dimensional characterization of biological samples, which is strongly required in cellular biology, has been successfully achieved either by confocal fluorescent microscopy (Gu 1996) or the nonlinear optical extension of fluorescence, two-photon excited fluorescent microscopy (Denk et al. 1990). While fluorescence-based spectroscopy can provide very bright and flexible signals, these are still indirect methods to investigate the constitution and conformation of the target molecules. Raman spectroscopy (Chalmers and Griffiths 2002) is an ideal complement to directly derive the molecular vibrations without staining with the indirect fluorescent molecules. However, Raman spectroscopy has been used not as "microscopy" but as "spectroscopy," because the main drawback of Raman process is its extreme inefficiency; e.g., Raman scattering cross section is 10^{-30} cm^2, fluorescent (absorption) cross section is 10^{-16} cm^2. Because of this major drawback, it has been almost impossible to apply Raman spectroscopy to microscopic use, in which the observing volume becomes small and fast image acquisition is required. However, owing to the recent advance of highly efficient detectors with extremely small background noises, such as liquid nitrogen–cooled charge-coupled-device (CCD) camera and avalanche photo diode with single photon sensitivity, Raman spectroscopy has been successfully developed as confocal Raman microscopy (Puppels et al. 1990), which also possesses three-dimensional resolution. In terms of nonlinear optical variation of Raman spectroscopy, in which we can expect higher sensitivity, coherent anti-Stokes Raman scattering (CARS) (Shen 1984) is one of the most promising nonlinear Raman spectroscopies. Actually, Duncan et al. (1982) reported scanning coherent anti-Stokes Raman microscopy before the successful report of confocal Raman microscopy. Different from fluorescent spectroscopy, Raman scattering is a coherent process, which requires momentum conservation during the instant scattering process; phase matching of two laser pulses has to be considered in CARS. Because of the limitation of this phase-matching condition in CARS, high numerical aperture (NA) objective lens was not used and, thus, two pulsed beams with proper wave vectors were overlapped both in time and space instead. This geometry actually gives three-dimensional spatial resolution and has been quite useful to directly investigate physically inaccessible positions such as explosion process characterization (Eckbreth 1996), but the spatial resolution was not high enough for biochemical applications. Recently, Zumbusch et al. (1999) and Hashimoto et al. (2000) reported CARS microscopy with collinear optical geometry using a high NA objective lens, in which they proved that the phase-matching condition could be relaxed in the small sample volume and automatically satisfied by the wide availability of wave vectors at the tight focusing spot. In this stage, the CARS microscope has become a competitive spectroscopic technique to two-photon fluorescence microscopy, and even

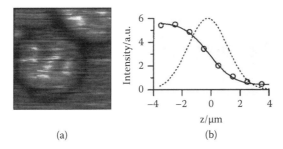

(a) (b)

FIGURE 10.1 CARS image of yeast cell in water observed at 1215 cm⁻¹. (b) Estimation of the *z*-axis resolution of the CARS microscopy system. Open circles, observed data; solid curve, result of curve fitting; dashed curve, differential of the solid curve.

more advantageous because CARS is one of the vibrational spectroscopic techniques, which allows for direct observation of targets without dye labeling and also because CARS uses anti-Stokes signals, which can be easily separable from autofluorescence. Figure 10.1a shows the CARS microscope image of a yeast cell in water observed at 1215 cm⁻¹, which corresponds to the amido III band of protein (Hashimoto et al. 2000). The lateral spatial resolution of <1 μm was achieved, and according to *z*-axis response of the CARS signals in Figure 10.1b, the vertical resolution of ~3.2 μm has been reported due to analogous nonlinear optical feature of two-photon excited fluorescence microscope (Denk et al. 1990). In the recent development of the CARS microscope, 230 nm lateral resolution and 750 nm vertical resolution have been successfully reported (Cheng and Xie. 2004), which is a bit better than the diffraction limit of an near-infrared (NIR) light source (~800 nm) used for CARS due to the nonlinear response of the materials (Kawata et al. 2001; Hell and Wichmann 1994; Takada et al. 2005). The future main issues of CARS microscopy, in order to replace one-photon or two-photon fluorescence microscopy for biochemical applications, are technical difficulty of optical arrangement and higher costs because of the necessity of two laser pulses. Moreover, since the spatial resolution of CARS microscope is generally the same as one-photon or two-photon fluorescence microscopes, in which the spatial resolutions are all determined by the diffraction limit of a light, CARS microscope has to show its wide applicability to use the main advantage of direct observation without staining over the rapid advances of dye molecules and site-specific labeling technologies of the dyes in fluorescence microscope. The alternative way to push CARS microscope ahead over fluorescence microscopes is to improve the spatial resolution. In this chapter, we discuss the improvement of the spatial resolution in CARS microscopy, which is based on near-field optical technique using a nano-sized metallic tip (Inouye and Kawata 1994; Fischer and Pohl 1989; Bachelot et al. 1995; Kawata 2001). This technique has been recently recognized as tip-enhanced spectroscopy, in particular, the combination with Raman spectroscopy, tip-enhanced Raman scattering (TERS) (Kawata and Shalaev 2007), because of the analogousness to surface enhanced Raman scattering (SERS) (Chang and Furtak 1981). In section II, we provide you with a simple sketch of how to generate a nanolight source in near-field, and then introduce TERS as a tool to enable visualizing the molecular vibrations of organic molecules beyond the diffraction limit of a light due to the photon confinement by the near-field optics. In Section III, we discuss the combination of tip enhancement and nonlinear vibrational spectroscopy, CARS, to further confine the photons owing to the nonlinear response of materials. Spatial resolution down to ~15 nm is successfully demonstrated in DNA-based adenine molecules. In section IV, we discuss future possibilities of the combination between near-field optics and other nonlinear optics for biochemical applications.

II. TIP-ENHANCED RAMAN SCATTERING

Near-field microscopy is characterized by its super-resolution capability. It can exceed the classical limit of spatial resolution, so-called diffraction limit, due to the wave nature of photons (Born and Wolf 1999). The imaging mechanism of near-field microscopy is different from classical optical microscopy; the light intensity is detected as a result of strong electromagnetic interaction between the probe and the sample structure in the near-field via evanescent photons, so that the system is not a linear, passive one but a more complex one. In this section, we show a simple way to confine the photons in the near-field as a nano light source for various spectroscopies. Then, we will introduce one of the spectroscopic applications, Raman spectroscopy in the near-field, that is, tip-enhanced Raman scattering (TERS) spectroscopy for biochemical sensing. As a demonstration of the capability of TERS, we focus on the identification of different organic molecules and different vibrational modes of single-walled carbon nanotubes by TERS.

A. PHOTON CONFINEMENTS IN THE NEAR-FIELD:
EVANESCENT FIELD AS A NANOLIGHT SOURCE

According to Fourier optics (Goodman 1996), a small aperture or a small object with a finite size can be regarded as an expansion of grating sets, which have broad distributions of lattice constants from much smaller to larger than the wavelength of a light. These concepts are illustrated in Figure 10.2. Small aperture (Figure 10.2a) and small object (Figure 10.2b) with the same size ($= 2a$, which is much smaller than the wavelength of light) become the same *Sinc* function by Fourier transformation. Among those gratings, the grating with a lattice constant larger than the wavelength can diffract the light to the proper angle that satisfies the diffraction condition, and thus the diffracted light will propagate to the specific direction, while the grating with smaller lattice constant than the light wavelength will not satisfy the propagation condition of the light, and thus the diffracted light becomes evanescent field, which cannot propagate from the grating structure and is localized at the structure (Kawata 2001). With the use of these evanescent field components, light field can be confined and localized either at the aperture or the structure, which results in the generation of a nano-light source in which the size corresponds to the diameter ($= 2a$) of the aperture and structure. This is because the evanescent light components have much higher wave vector, k, than the propagation light component as illustrated in Figure 10.2c. Each concept of Figure 10.2a and b has been realized as so-called

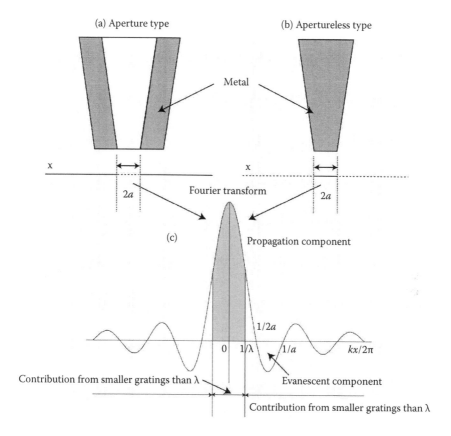

FIGURE 10.2 Concept of (a) aperture type and (b) apertureless type near-field microscopes; (c) is the Fourier transform function of (a) and (b), which shows propagation and evanescent light components.

"aperture type" near-field microscopes (Pohl et al. 1984; Harootunian et al. 1986) and "apertureless type" near-field microscopes (Inouye and Kawata 1994; Fischer et al. 1989; Bachelot et al. 1995; Zenhausern et al. 1994; Sugiura et al. 1997), respectively. Each typical near-field probe is illustrated in Figure 10.2a and b. Aperture type probes are generally made of sharpened optical fiber coated with metal in order to block the leakage of light from the taper except for the aperture, and apertureless probes are generally sharp silicon cantilever tips of atomic force microscope or sharpened pure metallic wires. Theoretically, the same spatial resolution can be expected both in aperture and apertureless near-field microscopes. Higher spatial resolution requires either smaller aperture of the optical fiber in Figure 10.2a or smaller diameter of the sharpened tip in Figure 10.2b. According to Figure 10.2c, the smaller the aperture or the diameter of the tip becomes (a becomes smaller), the greater the evanescent component is and the higher the wave vector becomes. However, practically, since a smaller aperture simply makes the near-field signal smaller, it would be impossible to detect the weak signals in the nanoscale. On the other hand, additional signal enhancement can be expected in the case of aperture-less type probes, especially when the probes are made of metal. This additional signal enhancement is specifically expected when the tip diameter becomes smaller, which corresponds to higher spatial resolution. In the next section, we discuss the mechanism of the field enhancement effect of a metallic probe tip.

B. A METALLIC PROBE TIP ENHANCES THE NANOLIGHT SOURCE: TIP-ENHANCEMENT EFFECT

When a metallic probe, which has a nanometric tip, is illuminated with an optical field, conductive free electrons collectively oscillate at the surface of the metal (Figure 10.3). The quantum of the induced oscillation is referred to as surface plasmon polariton (SPP) (Raether 1988). The electrons (and the positive charge) are concentrated at the tip apex and strongly generate an external electric field. Photon energy is confined in the local vicinity of the tip. Therefore, the metallic tip works as a photon reservoir.

The local electric field is used as a nanolight source to excite photon-matter interaction for various spectroscopic techniques, including fluorescence (Hayazawa et al. 1999; Azoulav et al. 1999; Hamann et al. 2000; Lessard et al. 2000), two-photon excited fluorescence (Sánchez et al. 1999), Raman scattering (Inouye et al. 1999; Hayazawa et al. 2000; Stöckle et al. 2000; Anderson 2000), and coherent anti-Stokes Raman spectroscopy (Hayazawa et al. 2004a; Ichimura et al. 2004a,b). In particular, the enhancement of Raman scattering by metallic structures has been widely investigated as SERS since the 1970s (Fleischmann et al. 1974; Jeanmaire and Van Duyne 1977; Albrecht and Creighton 1977). For the SERS effect, metallic nanoparticles and their aggregates are extensively used in various models to understand the physical behavior of the local electric field and the magnitude of the enhancement (Chang and Furtak 1981). Similarly, in the case of the tip enhancement, the local field has been calculated with numerical analysis in the same manner based on

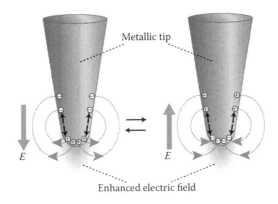

Metallic tip

E

E

Enhanced electric field

FIGURE 10.3 Quantized oscillation of electrons at the surface of a metallic probe tip. This is so called surface plasmon. As the charge distribution is confined tightly at the sharp tip end, the subsequent electric field at the tip is strongly enhanced.

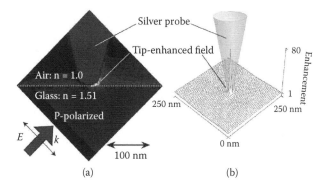

FIGURE 10.4 (a) Electric field intensity localized at a silver probe tip calculated by FDTD method. The three-dimensional calculated area was a 300 nm cube and was divided into $100 \times 100 \times 100$ pixels. P-polarized light with 488 nm wavelength is introduced at an incident angle of 45 degrees. (b) Electric field intensity distribution on the glass surface. The field is enhanced up to 80.

electromagnetic mechanism. Figure 10.4a shows the local field distribution near a metallic probe tip calculated via the finite difference time-domain (FDTD) method (Kunz and Luebbers 1999), which gives a numerical solution of Maxwell's equations (Furukawa and Kawata 1998). It can be seen that the optical field is highly confined into a tiny volume near the tip end. In addition to the spatial confinement, the optical field is strongly amplified by a factor of 80 at the maximum. Figure 10.4b shows a light intensity distribution on the dielectric substrate. Figure 10.4 shows the three-dimensional confinement of the optical field and amplification of the light intensity in the confined spot. The size of the spot is comparable to that of the probe tip (Martin et al. 2001; Novotny et al. 1997).

It has been shown that a metallic probe tip highly localizes and strongly amplifies the optical field through the resonance effect of the plasmon polaritons at the probe tip. This concept has made it possible to optically observe nanometric samples with a nanometric spatial resolution.

C. GENERAL TERS MICROSCOPY CONFIGURATION

The basic system configurations so far reported by several groups (Hayazawa et al. 2000, 2001, 2002, 2003; Hartschuh et al. 2003; Stöckle et al. 2000) are based on a combination of an atomic force microscope (AFM) and an inverted optical microscope as illustrated in Figure 10.5a. An expanded and collimated light field from a visible laser enters into the epi-illumination optics of an inverted optical microscope. A circular mask is inserted in the beam path of the illumination light and located at the conjugated plane of the pupil of the objective lens (NA = 1.4). The mask rejects the part of the beam corresponding to focusing angles that are less than NA = 1.0, while the transmitted light forms a focused spot that produces an evanescent field on the sample surface (Hayazawa et al. 1999). As the metallic tip is moved closer to the focused spot that is generated by a high numerical aperture (NA = 1.4) objective

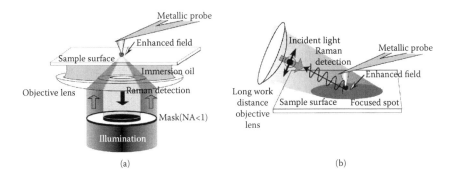

(a)   (b)

FIGURE 10.5 TERS configurations: (a) transmission mode; (b) reflection mode.

lens, a localized enhanced electric field is generated at the tip apex as described in section II B. The localized enhanced electric field at the tip as a nanolight source is scattered inelastically by the Raman active molecules, which corresponds to TERS. The TERS is collected by the same objective lens, and is directed to the spectrometer that is equipped with a liquid nitrogen–cooled CCD camera for TERS spectroscopy and with an avalanche photodiode for TERS imaging (Hayazawa et al. 2002). The avalanche photodiode is located after the exit slit of the spectrometer so that a specific Raman-shifted line can be detected. Excitation light or Rayleigh scattering is sufficiently rejected by a notch filter before the spectrometer. The distance between the sample and the metallic tip is regulated by AFM operation either in contact or noncontact mode, and the sample is scanned with piezoelectric transducers (PZT) in the X-Y plane. Scanning the XY-PZT sample stage while simultaneously detecting the Raman signal with the avalanche photodiode can perform TERS imaging at the specific Raman-shifted line.

It should be pointed out that the system configuration above has been recently recognized as "transmission mode" TERS (Hayazawa and Saito 2007) because the light needs to transmit through the sample. So, the applicability of the system is limited to very thin or transparent samples. In order to apply TERS to thick or opaque samples, several groups are developing so-called "reflection mode" TERS mainly for characterizing semiconductive silicon-based materials (Hayazawa et al. 2005; Saito et al. 2006; Sun and Shen 2003). This configuration is illustrated in Figure 10.5b. While the reflection mode is advantageous for opaque samples, the background signals are relatively higher than transmission mode because a lower NA objective lens with a long working distance is used due to the limited spatial clearance around the cantilever head with AFM scanner. All the experimental configurations discussed below are based on the transmission mode. Note that some system improvements for higher sensitivity in TERS have been recently reported such as the use of radial polarization illumination (Hayazawa et al. 2004b) or gap mode SPP excitation (Hayazawa et al. 2000; Hayazawa et al. 2001, 2002, 2007a; Pettinger et al. 2004) for the transmission mode TERS and depolarization detection configurations (Poborchii et al. 2005; Lee et al. 2007; Ossikovski et al. 2007; Hayazawa et al. 2007b) for the reflection mode TERS.

D. SELECTIVE DETECTION OF DIFFERENT ORGANIC MOLECULES BY TERS

In this section, we demonstrate the capability of direct molecular determination by TERS. Each molecule has its own characteristic molecular vibrations corresponding to different energies. Raman spectra can clearly distinguish the molecules, even nonfluorescent molecules, while most of the fluorescent spectra can easily overlap each other due to broad energy distribution in the room temperature. Aggregates of Rhodamine 6G (R6G) and Crystal Violet (CV) molecules were used as a sample for TERS spectroscopy and microscopy (Hayazawa et al. 2002). Samples are prepared by casting ethanol solution of R6G and dried on a cover slip covered with an 8-nm-thick silver film. The distribution of molecules for both samples was fairly inhomogeneous and was set to have a 1 nm average thickness of the layer of molecules. The silver film is one of very popular surface enhancer of SERS effect, and is effective not only in enhancement of Raman scattering cross section, but also in the reduction of the fluorescence due to the fluorescence energy transfer from the molecules to metal (Chance et al. 1978). Both R6G and CV have absorption at the excitation wavelength of 532 nm, and have fluorescence overlapped with Raman peaks. Consequently, the observed Raman spectra are due to surface enhanced resonant Raman scattering (SERRS). Figure 10.6 shows the TERS spectra of R6G (Figure 10.6a), CV (Figure 10.6b), and complex of R6G and CV, respectively

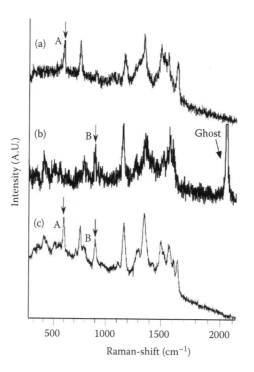

FIGURE 10.6 TERS spectra of (a) Rhodamine 6G, (b) Crystal Violet, and (c) aggregates of Rhodamine 6G and Crystal Violet. The Raman peaks at A and B were employed for TERS imaging.

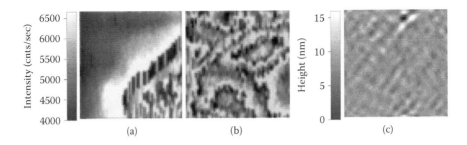

FIGURE 10.7 TERS images obtained at the specific vibration mode of (a) Rhodamine 6G and (b) Crystal Violet. The dimension of both images with 64×64 pixels is $1 \times 1 \ \mu m^2$. (c) The simultaneously obtained topographic image of the same area.

(Figure 10.6c). Several Raman-shifted lines of R6G and CV molecules are observed, and all these peaks are in good agreement with the former works by other authors (Nie and Emory 1997; Hildebrandt and Stockburger 1984; Sunder and Bernstein 1981; Watanabe and Pettingger 1982). In addition, Watanabe et al. recently reported the detailed analysis of vibration modes of R6G molecules in TERS spectroscopy (Watanabe et al. 2005).

To investigate the spatial resolving power of TERS spectroscopy and to identify molecular vibration distributions among different kinds of molecules, TERS images of an aggregated sample of R6G and CV molecules were obtained at characteristic Raman peaks of each molecule. Figure 10.7 shows TERS images at the same area of the sample. Figure 10.7a was obtained at 607 cm^{-1}, which corresponds to the Raman-shifted peak of the C-C-C in-plane bending vibration mode of R6G indicated by arrow "A" in Figure 10.6. Figure 10.7b was obtained at 908 cm^{-1}, which corresponds to the Raman-shifted peak of the C-H out-of-plane bending vibration mode of CV indicated by arrow "B" in Figure 10.6. Now we can selectively obtain the distributions of each vibration mode that we cannot distinguish in the topographic image shown in Figure 10.7c. The distributions of each vibration mode are quite different and show complicated structures reflecting the inhomogeneous distributions of both molecules. According to Figure 10.7a, R6G molecules are mainly localized at the lower right of the area. On the other hand, Figure 10.7b shows that CV molecules are randomly dispersed in the scanned area. In Figure 10.7c, the island structures of the silver film are observed in the topographic image because the average thickness of the aggregated molecular layer is at ~1 nm and much thinner than the silver film (average thickness: 8 nm). Accordingly, the distributions of both molecules are not clearly seen in the topographic image that reflects the pancake structure of the silver grains (30–50 nm in diameter and 8 nm in thickness). Note that without a metallic tip (far-field detection), it is impossible to obtain such high-resolution images because the far-field signal is averaged inside of the diffraction limited focused spot. In the case of the mixed sample used for imaging, both R6G and CV molecules are randomly dispersed on the silver-coated cover slip so that the far-field Raman signal from the focused spot is averaged to almost constant value. Figure 10.7 shows that the TERS images attain the molecular vibration distributions with a high sensitivity and high spatial resolution, even if the thickness of the molecular layer is 1 nm. Here, organic dye molecules adsorbed on silver films are used. While the silver films are necessary for dye molecules to quench the strong fluorescence, this configuration can work also as so-called "gap mode" (Xu et al. 1999; Hayashi and Konishi 2005) for the higher enhancement effect. With enough tip-enhancement effect without fluorescence we can carry out direct observation of molecular vibration without the aid of silver island films, e.g., adenine molecules, single-walled carbon nanotubes (Hayazawa et al. 2003). It should be noted that the signal enhancement discussed above was based on an electromagnetic mechanism due to tip-enhancement effect. However, chemical-enhancement effect (Watanabe et al. 2004; Hayazawa et al. 2006), which is observed in SERS (Otto et al. 1992; Kambhampati et al. 1998; Lu 2005), has also been reported in TERS for higher sensitivity and spatial resolution.

E. SELECTIVE DETECTION OF DIFFERENT VIBRATION MODES IN SINGLE-WALLED CARBON NANOTUBES

In section II D, we showed the sensitivity of TERS to distinguish different molecules by focusing on characteristic vibration modes of each molecule. In this section, we show another capability of TERS to characterize the structural difference of single-walled carbon nanotubes (SWNTs) as a sample.

Raman spectroscopy has been a promising tool to analyze SWNTs (Saito and Kataura 2001). The radial breathing mode (RBM), which appears in the low-frequency region, 100–400 cm^{-1}, in a Raman spectrum, is one of the most significant vibrational modes of SWNTs, because this mode has much stronger dependency on the diameter of the SWNTs than the other vibrational modes (Rao et al. 1997; Kataura et al. 1999). From the position of the RBM in a Raman spectrum, the diameter of SWNTs can be determined with an accuracy of an angstrom. However, due to the diffraction limit of light, conventional micro-Raman spectroscopy provides an averaged detection of Raman scattering from SWNTs lying inside the illumination focal spot (Jiang et al. 2003), and hence the information about diameter distribution within the diffraction-limited focal area is lost. In order to overcome this diffraction limit, TERS can be used. Several groups have observed individual isolated SWNTs by TERS (Hayazawa et al. 2003; Anderson et al. 2005).

Here, we introduce the structural sensitive detection of TERS on a bundle of SWNTs. In order to identify the diameter distribution within a bundle, diameter-selective TERS imaging of an isolated SWNT bundle, using three different frequencies of RBM preselected from a TERS spectrum of the bundle, was performed. The area surrounded by the dotted line in Figure 10.8a, which is a topographic image of the sample, indicates the bundle selected for this study. According to a diameter d dependence of the frequency of the RBM including the bundle effect (Henrard et al. 1999), i.e., d nm $= 232/(\omega\text{-}6.5)$ [cm^{-1}] (Alvarez et al. 2000), the spectrometer was fixed at frequencies 195 cm^{-1} (diameter: 1.23 nm), 244 cm^{-1} (diameter: 0.97 nm), and 278 cm^{-1} (diameter: 0.85 nm), and the TERS images were obtained, which are shown

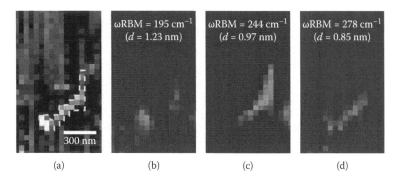

FIGURE 10.8 (a) An AFM image of a bundle of SWNTs. TERS images are obtained at the frequencies of (b) 195 cm⁻¹, (c) 244 cm⁻¹, and (d) 278 cm⁻¹.

in Figures 10.8b and d, respectively. Figure 10.8b reveals that the SWNTs having a diameter of 1.23 nm are localized at both edges of the bundle, Figure 10.8c shows that the SWNTs with a diameter of 0.97 nm are prominently distributed toward the central and upper part of the bundle, and Figure 10.8d indicates that the SWNTs with a diameter of 0.85 nm are distributed mainly toward the lower part of the bundle. With the use of TERS, it was clearly demonstrated that the structural difference of a single molecule could be differentiated into color variations (different Roman frequencies) that correspond to specific vibration energies of each structure.

In section II D, we introduced the utilization of chemical enhancement effect for higher sensitivity in TERS. Here, it should be pointed out that in addition to electromagnetic enhancement and chemical enhancement effects, physical deformation induced by tip-applied force showed extra enhancement effect in TERS on carbon materials such as SWNTs and fullerene molecules (Yano et al. 2005, 2006; Verma et al. 2006). This tip-pressurized effect is a unique feature of TERS and not observable in SERS. Since the spatial resolution of TERS with tip-pressurized effect is determined by the size of the very end of the metallic tip that has direct contact with the molecules, this is a very promising approach to improve the spatial resolution of the near-field microscope.

In section II, we introduced TERS for molecular vibrational identification in the nanoscale, which should be a promising technique for biochemical applications, since the strong requirement in near-field microscopes, including TERS, is to achieve higher spatial resolution. However, the higher the spatial resolution is, the smaller the signal level becomes. An improvement in sensitivity is indispensable to microscopic detections in the nanoscale. One of the ways to answer this requirement is to optimize the tip-enhancement effect. The ways to increase the field enhancement factor at the metallic coated tip apex include tip material and geometry modification (Sánchez et al. 2002; Ropers et al. 2007), excitation wavelength selection (or tuning plasmon resonance) (Nordlander and Le 2006; Cui et al. 2007), optical system (illumination and detection) optimization and proper use of probing light polarization (Novotny et al. 1998; Hayazawa et al. 2004b; Chen and Zhan 2007; Saito et al.

2005; Ichimura et al. 2007). Another way is to suppress the far-field background Raman signals from the light-illuminated spot by an objective lens, because when the observing volume becomes much smaller than the diffraction-limited focused spot, the far-field Raman signals from the entire focused spot may overwhelm the TERS signal: in particular, this background signal would be serious when observing bulk materials such as biological cells. The most efficient way to improve the signal-to-noise ratio (SNR) in TERS is to increase the field enhancement and suppress the background signal at the same time. The answer for these two requirements is to use nonlinear optical response of the materials. In section III, we introduce the combination of nonlinear optics and near-field optics toward higher sensitivity and resolution for biochemical samples.

III. NONLINEAR EFFECT IN TIP-ENHANCED MICROSCOPY AND SPECTROSCOPY

When tip-enhanced microscopy and spectroscopy are combined with high-order nonlinear optical effects, the spatial resolution can be essentially improved. Photons generated by the nonlinear optical effects are spatially confined in a smaller volume than the size of the tip-enhanced field of a fundamental frequency because the intensity of the nonlinear effects is proportional to the high-order powers (square, cube, etc.) of the excitation light intensity. The spatial distribution of the signal emission (harmonic signal, two-photon excited fluorescence, etc.) becomes narrower than the intensity distribution of the excitation field. This leads to the reduction of the effective volume of light-matter interaction. Higher-order optical effects give much finer spatial response (Figure 10.9a). This concept has been realized in far-field optical microscopy and optical fabrication (Denk et al. 1990; Kawata et al. 2001). In addition to the spatial confinement, because of the nonlinear responses, even a small enhancement of the excitation field could lead to a huge enhancement of the emitted signal, allowing for reduction of the far-field background.

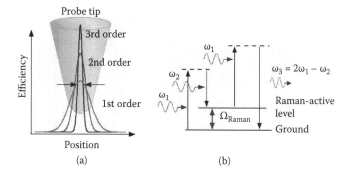

FIGURE 10.9 (a) Spatial confinement of the excitation efficiency of higher-order nonlinear effects. The first-order function represents the distribution of the tip-enhanced field intensity as shown in Figures 10.2 and 10.3. (b) Energy diagram of the CARS process.

A. Tip-Enhanced Coherent Anti-Stokes Raman Scattering

In order to realize molecular-vibration spectroscopy, coherent anti-Stokes Raman scattering (CARS) spectroscopy is employed, which is one of the most widely used nonlinear Raman spectroscopes (Shen 1984). CARS spectroscopy uses three incident fields including a pump field (ω_1), a Stokes field (ω_2; $\omega_2 < \omega_1$), and a probe field ($\omega_1' = \omega_1$), and induces a nonlinear polarization at the frequency of $\omega_3 = 2\omega_1 - \omega_2$, which is given in a scalar form by

$$P_{CARS}^{(3)}(\omega_3 = 2\omega_1 - \omega_2) = \chi^{(3)} E(\omega_1) E^*(\omega_2) E(\omega_1) \qquad (10.1)$$

where $\chi^{(3)}$ represents the third-order nonlinear susceptibility, $E(\omega_1)$ and $E(\omega_2)$ are the electric fields for excitation light. $E^*(\omega_2)$ denotes the complex conjugate of $E(\omega_2)$. The nonlinear susceptibility is expressed by vibration-resonant term ($\chi^{(3)}{}_R$) and non-resonant term ($\chi^{(3)}{}_{NR}$):

$$\chi^{(3)} = \chi^{(3)}_R + \chi^{(3)}_{NR} = \frac{A}{\Omega_{Raman} - (\omega_1 - \omega_2) - i\Gamma} + \chi^{(3)}_{NR} \qquad (10.2)$$

The coefficient of the fraction, A, denotes a constant related to the strength of the vibration, Ω_{Raman} denotes one of the specific molecular vibrational frequencies of a given sample, and Γ corresponds to the spectral bandwidth of the vibration mode. When the frequency difference of ω_1 and ω_2 ($\omega_1 - \omega_2$) coincides with Ω_{Raman}, the anti-Stokes Raman signal is resonantly generated. Figure 10.9b shows an energy diagram for the CARS process. $\chi^{(3)}_{NR}$ is a contribution from transition process, which does not undergo the vibrational state. In particular, the process, which undergoes the $2\omega_1$ state, may be resonant or pre-resonant to an electronic state, resulting in strong contribution to the susceptibility given by Equation (10.2).

One can obtain a CARS spectrum by plotting the CARS signal intensity with sweeping ω_2. The CARS spectrum gives essentially identical information with spontaneous Raman spectra (Levenson and Kano 1988). In ordinary CARS spectroscopy, the propagation angles of incident electric fields have to fulfill the phase-matching condition, $k_{CARS} = 2k_1 - k_2$, to induce CARS polarization (Shen 1984). However, when the CARS polarization is induced in a volume smaller than the wavelength of CARS field, the phase-matching condition is automatically satisfied. In the small volume, the induced polarization can oscillate in phase, and the wave vector of CARS field loses the relation with the incident excitation field. This concept has been commonly noticed in laser scanning CARS microscopy (Zumbusch et al. 1999; Hashimoto et al. 2000; Hashimoto and Araki 2001; Cheng et al. 2001), in which excitation beams are focused into a volume smaller than wavelengths by a high NA objective lens. Furthermore, surface enhanced CARS was also reported, in which CARS was amplified by isolated gold colloidal nanoparticles (Liang et al. 1994; Ichimura et al. 2003). These reports verified the possibility of the local enhancement of CARS by a metallic nanostructure. Based on this concept, one can observe CARS signals generated by the enhanced field at a metallic tip end of nanometric scale.

The nonlinear polarization of tip-enhanced CARS (TE-CARS) is expressed by

$$P^{(3)}_{\text{TECARS}}(\omega_3) = L(\omega_3)\{\chi^{(3)}[L(\omega_1)E(\omega_1)][L^*(\omega_2)E^*(\omega_2)][L(\omega_1)E(\omega_1)]\}$$
$$= L(\omega_3)L(\omega_1)L^*(\omega_2)L(\omega_1)\{\chi^{(3)}E(\omega_1)E^*(\omega_2)E(\omega_1)\}$$

(10.3)

where $L(\omega_i)$ ($i = 1,2,3$), is referred to as "local field factor" which represents the factor of enhancement at a given wavelength at a particular position, e.g., at the tip apex. Field enhancement is effective in each of the fields, including the CARS field. On the other hand, tip-enhanced normal Raman scattering can be expressed by

$$P^{(1)}_{\text{TERS}}(\omega_2) = L(\omega_2)\left\{\chi^{(1)}[L(\omega_1)E(\omega_1)]\right\}$$
$$= L(\omega_2)L(\omega_1) \times \left\{\chi^{(1)}E(\omega_1)\right\}$$

(10.4)

where $\chi^{(1)}$ is a linear susceptibility for normal Raman scattering. From Equations (10.3) and (10.4), the net enhancement factor for CARS has higher order dependence on the enhancement factors. Owing to the high-order nonlinearity of the CARS process, one can expect huge enhancement for CARS intensity, which could be much higher than that for normal Raman scattering. The high-order nonlinearity results in not only increase of the signal intensity but also relative reduction of far-field background, which is the CARS/Raman signal generated at molecules far from the tip-enhanced spot. At times, far-field background makes near-field optical images difficult to interpret, and was previously discussed in TERS imaging (Hayazawa et al. 2001; Mehtani et al. 2005). In TE-CARS microscopy, the near-field contribution becomes dominant to the far-field contribution, allowing one to interpret the obtained images in a simple way. In terms of spatial resolution, the CARS polarization can be further confined to the very end of the tip in comparison with normal Raman scattering (Figure 10.9a). Under the assumption that a spatial distribution of the local field factor to take a $1/r^3$ form, where r denotes the distance from the tip center (Bohren and Huffman 1983), spatial distribution of the TE-CARS polarization would have a $1/r^{12}$ dependence, while that of TERS polarization would have a $1/r^6$ dependence. This leads to an essential difference of spatial distribution of TE-CARS and TERS.

B. TE-CARS MICROSCOPY CONFIGURATION

Figure 10.10 shows the experimental system of TE-CARS microscopy (Ichimura et al. 2004a). As similar to the TERS system (Hayazawa et al. 2000), the system mainly consists of an excitation laser, an inverted microscope, an AFM using a silver-coated probe, and a monochromator. Two mode-locked Ti:sapphire lasers (pulse duration: 5 picoseconds [ps]; spectral band width: 4 cm^{-1}; repetition rate: 80 MHz) are used for the excitation of CARS. The ω_1 and ω_2 beams are collinearly combined in time and space, and introduced into the microscope with an oil-immersion objective lens (NA = 1.4) focused onto the sample surface. As the z-polarized component of the

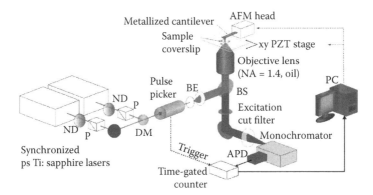

FIGURE 10.10 An experimental system of tip-enhanced CARS microscopy. See the text for detail. ND: neutral-density filter, P: polarizer, DM: dichroic mirror, BE: beam expander, BS: beam splitter, APD: avalanche photo diode.

electric field along the tip axis is dominant in the tip-enhanced local field, the CARS polarizations are induced along the z-direction. For effective coupling of incident fields and the local fields, the tip has to be at a position where the incident electric field in the z-direction is strong (Hayazawa et al. 2004b). An electro-optically modulated pulse picker controls the repetition rate of the excitation lasers. The backscattered CARS emission enhanced by the probe tip is collected with the objective lens and detected with an avalanche photodiode–based photon-counting module through an excitation-cut filter and the monochromator. The observing spectral width through the detection system is about 12 cm^{-1}. The pulse signals from the avalanche photo-diode (APD) are counted by a time-gated photon counter synchronously triggered with the pulse picker, which effectively reduces the dark counts down to almost 0 counts/s. Scanning the sample stage, while keeping the tip at the focused spot, can acquire two-dimensional TE-CARS images of a specific vibrational mode with a high spatial resolution.

C. DNA IMAGING BY TE-CARS

DNA imaging by TE-CARS microscopy was performed. We prepared two types of samples: (i) DNA clusters and (ii) DNA network. For preparation of DNA clusters, a poly(dA-dT) solution in water (250 μg/mL) was cast and dried on a cover slip in room temperature with a fixation time of ~24 hr. The dimensions of the clusters are typically ~20 nm in height and ~100 nm in width. For DNA network, DNA [poly(dA-dT)-poly(dA-dT)] dissolved in water (250 μg/mL) was mixed with MgCl$_2$ (0.5 μM) solution, then the DNA solution was cast on a cover slip and blow-dried after a fixation time of ~2 hr (Tanaka et al. 2001). Mg^{2+} has a role in the linkage between DNA and oxygen atoms of the glass surface. The DNA network consists of bundles of DNA double-helix filaments aligned parallel on the glass substrate. Since the diameter of single DNA double-helix filaments is ~2.5 nm, the height of

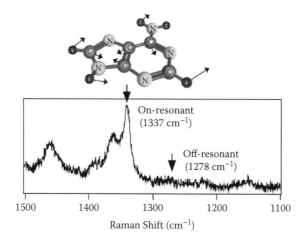

FIGURE 10.11 Tip enhanced spontaneous Raman spectrum of DNA sample (poly(dA-dT)-poly(dA-dT)). The vibration at 1337 cm^{-1} is employed as the target vibration. The corresponding vibration mode is illustrated in the figure.

the bundle structures is ~2.5 nm, and the width is from 2.5 nm (for single filaments) to a few tens of nanometers (for approximately 10 filaments). The frequency difference of the two excitation lasers was set to be 1337 cm^{-1} ($\omega_1 - \omega_2 = 1337$ cm^{-1}) corresponding to a Raman mode of adenine by tuning the excitation frequencies ω_1 and ω_2 to be 12,710 cm^{-1} (λ_1: 786.77 nm) and 11,373 cm^{-1} (λ_2: 879.25 nm), respectively. After the "on-resonant" imaging, the frequency of ω_2 was changed such that the frequency difference corresponds to none of the Raman-active vibrations ("off-resonant"). Figure 10.11 shows a normal Raman spectrum of the DNA in a part of the fingerprint region. The solid arrows on the spectrum denote the frequencies adopted for the "on-resonant" and "off-resonant" conditions in CARS imaging.

Figure 10.12 shows the TE-CARS images of the DNA clusters obtained by our system. Figures 10.12a and b are the TE-CARS image at the on-resonant frequency (1337 cm^{-1}) and the simultaneously acquired topographic AFM image. The DNA clusters of ~100 nm diameter are visualized in Figure 10.12a. The two DNA clusters with distance of ~160 nm are obviously distinguished by the tip-enhanced CARS imaging. This indicates that the CARS imaging successfully achieved super-resolving capability beyond the diffraction limit of light. At the off-resonant frequency (1278 cm^{-1}), the CARS signals mostly vanished in Figure 10.12c. Figures 10.12a and c verify that DNA molecules emit vibrationally resonant CARS at the specific frequency. However, there remains some slight signal increase at the clusters at the off-resonant frequency, as seen in Figure 10.12d, which is the same as Figure 10.12c but is shown with a different gray scale. This can be caused by both the frequency-invariant (nonresonant) component of the nonlinear susceptibility of DNA (Shen 1984) and the topographic artifact (Hecht et al. 1997). Figure 10.12e is a CARS image at the on-resonant frequency, which was obtained after removing the tip from the sample. The CARS signal was not detected in the CARS image

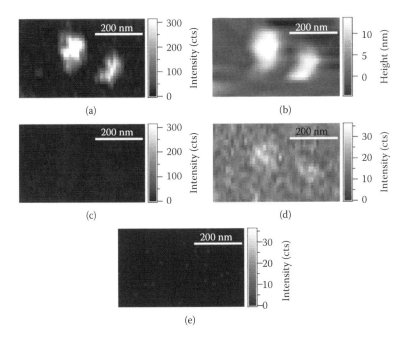

FIGURE 10.12 CARS images of the DNA clusters. (a) TE-CARS image at on-resonant frequency (1337 cm⁻¹), and (b) the simultaneously obtained topographic image. (c) TE-CARS image at the off-resonant frequency (1278 cm⁻¹). (d) The same image as (c) shown with a different gray scale. (e) CARS image of the corresponding area obtained without the silver tip. The scanned area is 500 nm × 300 nm. The number of photons counted in 100 ms was recorded for one pixel. The acquisition time was ~3 minutes for the image. The average powers of the ω_1 and ω_2 beams were 30 μW and 15 μW at the 800 kHz repetition rate.

without the silver tip, which confirms that the tip-enhanced field effectively induces the CARS polarization.

CARS images of the DNA network structure at the on- and off-resonant frequencies are shown in Figures 10.13a and b. The DNA bundles are observed at the resonant frequency in Figure 10.13a, while they cannot be visualized at the off-resonant frequency in Figure 10.13b. This indicates that the observed contrast is dominated by the vibrationally resonant CARS signals. Figure 10.13c shows one-dimensional profiles of the row indicated by solid arrows, which were acquired with a 5 nm step. The line profile acquired without the silver tip is also added for comparison. Only the CARS profile with the tip in the on-resonant condition has peaks at the positions of $x \sim 370$ nm and $x \sim 700$ nm where adenine molecules exist in the DNA double helix, while the other line profiles do not sense the existence of the molecules. The full width of half maximum of the right peak is found to be ~15 nm, as indicated in Figure 10.13c. This spatial resolution is much better than our previous works using spontaneous Raman scattering, which can be attributed to the nonlinearity of the CARS process.

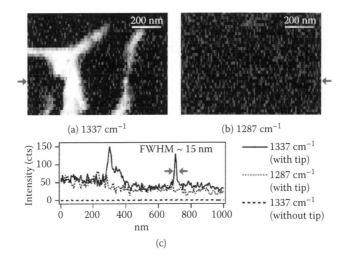

(a) 1337 cm^{-1} (b) 1287 cm^{-1}

FWHM ~ 15 nm

—— 1337 cm^{-1} (with tip)

········ 1287 cm^{-1} (with tip)

---- 1337 cm^{-1} (without tip)

(c)

FIGURE 10.13 CARS images of the DNA network. (a) TE-CARS image at on-resonant frequency (1337 cm^{-1}). (b) TE-CARS image at the off-resonant frequency (1278 cm^{-1}). (c) One-dimensional line profiles of the row indicated by the solid arrows. The scanned area is 1000 nm × 800 nm. The number of photons counted in 100 ms was recorded for one pixel. The acquisition time was ~12 minutes for the image. The average powers of the ω_1 and ω_2 beams were 45 μW and 23 μW at the 800 kHz repetition rate.

D. Nonlinear Emissions from Tip

It can be found that there exists background light in the presence of the tip, as is obvious from Figure 10.12. This background light is emitted from the silver-coated tip. The tip emits light at the same frequency as the CARS ($2\omega_1 - \omega_2$) by the third-order nonlinear susceptibility of silver, which is attributed to local nondegenerate four-wave mixing (NDFWM). In addition, noble metals such as gold and silver generate white light continuum (WLC), which is induced by multiphoton excited photoluminescence due to recombination radiation between electrons near Fermi level and photoexcited holes in the d band (Boyd et al. 1986; Wilcoxon and Martin 1998). These two components become background light and compete with the CARS process. Figure 10.14a is an experimental result of the nonlinear emission from a gold tip detected by a spectrometer with a liquid nitrogen–cooled CCD camera instead of the APD. Silver exhibits almost the same spectrum in this region. Both NDFWM emission at the peak of $2\omega_1 - \omega_2$ and the broad white continuum emission are seen in the spectrum. The laser frequencies of ω_1 and ω_2 are adjusted at 12,703.3 cm^{-1} (787.2 nm) and 11,396.0 cm^{-1} (877.5 nm), respectively. Accordingly, the NDFWM emission $2\omega_1 - \omega_2$ is observed at 14,010.6 cm^{-1} (713.7 nm). The band-pass filter used in the detection system rejects the light >720 nm and <690 nm in the results of Figure 10.14a; however, the WLC emission covers almost the whole visible wavelength region. In addition to the light emission due to NDFWM, metal tips also emit light at frequencies of $2\omega_1$, $2\omega_2$ (SHG) and $\omega_1 + \omega_2$ (sum-frequency generation [SFG])

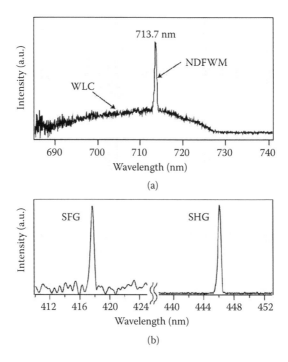

FIGURE 10.14 (a) A spectrum of non-degenerate four wave mixing (NDFWM) and white light continuum (WLC) emission from a gold tip. (b) A spectrum of second harmonic generation (SHG) and sum frequency generation (SFG) from a gold tip.

due to second-order nonlinear response at the metal surface (Zayats and Sandoghdar 2000; see also section IV A). Figure 10.14b shows a spectrum of the nonlinear emissions from a gold tip. For this measurement, the pump laser frequencies of ω_1 and ω_2 are adjusted at 12,733.3 cm^{-1} (785.3 nm) and 11,209.0 cm^{-1} (892.1 nm), respectively. In Figure 10.14b, strong peaks are observed at 417.7 nm and 446.1 nm, which exactly correspond to $\omega_1 + \omega_2$ and $2\omega_2$, respectively ($2\omega_1$ is out of range of the detection system). The nonlinear emissions can be used as a light source for optical nano-imaging at the blue to violet region.

In the experiments of Figures 10.12 and 10.13, the dominant background source is the NDFWM emission because the monochromator was used to selectively detect the signal at $2\omega_1 - \omega_2$. The background light can be seen at both the on-resonant and off-resonant frequencies, as they are independent of the molecular vibrations of the sample. Such light emission from a metallic tip degrades the image contrast and signal-to-noise ratio, and subsequently limits the smallest number of molecules that can be observed.

In Figure 10.12, the TE-CARS signal intensity largely surpasses the background because the number of molecules in the excited volume is enough to induce the large signal. It, however, depends on experimental conditions such as the number of molecules, excitation laser power, and optical density of samples. In other cases, the

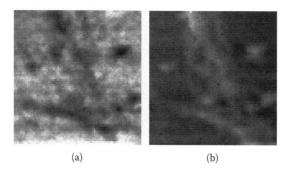

(a)   (b)

FIGURE 10.15 Inverted contrast between (a) an optical image and (b) a simultaneously obtained topographic image. The sample is SWNTs. The dimension of the images is 1 μm × 1 μm consisting of 80 pixels × 80 pixels.

vibrationally resonant CARS signal of the sample is sometimes overwhelmed by the nonresonant NDFWM emission from the tip. This is because when the average power or repetition rate is high, the local temperature of the observed sample increases so that the number of molecules at the vibrationally excited state increases, which results in the reduction or saturation of the vibrationally resonant CARS process, while the nonlinear emission from the tip is not saturated. Sensitivity of TE-CARS is quite varied due to the saturation effect of the CARS signal, which is strongly dependent on the sample species and the volume. In the case of Figure 10.13, the repetition rate is set to 800 kHz and the average laser power is adjusted to 45 μW for ω_1 and 23 μW for ω_2. Under these conditions, the saturation and the nonlinear emission are well suppressed.

Figure 10.15a is a typical near-field image reflecting the effect of the nonlinear emission from the tip (Hayazawa et al. 2004a). The sample is single-walled carbon nanotube (SWNT) bundles. Figure 10.15b shows a simultaneously observed topographic image. In this imaging, the repetition rate of the two Ti:sapphire lasers are set to 80 MHz while the peak power is the same as in Figure 10.12. As is clearly seen, the contrast of the optical image in Figure 10.15a is inverted from the topographic image in Figure 10.15b. The nonlinear emission from the tip is reduced when the tip is located at the SWNTs, while the emission is very strong when the tip is at the glass substrate (no sample). This can be attributed to the fact that the tip is retracted away from the glass substrate where the excitation light is tightly focused, which results in the lower evanescent excitation field, and accordingly the nonlinear emission is reduced.

IV. COMBINATION OF TIP-ENHANCEMENT AND SECOND-ORDER NONLINEAR SPECTROSCOPY

In Section III, we discussed the combination of near-field microscopy and third-order nonlinear spectroscopic technique, CARS. Since near-field microscopy is a surface sensing technique, even-order nonlinear processes, which require non-centrosymmetric conditions resulting in high surface sensitivity (Shen 1984), are the promising candidates for near-field nonlinear spectroscopy. In this section, we review mainly second-order nonlinear processes for biochemical applications in the nanoscale.

A. TIP-ENHANCED SHG MICROSCOPY TOWARD BIO-APPLICATIONS

Second harmonic generation (SHG) is one of the most intensively studied nonlinear optical effects that have ever been combined with near-field scanning optical microscopy (Shen et al. 2000; Zayats and Sandoghdar 2000; Zayats and Sandoghdar 2001; Takahashi and Zayats 2002). SHG, which is an even-order nonlinear process, is forbidden in centrosymmetric media under the dipole approximation (Shen 1984). Non-centrosymmetric molecules and lattices are allowed to exhibit SHG light. The second-order nonlinear polarization for SHG ($P^{(2)}_{SHG}$) is given in a scalar form by

$$P^{(2)}_{SHG}(\omega_2 = 2\omega_1) = \chi^{(2)}E(\omega_1)E(\omega_1) \tag{10.5}$$

where $\chi^{(2)}$ and $E(\omega_1)$ are the second-order nonlinear susceptibility for SHG and electric field at the incident frequency (ω_1), respectively. $\chi^{(2)}$ for a bulk of a centrosymmetric material is zero. However, even centrosymmetric materials possess nonzero values for $\chi^{(2)}$ at their surface because the symmetry is broken in the direction perpendicular to the surface. When a metallic nano-tip is in close vicinity of surface of a given sample, the tip locally enhances SHG. In a scalar form, the tip-enhanced (TE)-SHG polarization is given by

$$P^{(2)}_{TESHG}(\omega_2 = 2\omega_1) = L(\omega_2)\left\{\chi^{(2)}\left[L(\omega_1)E(\omega_1)][L(\omega_1)E(\omega_1)]\right]\right\}$$
$$= L(\omega_2)L(\omega_1)L(\omega_1)\left\{\chi^{(2)}E(\omega_1)E(\omega_1)\right\} \tag{10.6}$$

where $L(\omega_i)$ ($i = 1, 2$) is the local enhancement factor at ω_1 and ω_2. $L(\omega_i)$ is also dependent on the material and structure of the tip, and position on the surface. In addition to other nonlinear processes, TE-SH light can also be spatially confined in the very end of the tip owing to the nonlinear response to incident light intensity (Kawata et al. 1999).

A typical experimental system for TE-SHG microscopy basically consists of a pulsed laser, an AFM, and an inverted optical microscope. The system is not very different from the TERS and TE-CARS system described in previous sections. The only difference is the use of an optical filter that selectively passes the SH light.

One of the most important and interesting features of TE-SHG microscopy is that SH light is emitted not only from a sample but also from a metallic tip (Figure 10.14). Therefore, in order to correctly understand TE-SH images, magnitude correlation of susceptibilities of tip metal ($\chi^{(2)}_{tip}$) and sample ($\chi^{(2)}_{sample}$) has to be taken into consideration. When $\chi^{(2)}_{sample}$ is sufficiently larger than $\chi^{(2)}_{tip}$ ($\chi^{(2)}_{tip} < \chi^{(2)}_{sample}$), one can realize $\chi^{(2)}$ mapping on the given sample. So far, TE-SHG imaging of metal film surface (Zayats and Sandoghdar 2001) and ferroelectric surface (Mahieu-Williame et al. 2007) were realized and reported. In the opposite case ($\chi^{(2)}_{tip} > \chi^{(2)}_{sample}$), it is difficult to achieve SHG imaging of the sample. Instead, one can use the SH light generated by the tip as a nanosized light source to illuminate samples.

TE-SHG microscopy is a promising tool for nano-imaging of biomolecules. A variety of structural proteins have large susceptibilities and often compose well-ordered non-centrosymmetric secondary structures such as an α-helix. Collagen fibers take the form of triplet-twisted helix in rat tail tendon, and are one of the biomaterials that have been extensively observed by laser scanning far-field SHG microscopy (Chu et al. 2004). Molecular orientations of collagen fibers have also been observed by SHG microscopy. Other proteins including myosin and tubulin were also reported to be SHG active and were successfully observed by SHG microscopy (Campagnola et al. 2002; Boulesteix et al. 2004). Another advantage of SHG microscopy is that it can be used to measure the local static electric field across the membrane (membrane potential) because static field modifies $\chi^{(2)}$ and accordingly SH light intensity (Ben-Oren et al. 1999; Bouevitch et al. 1993). A promising study was reported by Peleg et al. in 1999, where the authors observed surface enhanced SHG from gold-sphere attached protein molecules on a living cell membrane (Peleg et al. 1999). This study guarantees TE-SHG microscopy to be capable of monitoring the membrane proteins with a high spatial resolution of tip size.

B. Tip-Enhanced SFG Microscopy

Sum-frequency generation (SFG) is also widely recognized as a powerful tool for intrinsically surface-specific spectroscopy (Shen 2000). As shown in Figure 10.16a, SFG spectroscopy, which requires two color input beams, is technically rather complicated, but one can expect additional features. For example, SFG is now expected to be a new probe for molecular chirality (Belkin et al. 2000; Champagne et al. 2000; Ji et al. 2006). From the viewpoint of biomolecular imaging, this feature is more attractive. Therefore, we hereinafter focus on sum-frequency (SF) chiral

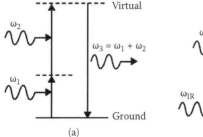
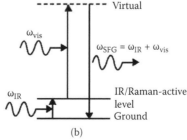

FIGURE 10.16 Energy diagrams of (a) SFG process and (b) vibrational SFG process.

spectroscopy. The dipole moment μ induced in a molecule placed in two electromagnetic fields with frequencies of ω_1 and ω_2 is given by

$$\mu_i = \mu_{0i} + \alpha_{ij} \cdot (E_j(\omega_1) + E_j(\omega_2)) +$$

$$1/2\beta_{ijk} : (E_j(\omega_1)E_k(\omega_1) + E_j(\omega_1)E_k(\omega_2) + E_j(\omega_2)E_k(\omega_2)), \quad (10.7)$$

where μ_0, α, and β are the permanent dipole moment, the polarizability tensor, and the hyperpolarizability tensor, respectively, and i, j, k refer to the molecular coordinates. The third term is responsible for the second-order nonlinear scattering; specifically, the $\beta{:}E(\omega_1)E(\omega_2)$ acts as a source of SFG. A nonlinear optical susceptibility of a bulk medium, $\chi^{(2)}$, is related to the molecular hyperpolarizability β by

$$\chi^{(2)}_{i'j'k'} \propto N \sum_{ijk} \langle (i' \cdot i)(j' \cdot j)(k' \cdot k) \rangle \beta_{ijl} \quad (10.8)$$

where N is the molecular density and i', j', k' refer to the laboratory coordinates. The angular brackets denote an average over the molecular orientation. Then, the nonlinear polarization as a source of coherent SFG from the bulk medium is described as follows:

$$P^{(2)}_{\mathrm{SFG}i'}(\omega_3 = \omega_1 + \omega_2) = \chi^{(2)}_{i'j'k'} : E_{j'}(\omega_1)E_{k'}(\omega_2). \quad (10.9)$$

As one can see, SHG is the specific case of SFG excited with $\omega_1 = \omega_2$. Note that molecular chirality is intrinsically a three-dimensional notion. Therefore, when $\omega_1 \neq \omega_2$, the $\chi^{(2)}$ for a chiral medium has non-vanishing elements even in isotropic liquid, meaning that SFG can be chirality specific. Importantly, chiral information is obtained as signal amplitude in SF spectroscopy. This is a great advantage for combination with tip-enhanced near-field spectroscopy because the enhancement effect is straightforward. Conversely, circular dichroism (CD), which has been more commonly used for chiral spectroscopy, is not suitable; chiral signal appears as small modulation of light polarization.

SF (chiral) spectroscopy can be easily modified for vibrational spectroscopy as shown in Figure 10.16b. This SF vibrational spectroscopy can benefit from plasmon resonances of a metallic tip by tuning the input beam frequency ω_{vis} and/or the output frequency ω_{SFG}. Therefore, spatial resolution of SF vibrational spectroscopy could reach to the order of several nanometers. Development of tip-enhanced SF chiral/vibrational spectroscopy is awaited for biomolecular imaging.

C. Tip-Enhanced Hyper-Raman Microscopy

SF radiation may be accompanied with *inelastic* nonlinear scattering, although the scattering cross section is extremely small, typically in the order of 10^{-65} cm^4 per molecule. Such an inelastic scattering effect in the second-order nonlinear optical

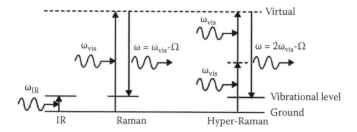

FIGURE 10.17 Energy diagram of IR absorption, Raman scattering, and hyper-Raman scattering.

effects, which can be understood by analogy of Raman scattering, is called hyper-Raman scattering (Denisov et al. 1987). Since hyper-Raman spectroscopy is usually used under the condition of $\omega_1 = \omega_2$, Equation (10.7) is simplified as follows:

$$\mu_i = \mu_{0i} + \alpha_{ij} \cdot E_j + 1/2\beta_{ijk} : E_j E_k \tag{10.10}$$

In Equation (10.10), the first-order expansion of the μ_0, α, β in the normal coordinate of vibration Q is, respectively, responsible for IR absorption, Raman scattering, and hyper-Raman scattering (Figure 10.17). According to the parity, one can easily understand that selection rules for hyper-Raman scattering are rather similar to those for IR (Cyvin et al. 1965; Christie and Lockwood 1971). Therefore, hyper-Raman spectroscopy can, in principle, be used as an alternative for IR spectroscopy and its spatial resolution is expected to be much better than IR microscopy (Shimada et al. 2006). Note that IR-mode detection by hyper-Raman spectroscopy is possible even in IR-opaque media, indicating a possible use of this technique for three-dimensional imaging of biological cells. In addition, hyper-Raman scattering can probe some silent modes, which are IR- and Raman-inactive vibrational modes (Ikeda and Vosaki 2008). Therefore, most of the vibrational information would be obtained from a combination of Raman and hyper-Raman spectroscopy. It is also possible to combine hyper-Raman spectroscopy with tip-enhanced near-field spectroscopy in a similar manner to Raman spectroscopy. Indeed, surface enhanced hyper-Raman scattering has already been reported in metal colloidal systems (Kneipp et al. 2006; Leng et al. 2006) or in artificial metal dimer systems (Ikeda et al. 2007). Tip-enhanced hyper-Raman microscopy could be a unique imaging tool for biomolecules.

D. Multipole Contributions in Tip-Enhanced Near-Field Spectroscopy

The above-mentioned nonlinear optical effects can be described by the perturbation of the electromagnetic field intensity under the electric dipole approximation. Actually, this approximation is broken in optical near-fields. Hence, a perturbation effect of multipole such as electric quadrupole or magnetic dipole should also be considered, although such a higher-order effect is normally negligible. Indeed, electric quadrupole contributions can be comparable with electric dipole contributions

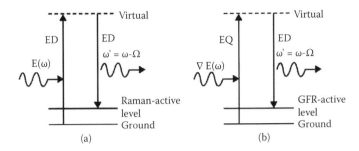

FIGURE 10.18 Comparison of (a) Raman scattering process and (b) gradient-field Raman (GFR) scattering process. ED and EQ denote electric dipole and electric quadrupole transitions, respectively.

near metallic surfaces, whereas magnetic dipole contributions remain small (Sass et al. 1981). By considering electric field gradients, the dipole moment induced in a molecule excited by an electromagnetic field is given by

$$\mu_i = \mu_{0i} + \alpha_{ij} \cdot E_j + 1/3 A_{ijk} : \nabla_j E_k + \cdots,$$ (10.11)

where ∇E is the electric field gradient and A is the dynamic molecular property tensor given by Buckingham et al. (1987). The first-order expansion of A in the Q is responsible for gradient-field Raman scattering (GFR). Raman scattering and GFR processes are illustrated in Figure 10.18. By comparison of the A in Equation (10.11) with the hyperpolarizability tensor β in Equation (10.10), one can notice that selection rules for GFR are the same as those for hyper-Raman scattering, meaning that IR-active modes can appear in Raman spectra if the magnitude of the GFR term becomes large enough. In the case of tip-enhanced near-field spectroscopy, the enhanced electric field is strongly localized near the tip end, and then GFR is expected to be observable. Indeed, there has already been reported an appearance of Raman-forbidden modes in aperture type near-field Raman spectroscopy (Ayars et al. 2001) because the evanescent field localized at the small aperture of a sharpened optical fiber probe also exhibits strong gradient field even though the field intensity is very weak due to the propagation loss in the optical fiber (see section II A). On the other hand, in the case of apertureless type, since the magnitude of field gradients is controllable and enhanced by incident field intensity and distance between a tip and sample surfaces, tip-enhanced GFR spectroscopy seems to be a promising tool for advanced near-field microscopy.

V. CONCLUSION

In this chapter, we showed the capability of near-field optical spectroscopy combined with vibrational spectroscopy and nonlinear optics for biochemical applications. The evanescent field localized at the nanoscale tip realized the extremely small light source for various spectroscopes in the near-field. Especially when the tip is made of metals, the evanescent field is strongly enhanced due to the excitation of surface plasmon polaritons, which can compensate for the possible small signal from the nanoscale region. We demonstrated the abilities of identifying different organic molecules and different vibration modes depending on Raman active molecular vibrations by TERS. For further improvement of the spatial resolution and sensitivity, we applied tip-enhancement effect to third-order nonlinear vibrational spectroscopy, TE-CARS. Owing to the nonlinear response of the materials, the photons are further confined at the tip apex, resulting in higher spatial resolution. As of now, 15-nm spatial resolution has been demonstrated in TE-CARS and the spatial resolution <10 nm is promising in the near future. In the last section, other possible nonlinear spectroscopic techniques such as SHG and SFG were discussed in terms of surface sensitive second-order nonlinear vibrational spectroscopy since near-field microscopes are also surface sensing techniques. Finally, we discussed the possibility of sensing IR-active vibration modes by high gradient field at the metallic tip apex. With this gradient-field Raman scattering, we will be able to detect both Raman active and IR-active modes at once. This interesting feature in the near-field is not possible in micro Raman measurements using a conventional objective lens.

In order to expand the number of users of TERS, TE-CARS, or other tip-enhanced spectroscopes in the field of biochemistry, the important issues are (i) the reproducibility of the tip-enhancement, which strongly depends on the tip shape, material, and light polarization to induce optimal plasmon modes; and (ii) the stability in the ambient conditions, because one of the biggest advantages of near-field microscopes is the feasibility of the measurements of room temperature, which causes undesirable fluctuations from circumstances such as vibrations and temperature. We believe that recent progress in nanoscale sciences could solve these difficulties soon because these are somewhat common problems among nanoscale researchers when pursuing higher spatial resolution.

REFERENCES

Albrecht, M. G., and Creighton, J. A. 1977. Anomalously intense Raman spectra of pyridine at a silver electrode. *J. Am. Chem. Soc.* 99:5215–17.

Alvarez, L., Righi, A., Guillard, T., Rols, S., Anglaret, E., Laplaze, D., and Sauvajol, J. L. 2000. Resonant Raman study of the structure and electronic properties of single wall carbon nanotubes. *Chem. Phys. Lett.* 316:186–90.

Anderson, M. S. 2000. Locally enhanced Raman spectroscopy with an atomic force microscope. *Appl. Phys. Lett.* 76:3130–32.

Anderson, N., Hartschuh, A., Cronin, S., and Novotny, L. 2005. Nanoscale vibrational analysis of single-walled carbon nanotubes. *J. Am. Chem. Soc.* 127:2533–77.

Ando, R., Hama, H., Hino, M. Y., Mizuno, H., and Miyawaki, A. 2002. An optical marker based on the UV-induced green-to-red photoconversion of a fluorescent protein. *Proc. Natl. Acad. Sci. USA* 99:12651–56.

Ayars, E. J., Jahncke, C. L., Paesler, M. A., and Hallen, H. D. 2001. Fundamental differences between micro- and nano-Raman spectroscopy. *J. Microsc.* 202:142–47.

Azoulav, J., Debarre, A., Richard, A., and Tchenio, T. 1999. Field enhancement and apertureless near-field optical spectroscopy of single molecules. *J. Microsc.* 194:486–90.

Bachelot, R., Gleyzes, P., and Boccara, A. C. 1995. Near-field optical microscope based on local perturbation of a diffraction spot. *Opt. Lett.* 20:1924–26.

Belkin, M. A., Kulakov, T. A., Ernst, K.-H., Yan, L., and Shen, Y. R. 2000. Sum-frequency vibrational spectroscopy on chiral liquids: A novel technique to probe molecular chirality. *Phy. Rev. Lett.* 85:4474–77.

Ben-Oren, I., Peleg, G., Lewis, A., Minke, B., and Loew, L. M. 1999. Infrared nonlinear optical measurements of membrane potential in photoreceptor cells. *Biophys. J.* 71:1616–20.

Bohren, C. F., and Huffman, D. R. 1983. *Absorption and scattering of light by small particles.* New York: Wiley Interscience.

Born, M., and Wolf, E. 1999. *Principles of optics,* 7th ed. Cambridge: Cambridge University Press.

Bouevitch, O., Lewis, A., Pinevsky, I., Wuskell, J. P., and Loew, L. M. 1993. Probing membrane-potential with nonlinear optics, *Biophys. J.* 65:672–79.

Boulesteix, T., Beaurepaire, E., Sauviat, M.-P., and Schanne-Klein, M.-C. 2004. Second-harmonic microscopy of unstained living cardiac myocytes: Measurements of sarcomere length with 20-nm accuracy. *Opt. Lett.* 29:2031–33.

Boyd, G. T., Yu, Z. H., and Shen, Y. R. 1986. Photoinduced luminescence from the noble metals and its enhancement on roughened surfaces. *Phys. Rev. B* 33:7923–36.

Buckingham, A. D. 1967. Permanent and induced molecular moments and long-range. intermolecular forces. *Adv. Chem. Phys.* 12:107–42.

Campagnola, P. J., Millard, A. C., Terasaki, M., Hoppe, P. E., Malone, C. J., and Mohler, W. A. 2002. Three-dimensional high-resolution second-harmonic generation imaging of endogenous structural proteins in biological tissues. *Biophys. J.* 82:493–508.

Chalmers, J. M., and Griffiths, P. R. 2002. *Handbook of vibrational spectroscopy.* Chichester: John Wiley & Sons.

Champagne, B., Fischer, P., and Buckingham, A. D. 2000. Ab initio investigation of the sum-frequency hyperpolarizability of small chiral molecules. *Chem. Phys. Letters* 331:83–88.

Chance, R. R., Prock, A., and Silbey, R. 1978. Molecular fluorescence and energy transfer near interfaces. *Adv. Chem. Phys.* 37:1–65.

Chang, R. K., and Furtak, T. E. 1981. *Surface enhanced Raman scattering.* New York: Plenum Press.

Chen, W., and Zhan, Q. 2007. Numerical study of an apertureless near field scanning optical microscope probe under radial polarization illumination. *Opt. Express* 15:4106–11.

Cheng, J.-X., Volkmer, A., Book, L. D., and Xie, X. S. 2001. An epi-detected coherent anti-Stokes Raman scattering (E-CARS) microscope with high spectral resolution and high sensitivity. *J. Phys. Chem. B* 105:1277–80.

Cheng, J.-X., and Xie, X. S. 2004. Coherent anti-Stokes Raman scattering microscopy: Instrumentation, theory and applications. *J. Phys. Chem. B* 108:827–40.

Christie, J. H., and Lockwood, D. J. 1971. Selection rules for three- and four-photon Raman interactions. *J. Chem. Phys.* 54:1141–54.

Chu, S.-W., Chen, S.-Y., Chern, G.-W., Tsai, T.-H., Chen, Y.-C., Lin, B.-L., and Sun, C.-K. 2004. Studies of $\chi^{(2)}/\chi^{(3)}$ tensors in submicron-scaled bio-tissues by polarization harmonics optical microscopy. *Biophys. J.* 86:3914–22.

Cui, X., Zhang, W., Yeo, B.-S., Zenobi, R., Hafner, C., and Erni, D. 2007. Tuning the resonance frequency of Ag-coated dielectric tips. *Opt. Express* 15:8309–16.

Cyvin, S. J., Rauch, J. E., and Decius, J. C. 1965. Theory of hyper-Raman effects (nonlinear inelastic light scattering): Selection rules and depolarization ratios for the second-order polarizability. *J. Chem. Phys.* 43:4083–95.

Denisov, V. N., Mavrin, B. N., and Podobedov, V. B. 1987. Hyper-Raman scattering by vibrational excitations in crystals, glasses and liquids, *Phys. Rep.* 151:1–92.

Denk, W., Strickler, J. H., and Webb, W. W. 1990. Two-photon laser scanning fluorescence microscopy. *Science* 248:73–76.

Duncan, M. D., Reintjes, J., and Manuccia, T. J. 1982. Scanning coherent anti-Stokes Raman microscope. *Opt. Lett.* 7:350–52.

Eckbreth, A. C. 1996. *Laser diagnostics for combustion temperature and species.* Amsterdam: Gordon & Breach Science.

Fischer, U. C., and Pohl, D. W. 1989. Observation of single-particle plasmons by near-field optical microscopy. *Phys. Rev. Lett.* 62:458–61.

Fleischmann, M., Hendra, P. J., and McQuillan, A. J. 1974. Raman spectra of pyridine adsorbed at a silver electrode. *Chem. Phys. Lett.* 26:163–66.

Furukawa, H., and Kawata, S. 1998. Local field enhancement with an apertureless near-field-microscope probe. *Opt. Commun.* 148:221–24.

Goodman, J. W. 1996. *Introduction to Fourier optics.* Singapore: McGraw-Hill.

Gu, M. 1996. *Principles of three-dimensional imaging in confocal microscopes.* Singapore: World Scientific, Inc.

Hamann, H. F., Gallagher, A., and Nesbitt, D. J. 2000. Near-field fluorescence imaging by localized field enhancement near a sharp probe tip. *Appl. Phys. Lett.* 76:1953–55.

Harootunian, A., Betzig, E., Isaacson, M., and Lewis, A. 1986. Super-resolution fluorescence near-field scanning optical microscopy. *Appl. Phys. Lett.* 49:674–76.

Hartschuh, A., Sánchez, E. J., Xie, X. S., and Novotny, L. 2003. High-resolution near-field Raman microscopy of single-walled carbon nanotubes. *Phys. Rev. Lett.* 90:095503.

Hashimoto, M., and Araki, T. 2001. Three dimensional coherent and optical transfer functions of coherent anti-stokes Raman scattering microscopy. *J. Opt. Soc. Am. A* 18:771–76.

Hashimoto, M., Araki, T., and Kawata, S. 2000. Molecular vibration imaging in the fingerprint region by use of coherent anti-Stokes Raman scattering microscopy with a collinear configuration. *Opt. Lett.* 25:1768–70.

Hayashi, S., and Konishi, T. 2005. Scanning near-field optical microscopic observation of surface-enhanced Raman scattering mediated by metallic particle-surface gap modes. *Jpn. J. Appl. Phys.* 44:5313–18.

Hayazawa, N., Ichimura, T., Hashimoto, M., Inouye, Y., and Kawata, S. 2004a. Amplification of coherent anti-Stokes Raman scattering by a metallic nanostructure for a high resolution vibration microscopy. *J. Appl. Phys.* 95:2676–81.

Hayazawa, N., Inouye, Y., and Kawata, S. 1999. Evanescent field excitation and measurement of dye fluorescence in a metallic probe near-field scanning optical microscope. *J. Microsc.* 194:472–76.

Hayazawa, N., Inouye, Y., Sekkat, Z., and Kawata, S. 2000. Metallized tip amplification of near-field Raman scattering. *Opt. Commun.* 183:333–36.

Hayazawa, N., Inouye, Y., Sekkat, Z., and Kawata, S. 2001. Near-field Raman scattering enhanced by a metallized tip. *Chem. Phys. Lett.* 335:369–74.

Hayazawa, N., Inouye, Y., Sekkat, Z., and Kawata, S. 2002. Near-field Raman imaging of organic molecules by an apertureless metallic probe scanning optical microscope. *J. Chem. Phys.* 117:1296–1301.

Hayazawa, N., Ishitobi, H., Taguchi, A., Ikeda, K., Tarun, A., and Kawata, S. 2007a. Focused excitation of surface plasmon polaritons for efficient field enhancement based on gap-mode in tip-enhanced spectroscopy. *Jpn. J. Appl. Phys.* 46:7995–99.

Hayazawa, N., Motohashi, M., Saito, Y., Ishitobi, H., Ono, A., Ichimura, T., Verma, P., and Kawata, S. 2007b. Visualization of localized strain of crystalline in nano-scale by tip-enhanced Raman spectroscope and microscope. *J. Raman Spectrosc.* 38:684–96.

Hayazawa, N., Motohashi, M., Saito, Y., and Kawata, S. 2005. Highly sensitive strain detection in strained silicon by surface enhanced Raman spectroscopy. *Appl. Phys. Lett.* 86:263114.

Hayazawa, N., and Saito, Y. 2007. Tip-enhanced spectroscopy for nano investigation of molecular vibrations. In *Applied scanning probe methods VI*, eds. B. Bhushan and S. Kawata, 257–85. Berlin: Springer.

Hayazawa, N., Saito, Y., and Kawata, S. 2004b. Detection and characterization of longitudinal field for tip-enhanced Raman spectroscopy. *Appl. Phys. Lett.* 85:6239–41.

Hayazawa, N., Watanabe, H., Saito, Y., and Kawata, S. 2006. Towards atomic site-selective sensitivity in tip-enhanced Raman spectroscopy. *J. Chem. Phys.* 125:244706.

Hayazawa, N., Yano, T., Watanabe, H., Inouye, Y., and Kawata, S. 2003. Detection of an individual single-wall carbon nanotube by tip-enhanced near-field Raman spectroscopy. *Chem. Phys. Lett.* 376:174–80.

Hecht, B., Bielefeldt, H., Inouye, Y., Pohl, D. W., and Novotny, L. 1997. Facts and artifacts in near-field optical microscopy. *J. Appl. Phys.* 81:2492–98.

Hell, S. W., and Wichmann, J. 1994. Breaking the diffraction resolution limit by stimulated emission: Stimulated-emission-depletion fluorescence microscopy. *Opt. Lett.* 19:780–82.

Henrard, L., Hernandez, E., Bernier, P., and Rubio, A. 1999. van der Waals interaction in nanotube bundles: Consequences on vibrational modes. *Phys. Rev. B* 60:R8521–24.

Hildebrandt, P., and Stockburger, M. 1984. Surface-enhanced resonance Raman spectroscopy of rhodamine 6G adsorbed on colloidal silver. *J. Phys. Chem.* 88:5935–44.

Ichimura, T., Hayazawa, N., Hashimoto, M., Inouye, Y., and Kawata, S. 2003. Local enhancement of coherent anti-Stokes Raman scattering by isolated gold nanoparticles. *J. Raman Spectrosc.* 34:651–54.

Ichimura, T., Hayazawa, N., Hashimoto, M., Inouye, Y., and Kawata, S. 2004a. Tip-enhanced coherent anti-Stokes Raman scattering for vibrational nano-imaging. *Phys. Rev. Lett.* 92:220801.

Ichimura, T., Hayazawa, N., Hashimoto, M., Inouye, Y., and Kawata, S. 2004b. Application of tip-enhanced microscopy for nonlinear Raman spectroscopy. *Appl. Phys. Lett.* 84:1768–70.

Ichimura, T., Watanabe, H., Morita, Y., Verma, P., Kawata, S., and Inouye, Y. 2007. Temporal fluctuation of tip-enhanced Raman spectra of adenine molecules. *J. Phys. Chem. C* 111:9460–64.

Ikeda, K., Takase, M., Sawai, Y., Nabika, H., Murakoshi, K., and Uosaki, K. 2007. Hyper-Raman scattering enhanced by anisotropic dimer plasmons on artificial nanostructures. *J. Chem. Phys.* 127:111103.

Ikeda, K., and Vosaki, K. 2008. Resonance hyper-Raman scattering of fullerene C60 microcrystals. *J. Phys. Chem. A* 112:790–93.

Inouye, Y., Hayazawa, N., Hayashi, K., Sekkat, Z., and Kawata, S. 1999. Near-field scanning optical microscope using a metallized cantilever tip for nanospectroscopy. *Proc. SPIE Int. Soc. Opt. Eng.* 3791:40–48.

Inouye, Y., and Kawata, S. 1994. Near-field scanning optical microscope using a metallic probe tip. *Opt. Lett.* 19:159–61.

Jeanmaire, D. I., and Van Duyne, R. P. 1977. Surface Raman spectroelectrochemistry. Part I. Heterocyclic, aromatic, and aliphatic amines adsorbed on the anodized silver electrode. *J. Electroanal. Chem.* 84:1–20.

Ji, N., Zhang, K., Yang, H., and Shen, Y. R. 2006. Three-dimensional chiral imaging by sum-frequency generation. *J. Am. Chem. Soc.* 128:3482–83.

Jiang, C., Zhao, J., Therese, H. A., Friedrich, M., and Mews, A. 2003. Raman imaging and spectroscopy of heterogeneous individual carbon nanotubes. *J. Phys. Chem. B* 107:8742–45.

Kambhampati, P., Child, C. M., Foster, M. C., and Campion, A. 1998. On the chemical mechanism of surface enhanced Raman scattering: Experiment and theory. *J. Chem. Phys.* 108:5013–26.

Kataura, H., Kumazawa, Y., Maniwa, Y., Umezu, I., Suzuki, S., Ohtsuka, Y., and Achiba, Y. 1999. Optical properties of single-wall carbon nanotubes. *Synth. Met.* 103:2555–58.

Kawata, S. 2001. *Near-field optics and surface plasmon polaritons.* New York: Springer.

Kawata, S., and Shalaev, V. M. 2007. *Tip enhancement.* Netherlands: Elsevier.

Kawata, S., Sun, H.-B., Tanaka, T., and Takada, K. 2001. Finer features for functional microdevices. *Nature* 412:697–98.

Kawata, Y., Xu, C., and Denk, W. 1999. Feasibility of molecular-resolution fluorescence near-field microscopy using multi-photon absorption and field enhancement near a sharp tip. *J. Appl. Phys.* 85: 1294–1301.

Kneipp, J., Kneipp, H., and Kneipp, K. 2006. Two-photon vibrational spectroscopy for biosciences based on surface enhanced hyper Raman scattering. *Proc. Natl. Acad. Sci. USA* 103:17149–53.

Kogure, T., Karasawa, S., Araki, T., Saito, K., Kinjo, M., and Miyawaki, A. 2006. A fluorescent variant of a protein from the stony coral Montipora facilitates dual-color single-laser fluorescence cross-correlation spectroscopy. *Nature Biotechnol.* 24:577–81.

Kunz, K. S., and Luebbers, R. J. 1999. *The finite difference time domain method for electromagnetics.* Boca Raton: CRC Press.

Lee, N., Hartschuh, R. D., Mehtani, D., Kisliuk, A., Maguire, J. F., Green, M., Foster, M. D., and Sokolov, A. P. 2007. High contrast scanning nano-Raman spectroscopy of silicon. *J. Raman Spectrosc.* 38:789–96.

Leng, W., Woo, H. Y., Vak, D., Bazan, G. C., and Kelley, A. M. 2006. Surface-enhanced resonance Raman and hyper-Raman spectroscopy of water-soluble substituted stilbene and distyrylbenzene chromophores. *J. Raman Spectrosc.* 37:132–41.

Lessard, T. J., Lessard, G. A., and Quake, S. R. 2000. An apertureless near-field microscope for fluorescence imaging. *Appl. Phys. Lett.* 76:378–80.

Levenson, M. D., and Kano, S. S. 1988. *Introduction to nonlinear optical spectroscopy.* Boston: Academic Press.

Liang, E. J., Weippert, A., Funk, J. M., Materny, A., and Kiefer, W. 1994. Experimental observation of surface-enhanced coherent anti-Stokes Raman scattering. *Chem. Phys. Lett.* 227:115–20.

Lu, H. P. 2005. Site-specific Raman spectroscopy and chemical dynamics of nanoscale interstitial systems. *J. Phys.: Condens. Matter* 17:R333–55.

Mahieu-Williame, L., Grésillon, S., Cuniot-Ponsard, M., and Boccara, C. 2007. Second harmonic generation in the near field and far field: A sensitive tool to probe crystalline homogeneity. *J. Appl. Phys.* 101:083111.

Martin, Y. C., Hamann, H. F., and Wickramasinghe, H. K. 2001. Strength of the electric field in apertureless near-field optical microscopy. *J. Appl. Phys.* 89:5774–78.

Mehtani, D., Lee, N., Hartschuh, R. D., Kisliuk, A., Foster, M. D., Sokolov, A. P., and Maguire, J. F. 2005. Nano-Raman spectroscopy with side-illumination optics. *J. Raman Spectrosc.* 36:1068–75.

Nie, S., and Emory, S. R. 1997. Surface-enhanced resonance Raman spectroscopy of rhodamine 6G adsorbed on colloidal silver. *Science* 275:1102–6.

Nordlander, P., and Le, F. 2006. Plasmonic structure and electromagnetic field enhancements in the metallic nanoparticle-film system. *Appl. Phys. B* 84:35–41.

Novotny, L., Bian, R. X., and Xie, X. S. 1997. Theory of nanometric optical tweezers. *Phys. Rev. Lett.* 79:645–48.

Novotny, L., Sánchez, E. J., and Xie, X. S. 1998. Near-field optical imaging using metal tips illuminated by higher-order Hermite-Gaussian beams. *Ultramicrosc.* 71:21–29.

Ossikovski, R., Nguyen, Q., and Picardi, G. 2007. Simple model for the polarization effects in tip-enhanced Raman spectroscopy. *Phys. Rev. B* 75:045412.

Otto, A., Mrozek, I., Grabhorn, H., and Akemann, W. 1992. Surface-enhanced Raman scattering. *J. Phys. Condens. Matter* 4:1143–1212.

Peleg, G., Lewis, A., Linial, M., and Loew, L. M. 1999. Non-linear optical measurement of membrane potential around single molecules at selected cellular sites. *Proc. Natl. Acad. Sci. USA* 96:6700–4.

Pettinger, B., Ren, B., Picardi, G., Schuster, R., and Ertl, G. 2004. Nanoscale probing of adsorbed species by tip-enhanced Raman spectroscopy. *Phys. Rev. Lett.* 92:096101.

Poborchii, V., Tada, T., and Kanayama, T. 2005. Subwavelength-resolution Raman microscopy of Si structures using metal-particle-topped AFM probe. *Jpn. J. Appl. Phys.* 44:L202–5.

Pohl, D. W., Denk, W., and Lanz, M. 1984. Optical stethoscopy: Image recording with resolution $\Omega/20$. *Appl. Phys. Lett.* 44:651–53.

Prasher, D. C., Eckenrode, V. K., Ward, W. W., Prendergast, F. G., and Cormier, M. J. 1992. Primary structure of the Aequorea victoria green-fluorescent protein. *Gene* 111:229–33.

Puppels, G. J., de Mul, F. F. M., Otto, C., Greve, J., Robert-Nicoud, M., Arndt-Jovin, D. J., and Jovin, T. M. 1990. Studying single living cells and chromosomes by confocal Raman microscopy. *Nature* 347:301–3.

Raether, H. 1988. *Surface plasmon polaritons on smooth and rough surfaces and on gratings.* Berlin: Springer.

Rao, A. M., Richter, E., Bandow, S., Chase, B., Eklund, P. C., Williams, K. A., Fang, S. et al. 1997. Diameter-selective Raman scattering from vibrational modes in carbon nanotubes. *Science* 275:187–91.

Ropers, C., Neacsu, C. C., Elsaesser, T., Albrecht, M., Raschke, M. B., and Lienau, C. 2007. Grating coupling of surface plasmons onto metallic tips. *Nano Lett.* 7:2784–88.

Saito, R., and Kataura, H. 2001. Optical properties and Raman spectroscopy of carbon nanotubes. In *Topics in applied physics: Carbon nanotubes*, eds. M. S. Dresselhaus, G. Dresselhaus, and P. Avouris, 213–46. Berlin: Springer.

Saito, Y., Hayazawa, N., Kataura, H., Tsukagoshi, K., Inouye, Y., and Kawata, S. 2005. Polarization measurements in tip-enhanced Raman spectroscopy applied to single-walled carbon nanotubes. *Chem. Phys. Lett.* 410:136–41.

Saito, Y., Motohashi, M., Hayazawa, N., Iyoki, M., and Kawata, S. 2006. Nanoscale characterization of strained silicon by tip-enhanced Raman spectroscope in reflection mode. *Appl. Phys. Lett.* 88:143109.

Sánchez, E. J., Krug, J. T., and Xie, X. S. 2002. Ion and electron beam assisted growth of nanometric SimOn structures for near-field microscopy. *Rev. Sci. Instrum.* 73:3901–7.

Sánchez, E. J., Novotny, L., and Xie, X. S. 1999. Near-field fluorescence microscopy based on two-photon excitation with metal tips. *Phys. Rev. Lett.* 82:4014–17.

Sass, J. K., Neff, H., Moskovits, M., and Holloway, S. 1981. Electric field gradient effects on the spectroscopy of adsorbed molecules. *J. Phys. Chem.* 85:621–23.

Shen, Y. R. 1984. *The principles of nonlinear optics*. New York: John Wiley & Sons.

Shen., Y. R. 2000. Surface nonlinear optics: A historical perspective, *IEEE J. Selected Top. Quantum Electron.* 6:1375–79.

Shen, Y. R., Swiatkiewicz, J., Winiarz, J., Markowicz, P., and Prasad, P. N. 2000. Second-harmonic and sum-frequency imaging of organic nanocrystals with photon scanning tunneling microscope. *Appl. Phys. Lett.* 77:2946–48.

Shimada, R., Kano, H., and Hamaguchi, H. 2006. Hyper-Raman microspectroscopy: A new approach to completing vibrational spectral and imaging information under a microscope. *Opt. Lett.* 31:320–22.

Shimomura, O., Johnson, F. H., and Saiga, Y. 1962. Extraction, purification and properties of aequorin, a bioluminescent protein from the luminous hydromedusan, Aequorea. *J. Cell. Comp. Physiol.* 59:223–40.

Stöckle, R. M., Suh, Y. D., Deckert, V., and Zenobi, R. 2000. Nanoscale chemical analysis by tip-enhanced Raman spectroscopy. *Chem. Phys. Lett.* 318:131–36.

Sugiura, T., Okada, T., Inouye, Y., Nakamura, O., and Kawata, S. 1997. Gold-bead scanning near-field optical microscope with laser-force position control. *Opt. Lett.* 22:1663–65.

Sun, W. X., and Shen, Z. X. 2003. Apertureless near-field scanning Raman microscopy using reflection scattering geometry. *Ultramicroscopy.* 94:237–44.

Sunder, S., and Bernstein, H. J. 1981. Resonance Raman spectrum of a deuterated crystal violet: [p(CH3)2N•C6D4]3C+Cl–. *Can. J. Chem.* 59:964–67.

Takada, K., Sun, H. B., and Kawata, S. 2005. Improved spatial resolution and surface roughness in photopolymerization-based laser nanowriting. *Appl. Phys. Lett.* 86:071122.

Takahashi, S., and Zayats, A. V. 2002. Near-field second-harmonic generation at a metal tip apex. *Appl. Phys. Lett.* 80:3479–81.

Tanaka, S., Cai, L. T., Tabata, H., and Kawai, T. 2001. Formation of two-dimensional network structure of DNA molecules on Si substrate. *Jpn. J. Appl. Phys.* 40:L407–9.

Tsien, R. Y. 1998. The green fluorescent protein. *Annu. Rev. Biochem.* 67:509–44.

Verma, P., Yamada, K., Watanabe, H., Inouye, Y., and Kawata, S. 2006. Near-field Raman scattering investigation of tip effects on C60 molecules. *Phys. Rev. B* 73:045416.

Watanabe, H., Hayazawa, N., Inouye, Y., and Kawata, S. 2005. DFT vibrational calculations of rhodamine 6G adsorbed on silver: Analysis of tip-enhanced Raman spectroscopy. *J. Phys. Chem. B* 109:5012–20.

Watanabe, H., Ishida, Y., Hayazawa, N., Inouye, Y., and Kawata, S. 2004. Tip-enhanced near-field Raman analysis of tip-pressurized adenine molecule. *Phys. Rev. B* 69:1–11.

Watanabe, T., and Pettinger, B. 1982. Surface-enhanced Raman scattering from crystal violet adsorbed on a silver electrode. *Chem. Phys. Lett.* 89:501–7.

Wilcoxon, J. P., and Martin, J. E. 1998. Photoluminescence from nanosize gold clusters. *J. Chem. Phys.* 108:9137–43.

Xu, H., Bjerneld, E. J., Käll, M., and Börjesson, L. 1999. Spectroscopy of single hemoglobin molecules by surface enhanced Raman scattering. *Phys. Rev. Lett.* 83:4357–60.

Yano, T., Inouye, Y., and Kawata, S. 2006. Nanoscale uniaxial pressure effect of a carbon nanotube bundle on tip-enhanced near-field Raman spectra. *Nano Lett.* 6:1269–73.

Yano, T., Verma, P., Inouye, Y., and Kawata, S. 2005. Diameter-selective near-field Raman analysis and imaging of isolated carbon nanotube bundles. *Appl. Phys. Lett.* 88:093125.

Zayats, A. V., and Sandoghdar, V. 2000. Apertureless scanning near-field second-harmonic microscopy. *Opt. Commun.* 178:245–49.

Zayats, A. V., and Sandoghdar, V. 2001. Apertureless near-field optical microscopy via local second-harmonic generation. *J. Microsc.* 2002:94–99.

Zenhausern, F., O'Boyle, M. P., and Wickramasinghe, H. K. 1994. Apertureless near-field optical microscope. *Appl. Phys. Lett.* 65:1623–25.

Zumbusch, A., Holton, G. R., and Xie, X. S. 1999. Three-dimensional vibrational imaging by coherent anti-Stokes Raman scattering. *Phys. Rev. Lett.* 82:4142–45.

Index

Symbols

α-perylene microcrystals, 67

A

Autofluorescence (AF), 34, 46
Axial resolution (Z), femtosecond fluorescence
    up-conversion microscope and, 59
Axon function, CARS and, 119–122

B

Barrier heights, 6–9, 23
Broadband femtosecond pulses, 171–174
Broadband multiphoton microspectroscopy,
    190–192

C

Calcium salt of dipicolinic acid (CaDPA), 186
Cancer, vibrational flow cytometry and, 157, 160
CARS. *See* Coherent anti-Stokes Raman
    scattering
CCD (charge-coupled device), 54
Characterization of broadband laser sources,
    177–179
Charge-coupled device (CCD), 54
Chemical kinetics, 4
Chloroplasts
    high-resolution imaging of, 91–93
    multicontrast nonlinear imaging of, 89–92
Coherent anti-Stokes Raman scattering (CARS),
    34, 39, 46, 103–124
    broadband microspectroscopy and, 152–155
    combining with THG, 119–122
    detection sensitivity of, improving,
        108–114
    endoscopy and, 114–119
    FE-CARS and, 227–235
    FM-CARS approach to, 109–114
    heterodyne detection and, 188–190
    interferometric imaging and, 109, 221
    laser source for, 104–108

microspectroscopy and, 167–196
molecular vibrations and, 169, 181, 185
multicontrast nonlinear microscopes
    and, 80
nonlinear Raman microspectroscopy and,
    141–145
nonlinear spectroscopic signals and, 169
nonresonant background in, 145–152
phase and, 213–215
protein crystallization and, 155–157
single-beam microspectroscopy/microscopy
    and, 179–190
spectral resonances and, 222–224
TE-CARS and, 252–259
TERS and, 242–251
Coherent control
    multiphoton processes and, 170
    single-beam CARS and, 181
Collagen molecules in solution, THG and, 137
Colloidal nanoparticles in solution, THG and,
    131
Confocal microscope, 59
Contrast mechanisms, nonlinear, 73–79

D

Dermis
    epidermis and, 33
    excitation of fluorescence and, 36
Difference electron densities, 11
Diffractive optics-based four-wave mixing with
    heterodyne detection, 17–19
Dihydrorhodamine (DHR), 43
Discrete states, 3–6
Dispersion
    femtosecond pulses and, 198–201
    GDD, 198, 200, 201–203
    GVD, 198
    measuring/eliminating, 200
    TOD, 199, 200
Dispersion compensation
    photobleaching and, 204–206
    TPM and, 203
DNA, imaged by TE-CARS, 254–257
Dynamics, theory of, 3–10

V

X